Resveratrol

State-of-the-Art
Science and Health Applications

Actionable Targets and
Mechanisms of Resveratrol

Other Related Titles from World Scientific

Carcinogens, DNA Damage and Cancer Risk: Mechanisms of Chemical Carcinogenesis
edited by Miriam C Poirier
ISBN: 978-981-3237-19-3

Inflammation: Fundamental Mechanisms
edited by Klaus Ley
ISBN: 978-981-3109-43-8

Nobel and Lasker Laureates of Chinese Descent: In Literature and Science
edited by Todd S Ing, Keith K Lau, Joseph M Chan, Hon-Lok Tang, Angela T Hadsell and Laurence K Chan
ISBN: 978-981-4704-60-1
ISBN: 978-981-4704-61-8 (pbk)

Resveratrol

State-of-the-Art Science and Health Applications

Actionable Targets and Mechanisms of Resveratrol

Editors

Joseph M Wu

Tze-chen Hsieh

New York Medical College, USA

World Scientific

NEW JERSEY • LONDON • SINGAPORE • BEIJING • SHANGHAI • HONG KONG • TAIPEI • CHENNAI • TOKYO

Published by

World Scientific Publishing Co. Pte. Ltd.

5 Toh Tuck Link, Singapore 596224

USA office: 27 Warren Street, Suite 401-402, Hackensack, NJ 07601

UK office: 57 Shelton Street, Covent Garden, London WC2H 9HE

Library of Congress Cataloging-in-Publication Data
Names: Wu, Joseph M., editor. | Hsieh, Tze-chen, editor.
Title: Resveratrol : state-of-the-art science and health applications : actionable targets and mechanisms of resveratrol / edited by Joseph M Wu, Tze-chen Hsieh.
Description: New Jersey : World Scientific, 2018. | Includes bibliographical references.
Identifiers: LCCN 2018020883 | ISBN 9789813270909 (hardcover : alk. paper)
Subjects: | MESH: Stilbenes--pharmacology | Antineoplastic Agents, Phytogenic | Anti-Inflammatory Agents, Non-Steroidal | Enzyme Inhibitors | Protective Agents
Classification: LCC RS431.A64 | NLM QV 269 | DDC 615.7/98--dc23
LC record available at https://lccn.loc.gov/2018020883

British Library Cataloguing-in-Publication Data
A catalogue record for this book is available from the British Library.

For any available supplementary material, please visit
http://www.worldscientific.com/worldscibooks/10.1142/11004#t=suppl

Typeset by Stallion Press
Email: enquiries@stallionpress.com

To my late parents who taught me the value of hard work, perseverance, service, and selfless considerations for others; these have been and remain the guiding light for my professional and personal activities.

To my children Amy and Mary for their belief in me; for reminding me to cheer on and appreciate life day to day.

To Dr. Tze-chen Hsieh, co-editor of this book for her unwavering commitment, lifelong support and devotion on studies of resveratrol; for maintaining equilibrium and a sense of optimism in all our joint daily events and adventures.

To all contributors for daring to think out-of-the-box and for tirelessly searching the uncanny dynamism of resveratrol.

JMW

This book is dedicated to the fond memory of my late father whose lifelong passion was about family and harmony.

TCH

Contents

Preface

When we were invited to serve as the editors of *Resveratrol: State-of-the-Art Science and Health Applications*, we were elated at the opportunity. We had worked in the field of chemoprevention and cardioprotection as a team for well over 24 years mostly as researchers but have had the opportunity to present our work as invited speakers. Being asked to have an active role in a new book on resveratrol is challenging as well as exciting, providing us with the opportunity to not only communicate our own studies, but also solicit participation and contribution from established and well-published expert colleagues, many of whom are leaders in the field.

This book is geared towards three audiences: resveratrol researchers, nutritional scientists and health care and government regulatory professionals. Each chapter covers topics using published and at times unpublished data to provide scholarly and comprehensive review and perspectives. Our aspiration is that the chapters in this volume will stimulate new, transcending ideas and provide collaborative opportunities. We also hope that busy health care practitioners will use this book as reliable, educational resources when administering clinical care or making decisions on the benefits, risks, or values of resveratrol in the prevention and treatment of diseases.

There is an undisputable historical record for disease preventative effects of the consumption of vegetables and fruits. Ample mention has been made of the usage of foods and their extracts as health beneficial and therapeutic tools in ancient and modern cultures.

Many clinical studies are ongoing concerning the health and clinical benefits of resveratrol. Patients today are exploring the use of dietary supplements to manage their diseases. It is challenging for health care practitioners and patients to keep up with the evolving and increasingly expanding scope of bioactivities of resveratrol, especially in weighing the perceived versus the scientifically validated benefits. We hope that this book will present a succinct and timely summary of recent advances and new research initiative in the field of resveratrol, and that the evidence and broadly based multidisciplinary efforts taken will provide a scientific perspective on the use of resveratrol in battling a variety of diseases.

This book is the result of many scientific minds who have shared a vision and aspiration of resveratrol. Gratitude is expressed to all who helped with the preparation of the book and illustrations, including the staff of World Scientific Publishing.

About the Editors

Joseph M. Wu, Ph.D., Professor, New York Medical College

Professor Wu received his B.S. (Honors) from McGill University in 1970, majoring in Biochemistry. He obtained his M.S. in Chemistry in 1972 and Biological Sciences in 1975, both from Florida State University. Following a two-year post-doctoral training in Biochemistry at Temple University, Dr. Wu became a faculty member at New York Medical College where he currently holds the rank of Professor of Biochemistry and Molecular Biology and is also a member of the Brander Cancer Research Institute.

Professor Wu is a biochemist/cell-molecular biologist with a long history of research on mechanism of gene regulation and expression in cancer and other chronic diseases. At New York Medical College, Dr. Wu has trained 11 Ph.D. and 65 M.S. students. He has authored 228 peer-reviewed scientific articles and reviews, and 21 book chapters and 139 abstracts. Since 1994, Dr. Wu has been studying the mechanism of action of resveratrol, retinoids, phytochemicals, and proprietary herbal formulations, in the context of chemoprevention and cardioprotection.

Tze-chen Hsieh, Ph.D., Research Associate Professor, New York Medical College

Dr. Hsieh is a Research Associate Professor in the Department of Biochemistry and Molecular Biology at New York Medical College. She is also a member of the Brander Cancer Research Institute at the same institution. Dr. Hsieh earned her B.S. in National

Chung Shing University in 1984 and received her M.S. and Ph.D. at Texas Tech University in 1988 and 1992. She started her training in the Department of Medicine at New York Medical College as a postdoctoral fellow in 1992. She joined the Department of Biochemistry and Molecular Biology as a Research Associate in 1994, and was promoted to the rank of Research Assistant Professor in 1997 and her current position in 2009.

Dr. Hsieh has been studying mechanism of cell proliferation, differentiation, and induction of apoptosis for the past ten years using prostate cancer as a model. She has authored 121 peer-reviewed scientific articles, 9 book chapters, 53 abstracts and has given 13 invited presentations since 1992. Although trained as a cell biologist, Dr. Hsieh has devoted a considerable amount of time and effort in learning and mastering the principles and applications of biochemical and molecular approaches and has applied this knowledge to her personal interest on studies of prostate cancer, particularly in the area of control of the tumor by resveratrol.

Chapter 1

Unraveling and Trailblazing Cardioprotection by Resveratrol

Tze-chen Hsieh and Joseph M. Wu

Biochemistry and Molecular Biology, New York Medical College, Valhalla, NY 10595, USA

Disease prevention by diet and nutrition has evolved from epidemiological studies to biochemical/cellular/molecular analyses, with mounting emphasis placed on the identification of specific dietary agents and elucidation of their actionable mechanisms. Resveratrol is a grape polyphenol that consistently shows preventative activity against coronary heart diseases (CHD) — a major cause of morbidity and mortality in the US. The role of resveratrol in cardioprotection stems from the "French paradox" reporting the association between low to moderate consumption of red wine and a reduced risk for CHD. The details of mechanisms remain to be elucidated.

This chapter begins with an overview of the studies performed in this laboratory and our perspective on bio-efficacy of resveratrol in conferring protection against CHD. The rationale, approaches and key findings from *in vitro* and animal studies on cardioprotection by resveratrol were summarized. Next, the strategy we used to isolate cellular target of resveratrol and the discovery of specific resveratrol targeting proteins (RTPs) were briefly introduced. The binding and interacting of resveratrol with its targeting protein was also reviewed. Lastly, we propose atherogenic microenvironment as a new frontier for future studies that may uncover new targets and mechanisms to further understand the cardioprotection by resveratrol.

1.1. Introduction

Resveratrol (3,5,4′-trihydroxystilbene) is a trihydroxylated stilbene abundant in red grape skins, peanuts, red wine and some other foods/drinks widely consumed in an average American diet.[1–3] This polyphenol has received considerable attention as a preventative agent against a variety of chronic diseases. Resveratrol first received scientific scrutiny as a protective agent against CHD based on the epidemiologic observation known as the French paradox.[4,5] Subsequent interest on chemoprevention by resveratrol was rooted in the report in 1997 by Pezzuto and coworkers that this polyphenol exhibits inhibitory activity on the major stages of carcinogenesis, respectively, initiation, promotion and progression.[2] At present, focus remains on the actionable mechanisms by which resveratrol exerts its cardioprotective and chemopreventive effects. In this chapter we review our approaches and *in vitro* and *in vivo* evidence from 1999–2017 directed at elucidating the role resveratrol plays in protection against coronary diseases (Figure 1.1).

1.2. *In vitro* and *in vivo* experiments confirm effect of resveratrol on cardioprotection

CHD is a major cause of morbidity and mortality which can result in loss of quality of life and premature death with a burdensome economic outcome to the public health system in the US. Atherosclerosis (AS) is a leading cause of CHD and it has witnessed a significant reduction in mortality over the past several decades, owing to knowledge gained in the understanding of its pathogenesis, efforts to modify life style factors that increase the risk of AS, and the emergence of new medications and treatments. Reduction in cigarette use and intake of cholesterol and saturated fat, and increased consumption of dietary fiber, whole grains, plant sterols, and phytochemicals all contribute to counteract risk of AS.

Cardioprotection by resveratrol as a grape-derived phytochemical is founded on results from a 1992 publication in *Lancet* frequently referred to as the "French paradox"[4]; namely, reporting

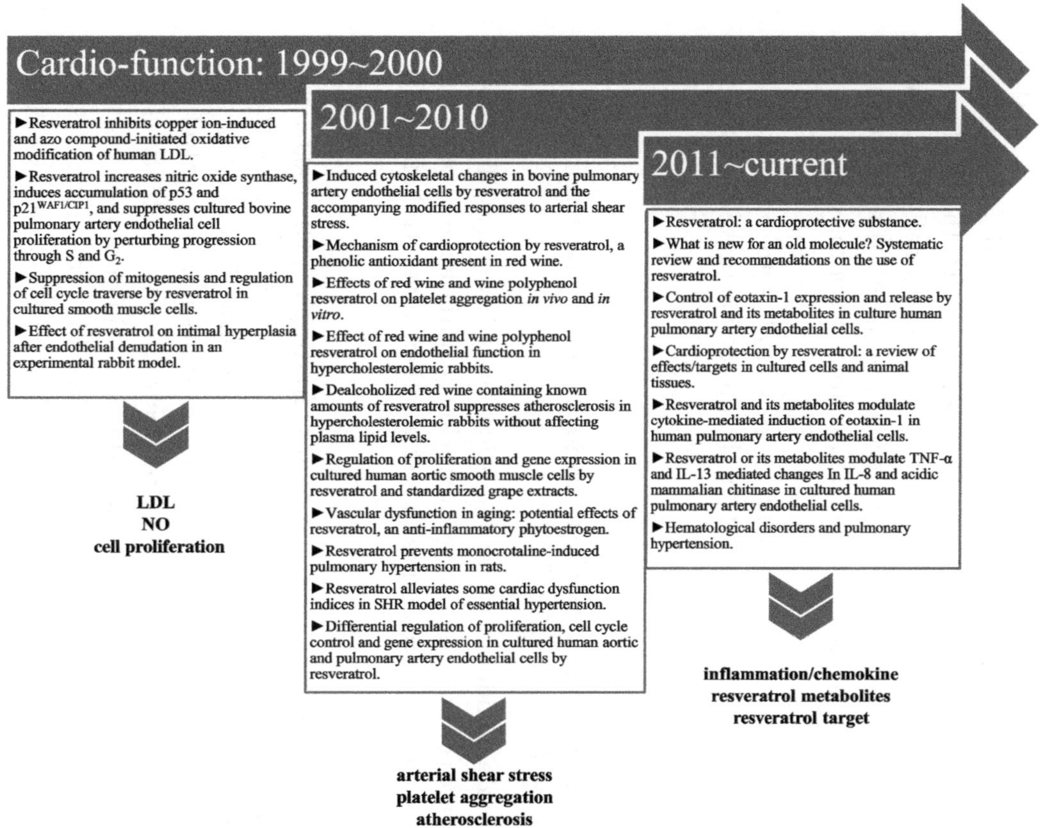

Fig. 1.1. A timeline that illustrates our plan of attack and published studies for dissecting the cardioprotective effects of resveratrol.

that the population residing in Southwest France had a relatively low incidence of heart disease, despite preference for a saturated fat rich diet. Important findings of this study are summarized. First, it validates epidemiological studies showing that the consumption of red wine is beneficial to human health.[6–8] Further, it reveals the potential link between intake of red wine and cardiovascular health,[4,9] providing support for the thesis that red wine may reduce clinical events in AS. Additionally, because of its relative abundance as a key ingredient in red wine, resveratrol may be hypothesized to contribute to cardioprotection by red wine.[10–12]

We performed *in vitro and in vivo* studies to investigate the various effects resveratrol exerts in the protection of AS. Key findings are summarized below.

1.2.1. *Resveratrol inhibits oxidative modification of human LDL*[13]

Rationale. Oxidative stress and oxidation of low density lipoprotein (LDL) are considered primary events in the initiation and progression of AS. We performed studies to test whether resveratrol affects LDL oxidation.

Experimental approach and key findings. LDL (500 μg/ml) purified from normolipidemic adults was oxidized using copper ions (40 μM) or azo compound (10 mM 2, 2′-azobis[2-amidinopropane dihydrochloride), with and without the addition of resveratrol. LDL oxidation was assayed by agarose electrophoretic mobility (REM) changes, and formation of thiobarbituric acid reactive substances (TBARS). Resveratrol ($\geq$50 μM) inhibited LDL oxidation as shown by reduction in REM (42.3%) and TBARS (70.5%), and by prolongation of lag phase in LDL oxidation.

Public health implications. Resveratrol protects against oxidized LDL-mediated initiation of AS, possibly by acting as a free radical scavenger.

1.2.2. *Resveratrol increases nitric oxide synthase, induces accumulation of p53 and p21WAF1/ CIP1, and suppresses cultured bovine pulmonary artery endothelial cell proliferation by perturbing progression through S and G_2*[14]

Rationale. Blood vessel endothelial lining could be susceptible to damage by oxidative stress resulting from oxidized LDL-elicited increase in reactive oxygen species (ROS) — an effect that can be counteracted by nitric oxide (NO) whose biosynthesis is under the control of endothelial nitric oxide synthase (eNOS). Whether and how resveratrol, as a red wine constituent, decreases the risk of CHD via control eNOS/NO reactions were tested.

Experimental approach and key findings. In cultured bovine pulmonary artery endothelial cells (BPAECs), resveratrol was found to induce eNOS. Since NO exerts multiple biological effects including vasodilation, platelet adhesion/aggregation, and cell proliferation and migration,[15–17] we further examined the control of BPAEC cell proliferation by resveratrol using flow cytometry. Resveratrol inhibited the cell proliferation by blocking cell progression through S and G_2 phases. Disruption in cell cycle progression is accompanied by an increase in the expression of tumor suppressor gene protein p53 and elevated level of cyclin-dependent kinase inhibitor p21(WAF1/CIP1).

Public health implications. Resveratrol delivered to endothelial cells (ECs) serves to inhibit EC proliferation to allow for rapid, efficient repair of damaged EC; this would effectively lower the probability of AS plaques formation and the development of CHD. Resveratrol affords cardioprotection by regulating NO biosynthesis via dynamic control of a functional eNOS.

1.2.3. *Suppression of mitogenesis and regulation of cell cycle traverse by resveratrol in cultured smooth muscle cells*[18]

Rationale. Atherosclerosis as a main cause of CHD involves the proliferation of smooth muscle cells (SMCs) in the intima. We investigated the effects of resveratrol on SMC proliferation.

Experimental approach and key findings. SMC proliferation in response to addition of mitogens, e.g. endothelin and platelet-derived growth factor (PDGF) was dose-dependently inhibited by resveratrol. The anti-mitogenic effects of resveratrol were attributed to a G_1 to S block in cell cycle traverse without induction of apoptosis.

Public health implications. The results further support the anti-AS effects of resveratrol.

1.2.4. *Effect of resveratrol on intimal hyperplasia after endothelial denudation in an experimental rabbit model*[19]

Rationale. Animal studies were performed to test whether resveratrol prevents hardening and narrowing of the arteries via inhibition of intimal thickening.

Experimental approach and key findings. Endothelial denudation was performed in the iliac artery of normal and hypercholesterolemic rabbits. Rabbits were also given resveratrol (2 and 4 mg/kg/day) intragastrically for 5 weeks beginning 1 week before denudation. Segments (2-cm) of the injured iliac artery from control and treated animals were excised and sections were stained for photomicrograph analysis. Intimal thickening occurred in denuded but not control artery sections. A significant increase in intimal area and decrease in cross-sectional area of lumen in the injured denuded arteries were observed, which was substantially reduced in the 4 mg/kg/day resveratrol treated groups. Comparison of the high dose resveratrol with the control group showed the following: (i) significant inhibition of the intimal proliferation index, (ii) increase in the relative luminal area, and (iii) reduction of SMC count in the thickened intima.

Public health implications. Resveratrol inhibits intimal hyperplasia of injured artery; this observation raises the possibility that resveratrol may be used to prevent and treat restenosis after angioplasty.

1.2.5. *Induced cytoskeletal changes in bovine pulmonary artery endothelial cells by resveratrol and the accompanying modified responses to arterial shear stress*[20]

Rationale. AS is initiated by inciting events leading to EC injury and an increase in EC permeability, we therefore asked whether resveratrol can affect EC integrity by modulating changes in EC cellular structure. Since blood flow induced shear stress on the EC of the blood vessel is a primary signal for arterial remodeling, we also tested whether resveratrol modulates EC structure in response to shear stress.

Experimental approach and key findings. Using BPAEC, we showed that treatment with resveratrol induced EC cell elongation. The EC morphologic changes were dependent on intracellular calcium and tyrosine kinase (not protein kinase C, PKC) activity, and were blocked by cytochalasin D and nocodazole suggesting the involvement of actin microfilaments and microtubule network in the resveratrol-induced cellular phenotype changes. Simulated arterial flow shear stress studies showed that resveratrol induced resistance to stress.[20]

Public health implications. Results of our studies show that resveratrol interacts with ECs *in vitro* and alters the structure/morphology of ECs in ways that render resistance to simulated arterial flow, making ECs less injury-prone to turbulent flow. The observed changes support the interpretation that resveratrol acts as an anti-AS agent.

1.2.6. *Effects of red wine and wine polyphenol resveratrol on platelet aggregation in vivo and in vitro*[21]

Rationale. Interactions between oxidized LDL and platelets leading to macrophage foam cell formation is the hallmark of early events in AS. Since we already demonstrated that resveratrol inhibits LDL oxidation, we asked whether resveratrol might reduce plaque formation via inhibition of platelet aggregation.

Experimental approach and key findings. The effects of resveratrol and red wine on platelet aggregation were investigated. Platelets isolated from healthy male volunteers were used for aggregation studies *in vitro*. Platelet aggregation rate (PAR) induced by collagen, thrombin, and ADP was dose-dependently inhibited by resveratrol. The effects of resveratrol on platelet aggregation in hypercholesterolemic rabbits were also studied. Animals were fed a high cholesterol diet, with and without concurrent gastric feeding of resveratrol (4 mg/kg/day). At 12 weeks post-treatment, the average PAR in platelets isolated from various animal groups was determined. PAR was significantly elevated in the high-cholesterol fed group compared to the normal group. This high fat diet-induced increase in PAR was reduced to control levels by resveratrol, when paired against the high cholesterol fed group. Interestingly, resveratrol treatment had no effect on serum lipid levels. Similar results were observed in animals that received intragastrically Chinese red wine (with or without alcohol, 4 ml/kg/day).

Public health implications. Both *in vitro* and *in vivo* experiments show that resveratrol inhibits platelet aggregation, reinforcing the notion that it exerts cardioprotective effects.

1.2.7. *Effect of red wine and wine polyphenol resveratrol on endothelial function in hypercholesterolemic rabbits*[22]

Rationale. In addition to platelet aggregation, the control of EC function by red wine and wine polyphenol resveratrol was investigated using hypercholesterolemic rabbits.

Experimental approach and key findings. In this study, flow-mediated dilation (FMD) in the femoral artery as a proxy for endothelial function was measured in rabbits fed a regular or high cholesterol diet. Compared to control group, rabbits fed a high-cholesterol diet showed a reduction in FMD while intragastric feeding of resveratrol (3 mg/kg/day), red wine (4 ml/kg/day), dealcoholized red wine (4 ml/kg/day) in hypercholesterolemic rabbits for 12 weeks significantly counteracted the reduction in FMD values.

That resveratrol and red wine improve endothelial function *in vivo* was also supported by significant elevation in plasma NO in parallel with decrease in endothelin-1 (ET-1) levels.

Public health implications. Resveratrol exerts cardioprotection by modulation of EC functions.

1.2.8. *Dealcoholized red wine containing known amounts of resveratrol suppresses atherosclerosis in hypercholesterolemic rabbits without affecting plasma lipid levels*[23]

Rationale. Red wine contains alcohol which could interact with resveratrol to elicit the observed cardioprotective effects. Whether resveratrol has alcohol-independent cardioprotective effects was investigated.

Experimental approach and key findings. Using a hypercholesterolemic rabbit model and resveratrol as a reference, cardioprotective properties of dealcoholized Chinese red wine were compared with alcohol-containing Chinese red wine with comparable amounts of resveratrol. Animals were fed a high cholesterol diet for a 12-week period and were simultaneously given water containing resveratrol (3 mg/kg/day) or regular or dealcoholized red wine (4 ml/kg/day) containing 3.98 mg/L and 3.23 mg/L resveratrol, respectively. Before and after the cholesterol challenge, total high density lipoprotein (HDL)- and LDL-cholesterol and triglyceride levels in the plasma were measured. Atherosclerotic plaques in the thoracic aorta and vascular/endothelial functions in the femoral artery were evaluated using histochemical methods and ultrasonographic image analysis. Compared to rabbits fed a regular diet, a significant increase in plasma levels of total, HDL- and LDL-cholesterol, but not triglycerides were observed in high cholesterol-fed animals. The cholesterol-induced lipid changes were also observed in animals concurrently fed dealcoholized or alcohol-containing red wine, or resveratrol. Atherosclerotic lesions were evident in specimens prepared from the thoracic aorta of high cholesterol-fed animals, but were significantly reduced in rabbits given dealcoholized red wine, red wine, or resveratrol. In addition, an ultrasound analysis of endothelial function showed a reduction of FMD in rabbits fed a high cholesterol diet compared to animals on control diet. This decrease was effectively prevented by the simultaneous exposure to dealcoholized red wine, red wine, or resveratrol.

Public health implications. Rabbits given dealcoholized red wine exhibited effects comparable to those of animals orally administered resveratrol. Accordingly, wine polyphenolics, rather than alcohol present in red wine, suffice to exert cardioprotection.

1.2.9. *Regulation of proliferation and gene expression in cultured human aortic smooth muscle cells by resveratrol and standardized grape extracts*[24]

Rationale. Can cardioprotective effects of resveratrol be reproduced by standardized grape extracts (SGE)?

Experimental approach and key findings. Comparison between resveratrol and SGE from the California Table Grape Commission shows similar growth inhibition and different gene expression profiles in cultured human aortic smooth muscle cells (HASMC), and differences in relaxation of rat carotid arterial rings.

Public health implications. SGE and resveratrol exhibit distinct mechanisms in the regulation of HASMC growth and gene expression.

1.2.10. *Resveratrol prevents monocrotaline-induced pulmonary hypertension in rats*[25]

Rationale. Whether resveratrol exerts protective effects on pulmonary vasculature, vis-à-vis, pulmonary hypertension was investigated.

Experimental approach and key findings. Pulmonary hypertension was established by injection with monocrotaline. Resveratrol treatment (25 mg/kg per day, oral administration, from day 1 post monocrotaline injection) attenuated right ventricular systolic pressure and pulmonary arterial remodeling, decreased expression of inflammatory cytokines including tumor necrosis factor-alpha (TNFα), interleukin 1-beta (IL-1β), IL-6, and platelet-derived growth factor-alpha/beta (PDGF-α/β). In addition, the proliferation of pulmonary arterial SMCs was inhibited by resveratrol. Treatment of rats with resveratrol also increased expression of endothelial NO synthase (eNOS), decreased oxidative stress, and improved endothelial function in small pulmonary arteries. The upregulation of NADPH oxidase associated with pulmonary arterial hypertension was significantly attenuated by resveratrol treatment.

Public health implications. Anti-inflammatory, antioxidant, and anti-proliferative effects exerted by resveratrol may contribute to the prevention of pulmonary hypertension.

1.2.11. *Differential regulation of proliferation, cell cycle control and gene expression in cultured human aortic and pulmonary artery endothelial cells by resveratrol*[26]

Rationale. To determine whether resveratrol exerts similar anti-mitogenic and gene modulatory activities in cells derived from different anatomical sites, using cultured human aortic and pulmonary artery endothelial cells, denoted HAECs and HPAECs, respectively.

Experimental approach and key findings. Time- and dose-dependent growth inhibition by resveratrol was observed in HAECs and HPAECs. Resveratrol induced growth inhibition was accompanied by disruption of cell cycle progression as assayed by flow cytometry. Real-time polymerase chain reaction (RT-PCR) results showed resveratrol inhibited eNOS and preproendothelin-1 (ppET-1) expression in both cell types. Resveratrol induced changes on mRNA levels were further analyzed to discover group gene alterations using human signal transduction pathway finder cDNA array. In HAECs, resveratrol treatment resulted in up regulation of cyclin-dependent kinase inhibitor p57, early growth response protein 1 (egr-1), forkhead box A2 and c-jun, and down regulation of bcl-xl, fibronectin-1, huntingtin interacting protein (HIP), mdm2, p53-inducible gene 3 (PIG3) and WSB1/SWIP-1. In HPAECs, resveratrol induced the expression of cathepsin D, intercellular adhesion molecule 1 (ICAM-1), c-jun and patched 1, and attenuated the expression of caudal type homeobox 1 (CDX1), engrailed homolog 1, fatty acid synthase (FASN), fibronectin-1, forkhead box A2, Hoxa-1, hsp27, PIG3, ELAM-1/E-selectin and WSB1/SWIP-1. Following resveratrol treatments, distinct and overlapping signaling pathways and mechanisms were observed in HAECs and HPAECs.

Public health implications. Cardioprotective effects of resveratrol may be contingent upon or influenced by the vascular bed of origin.

1.2.12. *Resveratrol: A cardioprotective substance*[7]

In this review, we summarize the cardioprotective activities of resveratrol based on our *in vitro, in vivo* and animal studies from 1999–2010.

The primary conclusions of our past studies are: (i) red wine consumption exhibits more CHD-prevention compared to other alcoholic drinks and (ii) cardioprotection may be attributed to grape-derived polyphenols, e.g., resveratrol, in red wine.

We also put forth the hypothesis that quantitative and qualitative differences in expression of resveratrol target protein, e.g., N-ribosyldihydronicotinamide:quinone oxidoreductase (NQO2) may underlie the multifaceted effects of resveratrol in various cell types in the vasculature. We propose a research roadmap as a framework to uncover a mechanism encompassing nongenomic and genomic effects and further insights on cardioprotection by resveratrol.

1.2.13. *What is new for an old molecule? Systematic review and recommendations on the use of resveratrol*[27]

The beneficial health effects of resveratrol supported by scientific data resulted in a set of published recommendations for its human use is the major outcome of the Resveratrol 2010 conference, held in Helsingør, Denmark.

Questions in search of recommendations include: (i) Can resveratrol be recommended in the prevention or treatment of human diseases? (ii) Are there observed "side effects" caused by the intake of resveratrol in humans? (iii) What is the relevant bio-efficacious dose of resveratrol? (iv) What valid data are available regarding an effect of resveratrol in various species of experimental animals? and (v) Which relevant (overall) mechanisms of action of resveratrol have been tested and confirmed?

Animal data showed promises of resveratrol in prevention of coronary heart diseases and other disorders such as cancer and diabetes. The published evidence is not sufficiently strong to justify a recommendation for the administration of resveratrol to humans, beyond the dose which can be obtained from dietary sources. There is a need for clinical trials, particularly regarding its mechanism of action and its safety and toxicology in human subjects.

1.2.14. *Control of eotaxin-1 expression and release by resveratrol and its metabolites in culture human pulmonary artery endothelial cells*[28]

Rationale. Inflammation plays a key role in CHD. However, anti-inflammation activity of resveratrol remains incompletely understood. We studied the effects of resveratrol in the modulation of inflammation.

Experimental approach and key findings. Pro-inflammatory cytokines IL-13 or TNF-α mediated induction of chemokine eotaxin-1 and its release in cultured human pulmonary artery endothelial cells, as confirmed by RT-PCR and eotaxin-1 gene promoter reporter assays were inhibited by resveratrol. The attenuated induction of eotaxin-1 gene expression was coordinated with the reduction of JAK-1, phosphorylated-STAT6 and p65 subunit of NF-κB. The inhibitory effect of resveratrol was also observed using its metabolite piceatannol.

Public health implications. Both resveratrol and its metabolite piceatannol exert anti-inflammatory activity as demonstrated by inhibition of transcription and expression of inflammation biomarker eotaxin-1.

1.2.15. *Resveratrol or its metabolites modulate TNF-α and IL-13 mediated changes in IL-8 and acidic mammalian chitinase (AMCase) in cultured human pulmonary artery endothelial cells*[29]

Rationale. Chronic inflammation is implicated in the etiology of CHD. Thus, the discovery of inflammation biomarkers, identification of agents with anti-inflammation properties, and studies of their cardioprotective mechanisms are of significant interest. We tested the effects of resveratrol and its metabolites in control of inflammatory chemokines IL-8 and AMCase.

Experimental approach and key findings. IL-8 mRNA level was increased following TNF-α treatment while AMCase mRNA was decreased in cells treated with TNF-α (100 ng/ml) or by the combination of IL-13 and TNF-α. Pre-treatment with resveratrol or its metabolite piceatannol significantly reduced the induction of IL-8 by combined treatment of IL-13/TNF-α. The reduced induction of IL-8 was coupled to inhibition of NF-κB p50 subunit mRNA by resveratrol, piceatannol, 4′-O- and 3-O-resveratrol-glucuronide, while NF-κB p65 subunit mRNA expression was only inhibited by 3-O-resveratrol-glucuronide. Cell fractionation and immunoblot analysis showed that combined IL-13/TNF-α induced nuclear translocation of NF-κB p50; the induced changes were down regulated by prior exposure to resveratrol.

Public health implications. The control of IL-8 and AMCase by pro-inflammatory cytokines in HPAEC may be considered as an *in vitro* model for screening and identifying dietary agents, including grape-derived polyphenols, with anti-inflammatory activities and cardioprotective potentials.

1.3. Identification and characterization of QR2 as a distinct RTP playing a role in mechanism of cardioprotection by resveratrol

LSP (ligand select proteomics) is a novel strategy we established to isolate and identify cell type and compartment specific RTPs, and elaboration of their potentials in cardioprotection. The affinity selection step accomplishes three objectives: First, with binding to resveratrol as the basis for inclusion in the analysis, there is a greater probability for a functional link to exist between proteins identified and cellular responses to resveratrol; second, LSP offers simplicity — via a single step concentration — for revealing low-abundance proteins that might go undetected in crude extracts; third, LSP can directly focus on the effects or responses at hand and hence more readily reveals the nature of proteins pertinent to the biology of resveratrol relevant to the question addressed. LSP in combination with MALDI-TOF-MS and cloning, have resulted in the identification of a RTP, denoted as QR2 or NQO2. The identification of RTP is significant; it lends credence to our hypothesis that resveratrol can interact with specific target proteins, which, directly or indirectly might contribute to its multitude of bioactivities including cardioprotection.

1.3.1. *QR2 in complex with resveratrol*

Analysis of crystal of QR2 in complex with resveratrol provides unequivocal evidence that resveratrol binds to the QR2 at the atomic level.[30] Overall, the x-ray crystallographic studies reveal crucial information on: (i) identity of binding domain in QR2 for resveratrol; (ii) the mechanism of interaction between resveratrol and QR2; (iii) residues in the QR2 crucial for effective binding to resveratrol.[30] Thus, resveratrol is shown to bind to the deep active site cavity, formed by hydrophobic side-chains from QR2 and the isoalloxazine ring of the bound co-factor flavin adenine dinucleotide (FAD). Interaction of resveratrol with QR2 is reinforced by hydrophobic and van der Waals forces, as well as by hydrogen bonding between the three hydroxyl groups of resveratrol and functional side chains of specific amino acid residues in QR2. These interactive features not only maintain the planar conformation of resveratrol, but also provide enthalpy for its high affinity binding to QR2. The deep and narrow indentation of the active site cavity fits perfectly the dimension required for binding by trans-resveratrol, and supports the hypothesis that a stilbenoid-binding domain may exist within QR2.

1.3.2. *Mechanism of cardioprotection by resveratrol*

Taking the published data obtained in the study of resveratrol on various aspects of the cardiovascular system as a whole, we propose that resveratrol act as a pleiotropic cellular modulator whose ability to confer cardioprotection may attribute to several possible mechanisms. First, resveratrol may simply be functioning as an antioxidant that attenuates LDL oxidation, and modulates ROS equilibrium to establish cellular redox homeostasis. This could be a significant dampening factor in the mitogenesis of SMC and EC; intracellular ROS such as superoxide and hydrogen peroxide are known to function as signaling molecules and may act as potent molecular drivers for proliferation of aforementioned cells that play an integral role in AS. Whether QR2 plays a role in CHD & AS remains to be established.

1.4. Tackling the atherogenic environment as future directions for studies of cardioprotective effects of resveratrol

1.4.1. *Maturation of AS lesions requires participation of multiple cell types and their dynamic interactions — possible existence of an atherogenic microenvironment*

Because its anatomical location places it in constant contact with chemicals and events with deleterious and/or protective potential, the endothelium is particularly vulnerable to injury, an event which is considered as the key to initiation of CHD. This concept was first articulated in the "response-to-injury" hypothesis by Ross.[31] Accompanying the injury to the endothelium are changes in the vessel architecture, blood components, hemodynamic flow, facilitated recruitment and attachment of leukocytes and platelets, migration of SMCs to the intima, increased assembly and deposition of extracellular matrix, culminating in inflammation.

One might envisage that injured/inflamed endothelium can induce the convergence of immune and other cells at the site of endothelial injury, by mediators and effectors elaborated by the cells *per se* or through their reciprocal interplay with ECs to generate over time an atherogenic microenvironment. Some of the cells involved may be physically juxtaposed to the endothelium thereby affording interaction; for other cells direct contact and engagement with the endothelium must require injury and may be driven by bioactive molecules secreted by the injured EC. This view of EC as a microcosm of interactive cells presents a model for understanding and elucidating the details of cell:cell communication, as pertaining to influence by cardioprotective dietary chemicals, e.g., resveratrol.

1.4.2. *The functional significance of cellular heterogeneity in the vascular system using single-cell profiling analysis*

EC and SMC are two major blood vessel cell types which are necessary for normal circulation and play an important role in various physiological processes, such as, angiogenesis, blood pressure control, coagulation, and inflammation. Thus the study of EC and SMC is crucial for understanding the proposed atherogenic microenvironment in the prevention of cardiovascular diseases.

The endothelial lining of blood vessels is known to show significant structural and functional heterogeneity.[32] Similarly SMCs are also phenotypically diverse depending on the type of vessel and organs in which they are located.[33] Therefore, analysis of EC and SMC isolated from tissues relevant to the research is essential for unraveling specific pathologies including atherogenic microenvironment in coronary artery disease.

Gene expression analysis as an approach to understanding of AS has been studied using bulk cDNA of pathologic lesions. Comparison of transcription profiles by measuring average of gene expression levels across the entire population of cells provide limited information since the interpretation of data must rely on the assumption that all cells of a given population are equal, or at least comparable, with respect to their gene expression repertoire. Such information may be predisposed to bias. As an example, two populations of 1,000 cells each may show a tenfold difference in the mRNA of a particular gene, but this change could result from every cell expressing ten times as much mRNA, or from a single cell expressing 10,000 times more of that mRNA. The difference between these two possible scenarios could have enormous phenotypic consequences in the context of pathology of the disease.

Limited data on AS are available with respect to gene expression at the single cell level. Individual cell approaches, compared to population cell studies, might easily distinguish between the scenarios mentioned above; moreover, since fewer cells are required the single cell approach can be used to interrogate the phenotypes of rare cells.

Whether inflamed subpopulation of ECs might be endowed with a specific gene expression profile repertoire, is unclear. Inflamed EC shows upregulation of adhesion molecules and pro-inflammatory mediators which are crucial for the pathogenesis of inflammatory vascular disease.[34] It is notable that vascular SMC also undergo phenotypical changes in response to external stimuli, resulting in a dedifferentiated state characterized by downregulation of contractile proteins and upregulation of pro-proliferative, pro-migratory and pro-inflammatory markers as well as enhanced matrix synthesis.[35,36] Dissection and understanding gene expression repertoire differences between healthy and diseased vascular cells will be essential for the design and development of preventive strategies selectively targeting the latter.

Experimentally, single cells may be isolated by laser capture microdissection, fluorescence-activated cell sorting, or microfluidic-based approaches. The acquired single cells can be immediately frozen and stored at −80°C until analysis. The analysis of tens to hundreds of genes in single cells may be extended to whole transcriptome assays using microarrays or next-generation sequencing.[37]

1.4.3. *Microparticles (MPs) and thymidine phosphorylase (TP) may also play a role in atherogenesis*

AS is initiated with injury to the endothelium where multiple cell types interact in an atherogenic microenvironment. These include macrophages, dendritic cells and T lymphocytes, which collectively elaborates the secretion of interferon (IFN)-α and β and pro-inflammatory mediators such as IFN-γ to facilitate the recruitment/attachment of macrophages to the endothelium for eventual migration into the intima.[38] During AS progression, the injured endothelium assumes a reactive state capable of changing the phenotype of neighboring cells. What are the molecular bases by which reactive endothelium forms, and how does cell:cell communication occur in the endothelial microenvironment? One possibility warranting investigation in the coordinated shedding of endothelial microparticles (EMPs) and angiogenic factor TP from the damaged EC.

Release of MPs is characteristic of all eukaryotic cells and has been described in several pathologies including AS.[39,40] MPs have also been implicated in the regulation of vascular functions.[41–43] EMP release is significantly correlated with acute coronary syndrome and AS.[39,40] Similarly, MPs can also be released by SMC.[44–46] Conceivably, because MPs are heterogeneous and are released by diverse cell types, it is possible that the phenotype of circulating MPs may depend on different pathological settings. Further, different MPs released by the endothelium may also exert differential pleiotropic effect on the vascular wall. To function as distinct, jettisoned particulate, cellular entities in AS lesions, shed EMPs are packaged with bioactive vascular effectors that can be transcellularly delivered to a multitude of cell types either directly juxtaposed with the injured endothelium or are recruited to the site of injury.[47–49] Released EMPs have been shown to regulate angiogenesis. Because TP has been identified as an important and potent angiogenic factor, as well as show overexpression in AS lesions,[50,51] we hypothesize that EMPs and TP play an integral role in affecting cellular events of both injured ECs and neighboring cells. This co-regulation of the interplay of these cell types may in part attributed to the multiple bioactive molecules present in EMPs shed by injured EC. As support, it is notable that direct injection of a plasmid vector encoding TP into ischemic canine myocardium or rabbit hind limb promotes angiogenesis in cardiac and skeletal muscle[52,53] and improves myocardial function and prevents hind limb necrosis. Thymidine phosphorylase has also been reported to inhibit vascular smooth muscle cell proliferation and migration.[54,55] Both laminar shear stress (5–20 dyne/cm^2) and resveratrol (20 μM) inhibit the release of EMPs via a sirtuin 1 (sirt1)-mediated increase in mitochondrial biogenesis.[56]

1.4.4. *Mast cell recruitment and activation of AS and prevention by resveratrol*

Inflammation plays an important role in the initiation and progression of AS. As mentioned, AS is initiated by the arterial EC in response to injury and ultimately manifested as atheromatous or fibro-fatty plaque. EC damage results in recruitment of multiple cell types and generation of inflammatory cytokines and chemokines. Recent evidence indicates that mast cells and their products play a key role in inflammation and AS.[57] Sirt1 has been reported to play a role in inflammation, for example, by negatively regulating FcεR1-mediated mast cell activation,[58] in allergy diseases. Pharmacologic doses of resveratrol may exert its cardioprotective effect by activating sirt1. Recent studies have provided support for therapeutic effects of resveratrol on allergic symptoms in both humans and rodents.[58] Details of mast cell recruitment and activation in AS and its prevention by resveratrol could be a fertile area for scientific investigation in the future.

1.4.5. *Cardiomyocyte regeneration using resveratrol and diet-based strategy*

Cardioprotection should encompass consideration of ways to curtail destruction of heart muscle cells known as cardiomyocytes. Cardiomyocytes play an integral role in normal and pathophysiological growth responses in the cardiovascular system, and are considered terminally differentiated shortly after formation of the heart. The adage that cardiomyocytes are restricted from entry into cell replication and mitosis in mammals, i.e., totally devoid of regeneration capabilities, has been a limiting factor in the development of strategies and drugs to combat myocardial pathologies. As such, prevention of cardiomyocyte cell death and discovery of ways to reactivate their silenced regeneration state, for example, by inducing the reentry into cell cycling, are both considered novel approaches since they are expected to enlarge the pool of living cells as well as afford the replacement of dead or dysfunctional cells.

Are cardiomyocytes totally and irreversibly prevented from undergoing cell regeneration? And how valid is the seemingly timeless biological dogma that the majority of human cells, once reaching specialization status, become permanently banned from cell division or acquisition of a different phenotype. In recent years, the viewpoint of total regeneration impotence for differentiated cells has undergone significant change and transformation, based on a number of new findings: first, through the discovery of endowed innate stem cells, damaged differentiated cells are now known to be capable of recreation, replacement and expansion; secondly, the apparent irrevocable dormant, locked-down state of specialized cells can be cued to re-enter the highly precise and dynamically regulated process of cell cycling, division, and proliferation. These hitherto unanticipated cornerstone features of tissue/cell regeneration have also been realized and accepted in cardiovascular and heart research, thus offering increasingly larger windows of hope that disease of the heart, involving an organ regarded as absolutely refractory to regeneration, may now be open to therapeutic intervention, rejuvenation and healing. An area warranting further testing is whether grape bioactive components, acting alone or in combination, have the potential to induce the regeneration of cardiomyocytes. This novel area of investigation is supported by resveratrol and grape constituents fulfilling several key aspects required for cardiomyocyte mitotic division in the new niche of the regenerating heart. First, the elegant studies of Engel and Keating[59,60] pioneered the notion that the restrictive ability of cardiomyocytes to undergo proliferation and hence regeneration may be readily reversed by inhibiting an enzyme known as p38 MAP kinase, which suppresses cardiomyocyte replication. Of note, resveratrol, acting as an agonist of sirt1, has been reported to inhibit p38 MAP kinase in neonatal rat cardiomyocytes.[61] Secondly, improvements in both cardiomyocyte

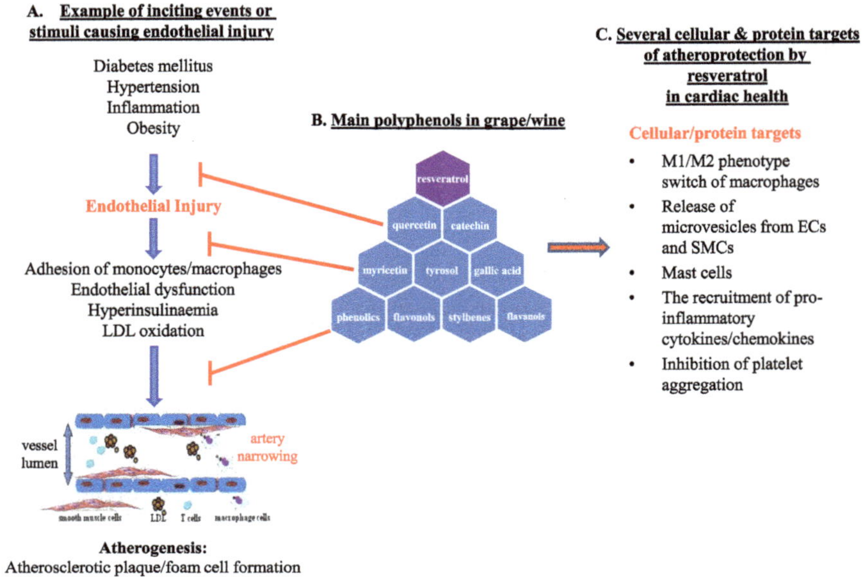

Fig. 1.2. **A**. A scheme showing several initiating stimuli or events leading to endothelial injury and the multitude of endothelial injury, AS-promoting changes that culminate in the establishment of atherogenesis. **B**. Atherogenesis and its sequence are postulated to be counteracted by grape polyphenol resveratrol, alone and in combination with other grape or wine phytochemicals, conferring cardioprotection. **C**. A summary of cellular events and protein targets for future investigation on atheroprotection by resveratrol in cardiac health.

proliferation and heart function are predicated on cardiomyocyte replication occurring in concomitance with increases in telomerase activity. It is noteworthy that resveratrol has been found to increase telomerase activity,[62–64] a key enzyme involved in cell and tissue regeneration.

1.5. Conclusion

Figure 1.2 summarizes the events leading up to CHD and prevention by resveratrol. First and foremost, CHD is a major yet preventable cause of death; a progressive reduction in CHD-related mortalities has been observed in the U.S over the past several decades.

Secondly, a significant inverse relationship between low/moderate consumption of red wine and the high risk for CHD reported in France, known as the French paradox, may be explained by polyphenols and phytochemicals contained in red wine, notably, trans-resveratrol.

Moreover, a novel affinity selection proteomic (LSP) approach to detect, select, and characterize cellular targets of resveratrol, using cultured HAEC and SMC, both of which have shown an integral role in the development of CHD, has identified QR2 as a distinct RTP.

Lastly, the mechanism of cardioprotection by resveratrol spans multiple cell types — platelets, endothelial and smooth muscle cells, whose dysfunctions contribute to clinical CHD. We show that resveratrol inhibits LDL oxidation, platelet aggregation, suppress proliferation of SMC and pulmonary aortic EC, and induces eNOS levels. Given orally, resveratrol also significantly reduced intimal thickening of denuded endothelium in rabbits. Further, its effects persisted in rabbits given alcohol-free red wine showing that wine polyphenolics *per se* in red wine accounted for much of the observed cardioprotective effects of resveratrol. We also propose areas of future investigation that may lead to new hypothesis-driven animal model studies and human clinical trials.

Acknowledgments

Support for cardioprotection by resveratrol: California Table Grape Commission, Phillip Morris USA Inc. and Phillip Morris International (JMW), and NYMC Intramural Research Award (TCH).

References

1. Soleas GJ, Diamandis EP, Goldberg DM. Resveratrol: a molecule whose time has come? And gone? *Clin Biochem* 1997; **30**: 91–113.
2. Jang M, Cai L, Udeani GO, Slowing KV, Thomas CF, Beecher CW, Fong HH, Farnsworth NR, Kinghorn AD, Mehta RG, Moon RC, Pezzuto JM. Cancer chemopreventive activity of resveratrol, a natural product derived from grapes. *Science* 1997; **275**: 218–20.
3. Soleas GJ, Diamandis EP, Goldberg DM. The world of resveratrol. *Adv Exp Med Biol* 2001; **492**: 159–82.
4. Renaud S, de Lorgeril M. Wine, alcohol, platelets, and the French paradox for coronary heart disease. *Lancet* 1992; **339**: 1523–6.

5. Wu JM, Wang ZR, Hsieh TC, Bruder JL, Zou JG, Huang YZ. Mechanism of cardioprotection by resveratrol, a phenolic antioxidant present in red wine (Review). *Int J Mol Med* 2001; **8**: 3–17.
6. Bertelli AA, Das DK. Grapes, wines, resveratrol, and heart health. *J Cardiovasc Pharmacol* 2009; **54**: 468–76.
7. Wu JM, Hsieh TC. Resveratrol: a cardioprotective substance. *Ann N Y Acad Sci* 2011; **1215**: 16–21.
8. Biagi M, Bertelli AA. Wine, alcohol and pills: What future for the French paradox? *Life Sci* 2015; **131**: 19–22.
9. Criqui MH, Ringel BL. Does diet or alcohol explain the French paradox? *Lancet* 1994; **344**: 1719–23.
10. Li H, Xia N, Förstermann U. Cardiovascular effects and molecular targets of resveratrol. *Nitric Oxide* 2012; **26**: 102–10.
11. Zordoky BN, Robertson IM, Dyck JR. Preclinical and clinical evidence for the role of resveratrol in the treatment of cardiovascular diseases. *Biochim Biophys Acta* 2015; **1852**: 1155–77.
12. Raj P, Zieroth S, Netticadan T. An overview of the efficacy of resveratrol in the management of ischemic heart disease. *Ann N Y Acad Sci* 2015; **1348**: 55–67.
13. Zou JG, Huang YZ, Chen Q, Wei EH, Hsieh TC, Wu JM. Resveratrol inhibits copper ion-induced and azo compound-initiated oxidative modification of human low density lipoprotein. *Biochem Mol Biol Int* 1999; **47**: 1089–96.
14. Hsieh TC, Juan G, Darzynkiewicz Z, Wu JM. Resveratrol increases nitric oxide synthase, induces accumulation of p53 and p21(WAF1/CIP1), and suppresses cultured bovine pulmonary artery endothelial cell proliferation by perturbing progression through S and G2. *Cancer Res* 1999; **59**: 2596–601.
15. Gauthier TW, Scalia R, Murohara T, Guo JP, Lefer AM. Nitric oxide protects against leukocyte-endothelium interactions in the early stages of hypercholesterolemia. *Arterioscler Thromb Vasc Biol* 1995; **15**: 1652–9.
16. Radomski MW, Palmer RM, Moncada S. The role of nitric oxide and cGMP in platelet adhesion to vascular endothelium. *Biochem Biophys Res Commun* 1987; **148**: 1482–9.
17. Zeiher AM, Fisslthaler B, Schray-Utz B, Busse R. Nitric oxide modulates the expression of monocyte chemoattractant protein 1 in cultured human endothelial cells. *Circ Res* 1995; **76**: 980–6.
18. Zou J, Huang Y, Chen Q, Wang N, Cao K, Hsieh TC, Wu JM. Suppression of mitogenesis and regulation of cell cycle traverse by resveratrol in cultured smooth muscle cells. *Int J Oncol* 1999; **15**: 647–51.
19. Zou J, Huang Y, Cao K, Yang G, Yin H, Len J, Hsieh TC, Wu JM. Effect of resveratrol on intimal hyperplasia after endothelial denudation in an experimental rabbit model. *Life Sci* 2000; **68**: 153–63.
20. Bruder JL, Hsieh T, Lerea KM, Olson SC, Wu JM. Induced cytoskeletal changes in bovine pulmonary artery endothelial cells by resveratrol and the accompanying modified responses to arterial shear stress. *BMC Cell Biol* 2001; **2**: 1.

21. Wang Z, Huang Y, Zou J, Cao K, Xu Y, Wu JM. Effects of red wine and wine polyphenol resveratrol on platelet aggregation in vivo and in vitro. *Int J Mol Med* 2002; **9**: 77–9.
22. Zou JG, Wang ZR, Huang YZ, Cao KJ, Wu JM. Effect of red wine and wine polyphenol resveratrol on endothelial function in hypercholesterolemic rabbits. *Int J Mol Med* 2003; **11**: 317–20.
23. Wang Z, Zou J, Cao K, Hsieh TC, Huang Y, Wu JM. Dealcoholized red wine containing known amounts of resveratrol suppresses atherosclerosis in hypercholesterolemic rabbits without affecting plasma lipid levels. *Int J Mol Med* 2005; **16**: 533–40.
24. Wang Z, Chen Y, Labinskyy N, Hsieh TC, Ungvari Z, Wu JM. Regulation of proliferation and gene expression in cultured human aortic smooth muscle cells by resveratrol and standardized grape extracts. *Biochem Biophys Res Commun* 2006; **346**: 367–76.
25. Csiszar A, Labinskyy N, Olson S, Pinto JT, Gupte S, Wu JM, Hu F, Ballabh P, Podlutsky A, Losonczy G, de Cabo R, Mathew R, *et al.* Resveratrol prevents monocrotaline-induced pulmonary hypertension in rats. *Hypertension* 2009; **54**: 668–75.
26. Hsieh TC, Lu X, Guo J, Wu JM. Differential regulation of proliferation, cell cycle control and gene expression in cultured human aortic and pulmonary artery endothelial cells by resveratrol. *Int J Mol Med* 2010; **26**: 743–9.
27. Vang O, Ahmad N, Baile CA, Baur JA, Brown K, Csiszar A, Das DK, Delmas D, Gottfried C, Lin HY, Ma QY, Mukhopadhyay P, *et al.* What is new for an old molecule? Systematic review and recommendations on the use of resveratrol. *PLoS One* 2011; **6**: e19881.
28. Yang CJ, Lin CY, Hsieh TC, Olson SC, Wu JM. Control of eotaxin-1 expression and release by resveratrol and its metabolites in culture human pulmonary artery endothelial cells. *Am J Cardiovasc Dis* 2011; **1**: 16–30.
29. C.J. Y, Hsieh TC, S.C. O, Wu JM. Resveratrol or its metabolites modulate TNF-alpha and IL-13 mediated changes in IL-8 and acidic mammalian chitinase in cultured human pulmonary artery endothelial cells, vol. 1 Journal of Cardiology and Vascular Medicine, 2013.
30. Buryanovskyy L, Fu Y, Boyd M, Ma Y, Hsieh TC, Wu JM, Zhang Z. Crystal structure of quinone reductase 2 in complex with resveratrol. *Biochemistry* 2004; **43**: 11417–26.
31. Ross R, Glomset J, Harker L. Response to injury and atherogenesis. *Am J Pathol* 1977; **86**: 675–84.
32. Aird WC. Endothelial cell heterogeneity. *Cold Spring Harb Perspect Med* 2012; **2**: a006429.
33. Zakharova IS, Zhiven' MK, Saaya SB, Shevchenko AI, Smirnova AM, Strunov A, Karpenko AA, Pokushalov EA, Ivanova LN, Makarevich PI, Parfyonova YV, Aboian E, *et al.* Endothelial and smooth muscle cells derived from human cardiac explants demonstrate angiogenic potential and suitable for design of cell-containing vascular grafts. *J Transl Med* 2017; **15**: 54.

34. Huo Y, Ley K. Adhesion molecules and atherogenesis. *Acta Physiol Scand* 2001; **173**: 35–43.
35. Owens GK. Regulation of differentiation of vascular smooth muscle cells. *Physiol Rev* 1995; **75**: 487–517.
36. Mack CP. Signaling mechanisms that regulate smooth muscle cell differentiation. *Arterioscler Thromb Vasc Biol* 2011; **31**: 1495–505.
37. Stegle O, Teichmann SA, Marioni JC. Computational and analytical challenges in single-cell transcriptomics. *Nat Rev Genet* 2015; **16**: 133–45.
38. Ilhan F, Kalkanli ST. Atherosclerosis and the role of immune cells. *World J Clin Cases* 2015; **3**: 345–52.
39. Mallat Z, Hugel B, Ohan J, Lesèche G, Freyssinet JM, Tedgui A. Shed membrane microparticles with procoagulant potential in human atherosclerotic plaques: a role for apoptosis in plaque thrombogenicity. *Circulation* 1999; **99**: 348–53.
40. Mallat Z, Benamer H, Hugel B, Benessiano J, Steg PG, Freyssinet JM, Tedgui A. Elevated levels of shed membrane microparticles with procoagulant potential in the peripheral circulating blood of patients with acute coronary syndromes. *Circulation* 2000; **101**: 841–3.
41. Diamant M, Tushuizen ME, Sturk A, Nieuwland R. Cellular microparticles: new players in the field of vascular disease? *Eur J Clin Invest* 2004; **34**: 392–401.
42. Brodsky SV, Zhang F, Nasjletti A, Goligorsky MS. Endothelium-derived microparticles impair endothelial function in vitro. *Am J Physiol Heart Circ Physiol* 2004; **286**: H1910-5.
43. Martínez MC, Tesse A, Zobairi F, Andriantsitohaina R. Shed membrane microparticles from circulating and vascular cells in regulating vascular function. *Am J Physiol Heart Circ Physiol* 2005; **288**: H1004-9.
44. Brisset AC, Terrisse AD, Dupouy D, Tellier L, Pech S, Navarro C, Sié P. Shedding of active tissue factor by aortic smooth muscle cells (SMCs) undergoing apoptosis. *Thromb Haemost* 2003; **90**: 511–8.
45. Essayagh S, Brisset AC, Terrisse AD, Dupouy D, Tellier L, Navarro C, Arnal JF, Sié P. Microparticles from apoptotic vascular smooth muscle cells induce endothelial dysfunction, a phenomenon prevented by beta3-integrin antagonists. *Thromb Haemost* 2005; **94**: 853–8.
46. Jia LX, Zhang WM, Li TT, Liu Y, Piao CM, Ma YC, Lu Y, Wang Y, Liu TT, Qi YF, Du J. ER stress dependent microparticles derived from smooth muscle cells promote endothelial dysfunction during thoracic aortic aneurysm and dissection. *Clin Sci (Lond)* 2017; **131**: 1287–99.
47. Ahn YS. Cell-derived microparticles: 'Miniature envoys with many faces'. *J Thromb Haemost* 2005; **3**: 884–7.
48. Abid Hussein MN, Nieuwland R, Hau CM, Evers LM, Meesters EW, Sturk A. Cell-derived microparticles contain caspase 3 in vitro and in vivo. *J Thromb Haemost* 2005; **3**: 888–96.

49. Morel O, Toti F, Hugel B, Freyssinet JM. Cellular microparticles: a disseminated storage pool of bioactive vascular effectors. *Curr Opin Hematol* 2004; **11**: 156–64.
50. Boyle JJ, Wilson B, Bicknell R, Harrower S, Weissberg PL, Fan TP. Expression of angiogenic factor thymidine phosphorylase and angiogenesis in human atherosclerosis. *J Pathol* 2000; **192**: 234–42.
51. Brill A, Dashevsky O, Rivo J, Gozal Y, Varon D. Platelet-derived microparticles induce angiogenesis and stimulate post-ischemic revascularization. *Cardiovasc Res* 2005; **67**: 30–8.
52. Li W, Tanaka K, Ihaya A, Fujibayashi Y, Takamatsu S, Morioka K, Sasaki M, Uesaka T, Kimura T, Yamada N, Tsuda T, Chiba Y. Gene therapy for chronic myocardial ischemia using platelet-derived endothelial cell growth factor in dogs. *Am J Physiol Heart Circ Physiol* 2005; **288**: H408-15.
53. Yamada N, Li W, Ihaya A, Kimura T, Morioka K, Uesaka T, Takamori A, Handa M, Tanabe S, Tanaka K. Platelet-derived endothelial cell growth factor gene therapy for limb ischemia. *J Vasc Surg* 2006; **44**: 1322–8.
54. Li W, Tanaka K, Morioka K, Uesaka T, Yamada N, Takamori A, Handa M, Tanabe S, Ihaya A. Thymidine phosphorylase gene transfer inhibits vascular smooth muscle cell proliferation by upregulating heme oxygenase-1 and p27KIP1. *Arterioscler Thromb Vasc Biol* 2005; **25**: 1370–5.
55. Yue H, Tanaka K, Furukawa T, Karnik SS, Li W. Thymidine phosphorylase inhibits vascular smooth muscle cell proliferation via upregulation of STAT3. *Biochim Biophys Acta* 2012; **1823**: 1316–23.
56. Kim JS, Kim B, Lee H, Thakkar S, Babbitt DM, Eguchi S, Brown MD, Park JY. Shear stress-induced mitochondrial biogenesis decreases the release of microparticles from endothelial cells. *Am J Physiol Heart Circ Physiol* 2015; **309**: H425–33.
57. Conti P, Shaik-Dasthagirisaeb Y. Atherosclerosis: a chronic inflammatory disease mediated by mast cells. *Cent Eur J Immunol* 2015; **40**: 380–6.
58. Li X, Lee YJ, Jin F, Park YN, Deng Y, Kang Y, Yang JH, Chang JH, Kim DY, Kim JA, Chang YC, Ko HJ, *et al.* Sirt1 negatively regulates FcεRI-mediated mast cell activation through AMPK- and PTP1B-dependent processes. *Sci Rep* 2017; **7**: 6444.
59. Engel FB, Schebesta M, Duong MT, Lu G, Ren S, Madwed JB, Jiang H, Wang Y, Keating MT. p38 MAP kinase inhibition enables proliferation of adult mammalian cardiomyocytes. *Genes Dev* 2005; **19**: 1175–87.
60. Engel FB, Hsieh PC, Lee RT, Keating MT. FGF1/p38 MAP kinase inhibitor therapy induces cardiomyocyte mitosis, reduces scarring, and rescues function after myocardial infarction. *Proc Natl Acad Sci U S A* 2006; **103**: 15546–51.
61. Ruan Y, Dong C, Patel J, Duan C, Wang X, Wu X, Cao Y, Pu L, Lu D, Shen T, Li J. SIRT1 suppresses doxorubicin-induced cardiotoxicity by regulating the oxidative stress and p38MAPK pathways. *Cell Physiol Biochem* 2015; **35**: 1116–24.

62. Pearce VP, Sherrell J, Lou Z, Kopelovich L, Wright WE, Shay JW. Immortalization of epithelial progenitor cells mediated by resveratrol. *Oncogene* 2008; **27**: 2365–74.
63. Wang XB, Zhu L, Huang J, Yin YG, Kong XQ, Rong QF, Shi AW, Cao KJ. Resveratrol-induced augmentation of telomerase activity delays senescence of endothelial progenitor cells. *Chin Med J (Engl)* 2011; **124**: 4310–5.
64. Yamashita S, Ogawa K, Ikei T, Udono M, Fujiki T, Katakura Y. SIRT1 prevents replicative senescence of normal human umbilical cord fibroblast through potentiating the transcription of human telomerase reverse transcriptase gene. *Biochem Biophys Res Commun* 2012; **417**: 630–4.

Chapter 2

Interrogating and Mining Mechanisms and Targets of Chemoprevention by Resveratrol

Tze-chen Hsieh, Barbara B. Doonan and Joseph M. Wu

Biochemistry and Molecular Biology,
New York Medical College,
Valhalla, NY 10595, USA

Resveratrol is a grape polyphenol that consistently shows preventative activity against malignant diseases, a major cause of morbidity and mortality in the US. We embarked on studies of chemoprevention of prostate cancer (CaP) by resveratrol, stemming from the seminal observation by Pezzuto and coworkers showing that this polyphenol exhibits inhibitory activities spanning the initiation, promotion, and progression stages of carcinogenesis.

This chapter begins with an overview and perspective on our strategy and approaches to provide evidence for bioefficacy of resveratrol in conferring protection against malignant diseases. Results of tissue culture experiments, animal studies, and exploratory human investigations that address issues of dose response, bioavailability and structure-function relationship, for correlation with biological effects and therapeutic potential of resveratrol are summarized. Next, the rationale for our hypothesis that resveratrol exerts its bioactivities via interaction with distinct cellular targets is discussed. The ability of resveratrol to bind and interact with specific proteins, denoted resveratrol targeting proteins (RTPs), in responsive cells is supported by its chemical structure — a stilbene core with defined hydrophilic side groups. Since distinct RTPs may reside in different cellular compartments and display quantitative and qualitative variations in their intracellular distribution, they could help explain the highly variable dose-dependent and cell type specific effects resveratrol reportedly exerts in different cell types and systems.

We then focus on our studies designed to identify and purify RTPs, by using an affinity approach whereby resveratrol is covalently linked to epoxy-activated agarose, generating a biospecific platform for enriching, purifying, and profiling RTPs. Promising results have been obtained using the resveratrol affinity chromatography approach, as examplified by the discovery of quinone reductase 2 (QR2) as a confirmed RTP isolated from various types of cultured cancer cells.

To show that QR2 is mechanistically linked to the chemopreventive attributes of resveratrol, we resorted to use of QR2-knockdown and over-expression tissue culture cells, and purification and characterization of QR2 interaction proteins and co-regulators. Lastly, we present the thesis that QR2, in coordination with quinone reductase 1 (QR1), controls the tumor suppressor protein p53. We postulate that resveratrol, through ability to complex QR2 and induce QR1, modulates the stability of p53, thereby contributing to a p53-dependent role in growth suppression and prostate specific antigen (PSA) regulation in CaP cells. Complex formation between resveratrol and QR2 may favor a reductive cellular environment restrictive to the progression of carcinogenesis.

2.1. Introduction

Despite progress made on detection and diagnosis of cancer in its early stages and some success in the treatment of blood-borne cancers, there has been no significant change in age-adjusted death rates for most cancers since the 1960s. In the United States cancer may have surpassed heart disease as the leading cause of death among the elderly; moreover, cancer has become an escalating cause of human deaths worldwide. In clinical and basic science circles, however, it is generally accepted that no "magic bullet" exists for cancer eradication. Therefore, in concept and practice, a prevailing goal is to manage cancers to enable patients to die with instead of from the disease. The trend, statistics and scientific perception clearly underscore the importance of cancer prevention; equally imperative is a readily implementable disease-preventative strategy.

Use of diet-based approaches and discovery of chemopreventive agents are founded on the pioneering work of Sporn in the 1970s, broadly defined as the use of natural or synthetic substances to reduce the risk of cancer development or the appearance of an incurable cancerous state, by inhibiting, reversing, and delaying the multistep carcinogenic process.[1–4] In the ensuing decades numerous chemopreventive agents have been identified, isolated and purified from plants, fruits and vegetables. Resveratrol is a grape-derived phytochemical that has received paramount attention as a preventative agent against cancer. Concerted research efforts have demonstrated that resveratrol utilizes multiple mechanisms to affect the reduction in cancer risks; however, significant open questions and gaps in its mechanism of action remain.

2.2. Resveratrol and chemoprevention

Our laboratory embarked on testing resveratrol as a chemopreventive agent in CaP stemming from the report in 1997 by Pezzuto and coworkers showing that this polyphenol exhibits inhibitory activities on initiation and promotion of hydrocarbon-induced skin tumor, and the progression of breast cancer in mice.[5] Publication of this seminal study was followed by many studies showing chemopreventive effects of resveratrol in various cancer cell lines.

Exposure to resveratrol results in differentiation of human promyelocytic leukemia cells.[6] Resveratrol induces apoptosis in numerous cancer cell types, including hormone-sensitive LNCaP,[7,8] hormone-insensitive DU145[9,10] CaP cells, mouse myeloid cells,[11] human B-cell leukemia cells,[12] and cells representing cancer of various anatomical origin, e.g., MCF-7, SW-480, HCE-7, Seg-1, HL-60,[13] human U251[14] and rat C6 glioma cells.[15] Induction of apoptosis by resveratrol was shown to involve multiple mechanisms including loss of mitochondrial function, release of cytochrome C, altered expression of the bcl-2 protein family, and activation of caspases, as validated by studies using U-937,[16] pancreatic carcinoma,[17] mouse myeloid leukemia,[11] normal and leukemia,[12] and Caco-2 colon carcinoma cells.[18] Moreover, both p53-independent mechanism, as evident by studies in HCT116 colon carcinoma cells,[19] and p53-dependent mechanism, as shown in JB6 epidermal[20] and HepG2 cells,[21] contribute to the induction of apoptosis by resveratrol. Resveratrol also blocks cell invasion[22] by mechanisms distinct from its anti-proliferative activities.[23] Resveratrol was also reported to suppress angiogenesis and glioma growth in rats.[24] Furthermore, dietary administration of resveratrol was found to suppress 7,12-Dimethylbenz[a]anthracene (DMBA)-induced mammary cancer in rats[25] while oral ingestion of resveratrol reduced the incidence and tumor size of premalignant and malignant colon lesions.[26,27] The combination of *in vitro* data and *in vivo* evidence in animals generally support the view that resveratrol, as a specific dietary ingredient, can function as a deterrent for malignant diseases. Since resveratrol is present in a variety of foods consumed by the public at large,[5,28,29] its feasibility and efficacy as a cancer preventive agent in human populations was investigated and supported by the studies of Brown and coworkers testing and showing that oral feeding of resveratrol to colon cancer patients significantly inhibited tumor cell proliferation in resected colon tissues. Moreover, these effects were correlated statistically with HPLC/UV/mass spectrometry authenticated resveratrol and 3-O-/4′-O-glucuronide and sulfate metabolites, at levels comparable to its anti-tumorigenic activity in tissue culture studies.[30] These results collectively lend support that resveratrol is a chemopreventive agent for colorectal and possibly other cancers.

In this chapter we will review our approaches and perspectives from 1999–2017 to elucidate mechanisms and targets and the translational implication of resveratrol in cancer prevention and therapy (Figure 2.1). We will focus on three areas: (i) early efforts on activity of resveratrol against several cancer types, (ii) identification of resveratrol target proteins, and (iii) building an integrative platform on studies of resveratrol as exemplified by functional activity network (FAN) analysis.

2.3. Chemopreventive studies of resveratrol from this laboratory: Early years

Anti-cancer activity elicited by resveratrol was first demonstrated[5] on skin and breast cancer in 1997. Our first 5 year efforts were focused on screening the anti-cancer activity of resveratrol in different cancer types. In 1999, we first reported the anti-CaP activity of resveratrol; currently, this study has received ~400 citations. We

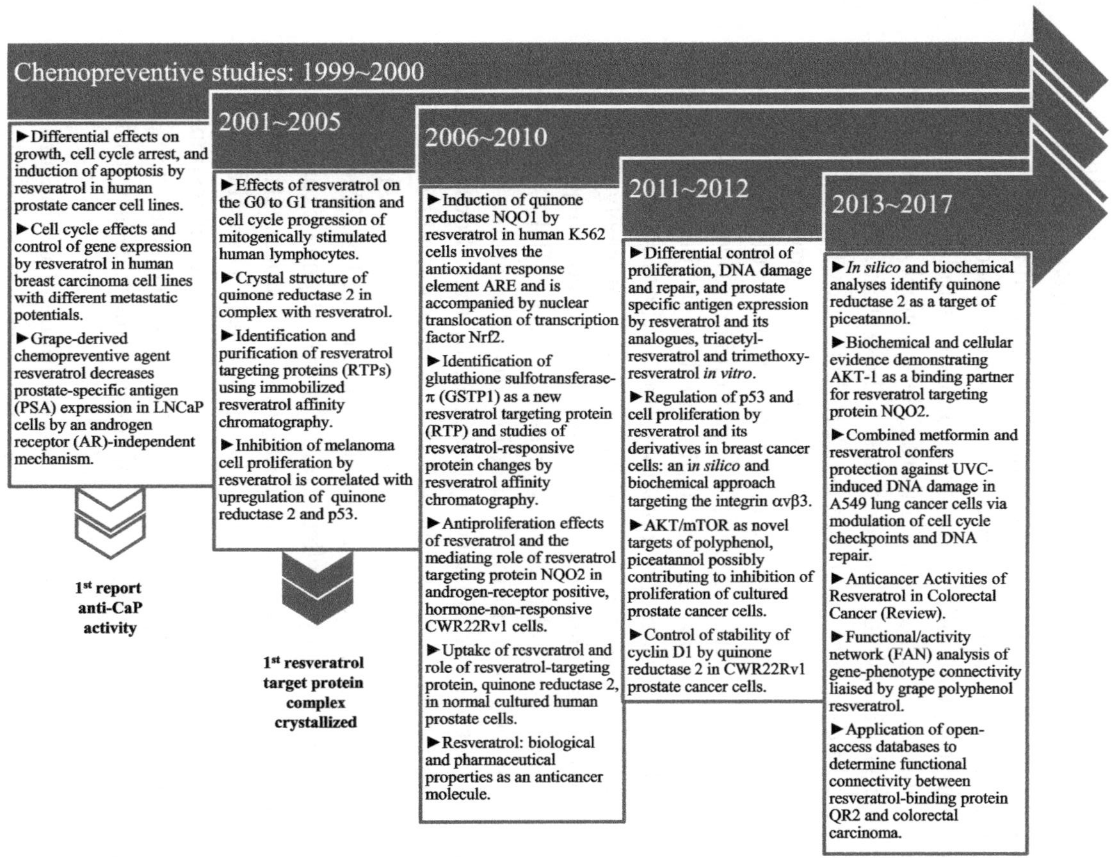

Fig. 2.1. A timeline of our published studies on chemoprevention by resveratrol.

showed that resveratrol suppressed both AD (androgen-dependent)-LNCaP and AI (androgen-independent, DU145 and PC-3) cell growth.[9] Growth inhibition by resveratrol was accompanied by partial disruption of G_1/S transition in AI cell lines. In LNCaP cells, resveratrol suppressed mitogenesis, induced apoptosis, and significantly lowered both intracellular and secreted PSA levels.[8,9] At $\leq 25\ \mu M$, the PSA-lowering effects of resveratrol occurred by an AR (androgen receptor)-independent mechanism, similar in magnitude to that observed using bioengineered AR-monoclonal antibody or AR-targeted ribozyme[24] approaches. These results suggest that resveratrol negatively modulates CaP cell growth, by affecting mitogenesis as well as inducing apoptosis, in a prostate cell-type-specific manner. Importantly, resveratrol also regulates PSA gene expression by an AR-independent mechanism. Because CaP has a slow natural history with a significant latency before it reaches end-stage HRPC (hormone refractory CaP), these results raise the possibility that resveratrol may be considered as an easily compliant, broad-based dietary deterrent that blocks one or more requisite genotypic and phenotypic alterations during AD→AI→HRPC transition, maintaining CaP in a hormone ablation treatable state.

Anti-breast cancer activity by resveratrol was shown in 1997.[5] Whether resveratrol exerts different effects in malignant cells with a common anatomical origin while displaying different invasive characteristics remained unknown at the time of our studies. Therefore we investigated the cell cycle effects and control of gene expression by resveratrol using human breast carcinoma cell lines with different metastatic potentials, based on the hypothesis that the intrinsic metastatic potential of cancer cells may affect their responses to resveratrol. Our results, reported in 1999, showed that resveratrol ($25\ \mu M$, 3 days) reduced growth of breast cancer cells with low or high metastatic potential by 60–80%.[31]

Resveratrol is widely consumed as a component of wine and many foods; thus, it is important to have knowledge about its effects on normal cell proliferation. In 2002, we reported our finding on the effects of resveratrol on normal, mitogenically stimulated lymphocytes using peripheral blood human lymphocytes.[32] At $\leq 50\ \mu M$

resveratrol, no detectable effects on resting lymphocytes were observed; however, resveratrol elicited dose- and time-dependent responses in phytohemagglutin (PHA) stimulated lymphocytes and prevented the lymphocyte progression through the cell cycle. In addition, the activation-induced apoptosis was reduced by ≥ 25 μM resveratrol. These data suggest that studies designed to test the efficacy of resveratrol as a chemopreventive agent *in vivo* should consider evaluating its immunomodulatory effect as revealed by the effect on cell cycle progression of the proliferating cell population, e.g., stimulated lymphocytes. In addition, since G_0 non-stimulated lymphocytes resemble quiescent stem cells, they may be considered a model that allows study of effect of agents including resveratrol on the control of G_0–G_1 transition, the stage that characterizes the induction of proliferation of stem cells.

A survey of published data on resveratrol chemopreventive activities showed that it has broad and highly variable cell type- and dose-dependent effects. What accounts for the diverse dose-effect relationship of resveratrol in various cell lines? To investigate the molecular basis for why cellular responsiveness to resveratrol shows large variations in dose dependence, we propose the "target-specific/dependent" hypothesis. In this hypothesis, resveratrol is postulated to interact with specific cellular targets in responsive cells.

In the next section, we review how we identify RTPs and the implication of our finding in explaining the issue of bioavailability of resveratrol *in vivo*.

2.4. Formulation of the resveratrol targeting protein (RTP) hypothesis — discovery of quinone reductase 2 (QR2) as a distinct RTP using resveratrol-capture bioaffinity chromatography and ligand-selection proteomics

The chemical structure of resveratrol contains an aromatic stilbene core with three hydrophilic side groups (Figure 2.2A) implicating its interaction with and modulation of target proteins. We hypothesize that resveratrol elicits its multiple bioactivities via binding to

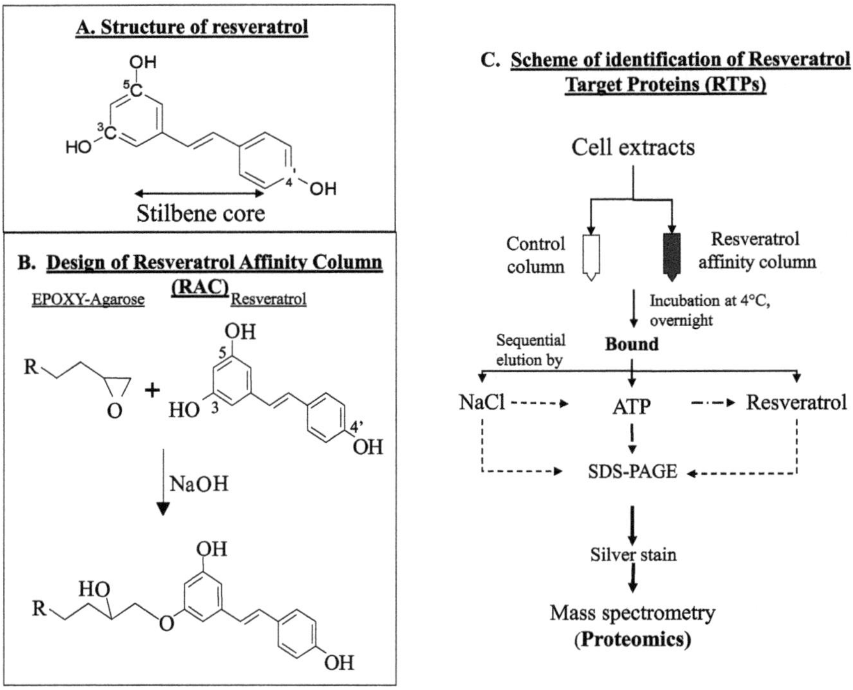

Fig. 2.2. **A.** Structure of resveratrol (trans-3,4′,5-trihydroxystibene). Its aromatic stilbene core structure and hydrophilic side groups support the hypothesis that the chemopreventive potential of resveratrol may in part lie in its ability to interact with and modulate cellular targets in responsive cells. **B.** Activation of carbohydrate-based resins with 12 atom-linker epoxy groups enables its derivatization with resveratrol, ligand of interest. **C.** Hypothesis: ligand (resveratrol)-immobilized agarose columns could "capture" proteins with a high affinity for resveratrol. The flow chart for isolation and purification of resveratrol targeting protein from cultured mammalian cell extracts by resveratrol affinity chromatography (RAC). The column is sequentially eluted with varying ionic strength buffers, or with nucleotides such as ATP to displace proteins with modest affinity for resveratrol. Proteins binding most tightly to the affinity column, denoted RTPs, are eluted with resveratrol. We consider these to be RTPs with resveratrol-relevant biological activities (modified from[33]).

and interacting with its targeting proteins present in resveratrol-responsive cells. According to this hypothesis, cell-type specific RTPs and quantitative and qualitative differences in their expression, coupled with differential uptake and metabolism of resveratrol may explain the large dose spread of resveratrol required for efficacy *in vitro*, and provide a rationale framework for the selective activity of resveratrol between normal and tumor cells. This hypothesis implies that once an RTP is identified, it can be characterized and evaluated for its participation or contribution in chemoprevention by resveratrol. The RTP hypothesis may also circumvent the seemingly irreconcilable dosage-bioavailability issue of resveratrol by becoming the focus of research.

2.4.1. *Testing the novel target hypothesis of resveratrol*

To test the presence of RTPs and to facilitate their identification and purification from cultured cell extracts, we developed a specific resveratrol affinity chromatography to isolate and purify RTPs.[33] This is adopted from a widely used classical, biochemical approach that separates and purifies target proteins by exploiting their specificity and affinity for an immobilized ligand, instead of the more general features of size, surface charge and hydrophobicity used in protein separation and purification.

2.4.1.1. *Synthesis of resveratrol affinity column (RAC) for purification and characterization of RTPs*

Resveratrol was immobilized on a solid matrix using epoxy-activated agarose beads through one of the hydrophilic side groups (Figure 2.2B). Evidently, the immobilization could occur at any one of the three hydroxyl positions. Earlier structure-function studies showed that modifications at 3- and 5-hydroxyls do not significantly affect the biological activity of resveratrol.[7,34,35] Therefore, we assumed that the majority of the immobilized resveratrol (statistically 67%) should retain ability to bind genuine RTPs similar to that of free resveratrol. The method for the covalent attachment of resveratrol to epoxy-activated agarose beads was detailed in our publication.[33]

To test the application of RAC, we used extracts prepared from various tumor cells.[33] The purification and characterization of RTPs using RAC is illustrated in Figure 2.2C. In brief, RTPs are resolved by changing NaCl concentrations and nature of the eluting ligands. Tightly bound RTPs are displaced from RAC by resveratrol. Differentially eluted proteins are checked for purity by SDS-PAGE and silver staining, and further characterized by mass spectrometry. Subsequently, QR2 in complex with resveratrol was crystalized by our group in 2004.[36]

2.4.1.2. *Attributes and limitations of RAC*

The utility of RAC is based on the specificity and affinity of resveratrol to select molecules relevant to its affect in cancer and other biological processes. As such, RAC captures targeting proteins possibly with direct or indirect roles in the bioactive effects of resveratrol.

Innovative features of this ligand (resveratrol)-affinity chromatography (RAC) approach are as follows: (i) it focuses on few proteins and hence reduces the complexity of the analysis. Proteins being analyzed are more likely to have a direct link to the biological response at hand; (ii) it detects and analyzes low abundance proteins that might be difficult to visualize and characterize using more conventional biochemical methods; (iii) it is easy to operate and therefore allows large numbers of samples to be processed quickly; (iv) it provides an unbiased method for testing and confirming similarities and differences in RTPs between different cell types; (v) it could generate RTP profiles that characterize different forms/stages of CaP. Such profiles may provide information applicable to drug design and discovery. A caveat of the RAC strategy is that while its affinity prefabrication feature offers a more focused analytical scope, it could render bias against proteins with low to no binding affinity for resveratrol, yet play an essential, though indirect role in its chemopreventive activity.

2.4.1.3. *Discovery of QR2 as a distinct resveratrol targeting protein RTP by RAC*

Extracts prepared from different CaP cells were fractionated by RAC and protein denoted RTP-22 was shown to be tightly bound. Resveratrol-eluted RTP-22 was characterization by mass spectrometry and further identified as quinone reductase 2 (QR2 or NQO2).[33] To explore the role, if any, QR2 might have in prostate carcinogenesis and CaP prevention by resveratrol, RAC was used to fractionate extracts prepared from cells representing early (LNCaP), intermediate, transition (CWR22Rv1) and late, hormone-refractory (DU145 and PC-3) stages of CaP. The expression of QR2 in these cell lines was monitored by RAC and western blot analysis. The silver stained protein profiles show QR2 to be the major protein retained on RAC in all cell extracts tested. QR2 expression was most pronounced in CWR22Rv1 cells using both assays. These findings suggest that QR2 may be involved in CaP progression and may also be considered a marker for CaP staging. The utility of RAC is additionally confirmed by the retention of ERα, ERβ, GSTP1, and AKT as putative targets of resveratrol.[37–40]

2.4.1.4. *Discovery of RTP has implications on the issue of bioavailability of resveratrol*

Although resveratrol has been shown to suppress cell growth and induce apoptosis in various cancer cell types, it suffers from the drawback that high doses are required for the observed effects *in vitro*. Moreover, low bioavailability of resveratrol is found in human and rodent studies.[41–43] A plasma peak level of 1–2 μM "total" resveratrol appeared 30–60 min after ingestion in humans.[41–45] The peak plasma levels of resveratrol administered at 10 mg/kg is estimated to be 1 μM in rodents.[43] Intragastric gavage of resveratrol, at doses of 2, 20 and 50 mg/kg, resulted in peak plasma values of 2, 1.2 and 6.6 μM, respectively.[43,46–48] Fresh grape skin contains approximately 30 μg/g of resveratrol,[49] while its concentration in red wine can range from 1.5 to 5 mg/L depending on the method of preparation.[50,51] The discovery of RTP exemplified by QR2 showed that resveratrol is a potent competitive inhibitor with a K_D of 35±15 nM.[36] Even with poor systemic bioavailability of resveratrol, it may be surmised that the concentration of resveratrol reached in the responsive tissues is sufficient to effectively modulate the QR2 activity and/or function.

2.4.1.5. *The role of RAC-discovered QR2 in chemoprevention of resveratrol*

The discovery of QR2 as RTP allowed us to explore and elucidate its role in resveratrol mediated anti-cancer activity. Several key findings are reviewed as part of the complete list of our studies on chemoprevention by resveratrol shown in Figure 2.1.

Example 1. Inhibition of melanoma cell proliferation by resveratrol is correlated with upregulation of QR2 and p53.[52]

Suppression of skin cancer by resveratrol was first reported in 1997;[5] to our knowledge hitherto it was not known whether resveratrol exerted similar anti-proliferative activities in melanoma cells with different metastatic potential. Thus, we tested whether comparable inhibition of growth and colony formation resulted from resveratrol treatment in both weakly metastatic Line IV clone 3 and on autologous, highly metastatic Line IV clone 1 cultured melanoma cells. A significant increase in p53 expression by resveratrol was observed in both cell lines. Resveratrol affinity chromatography revealed that the basal expression of QR2 was higher in clone 1 than clone 3 cells; however, the level of expression of QR2 was similarly and dose-dependently increased by resveratrol in both cells. We propose that induction of QR2 may relate to the observed increased expression of p53 that, in turn, contributes to the observed suppression of cell growth in both melanoma cell lines.

Example 2. Antiproliferation effects of resveratrol and the mediating role of resveratrol targeting protein NQO2 in androgen-receptor positive, hormone-non-responsive CWR22Rv1 cells.[53]

As mentioned previously, anti-CaP activities of resveratrol were evident in both androgen dependent LNCaP and androgen independent DU145 and PC-3 CaP cells. However, effects of resveratrol had not been investigated in AR-positive hormone-non-responsive CWR22Rv1 cells, which serve to mimic the transition stage of prostate carcinoma. Thus, both control and QR2 siRNA silenced CWR22Rv1 cells were used to investigate the anti-proliferative effects of resveratrol. Dose-dependent growth inhibition of QR2 expressing CWR22Rv1 resulted from exposure to resveratrol, and was accompanied by a reduction in NF-κB p65 level of expression. The suppression of NF-κB p65 expression was abrogated in QR2 siRNA silenced CWR22Rv1 cells. This study represented the first demonstration that resveratrol targeting protein QR2 plays a role in resveratrol-induced changes of NF-κB p65, suggesting that QR2 participates or contributes to the anti-CaP activity of resveratrol.

Example 3. Uptake of resveratrol and role of resveratrol-targeting protein, quinone reductase 2, in normal cultured human prostate cells.[54]

The accessibility and cellular distribution of resveratrol in non-cancerous prostate cells have not been investigated. We studied the uptake and intracellular trafficking of resveratrol in normally cultured prostate stromal (PrSCs) and epithelial cells (PrECs). Results showed that pre-treatment with resveratrol modulated its uptake and selective partitioning to membrane and organelle compartments in both PrSCs and PrECs. Resveratrol treatment also resulted in growth suppression accompanied by blockade of G_1-phase and a concomitant increase in S and G_2M-phases of the cell cycle. Resveratrol affinity column chromatography studies revealed the presence of QR2 in PrSCs. QR2 may play a role in resveratrol-mediated inhibition in PrSC growth and alteration in cell cycle distribution.

Example 4. Control of stability of cyclin D1 by quinone reductase 2 in CWR22Rv1 CaP cells.[55]

Our continuing efforts to elucidate anti-CaP activity of resveratrol next focused on the role of QR2 by employing and comparing QR2 expressing (sh08) with QR2 knockdown (sh25) CWR22Rv1 cells. Sh08 cells displayed slower proliferation, G_1 accumulation, significant decrease in the phosphorylation of retinoblastoma (Rb) and cyclin D1 and an increase in AKT activity paralleled by an elevation in phosphorylation and reduction/deactivation of GSK-3α/β at Ser21/9. Our results are summarized as follows: In QR2 expressing sh08 cells, QR2 binds AKT; this lowers AKT kinase activity accompanied by a more active GSK-3β which, in turn, increases cyclin D1 T286 phosphorylation, effectively shortening cyclin D1 $t_{1/2}$ via an increase in proteasome mediated degradation. In QR2 silenced sh25 cells, less QR2 is available for binding AKT. The high AKT increases GSK3β phosphorylation and decreases its activity, which, in turn, decreases cyclin D1 T286 phosphorylation, effectively prolonging cyclin D1 $t_{1/2}$ via a decrease in proteasome mediated degradation. This study reveals a hitherto never reported role of QR2 in the control of AKT/GSK-3β/cyclin D1, and highlights the involvement of QR2 in degradation of cyclin D1, as contributing to mechanism of chemoprevention by resveratrol.

Example 5. Biochemical and cellular evidence demonstrating AKT as a binding partner for resveratrol targeting protein.[40]

We recently reported that QR2 inhibited AKT activity.[55] To determine how resveratrol and its target protein QR2 exert control of AKT, we performed resveratrol affinity chromatography to obtain information on the interplay between resveratrol and AKT. We found that AKT was bound to the column with high affinity suggesting that AKT may be a genuine RTP of resveratrol. Molecular modeling was used to investigate QR2:AKT interaction. The pleckstrin homology (PH) and kinase domains of AKT were shown to bind to chains A and B of QR2 using pull-down and deletion assays. Furthermore, access to PIP3 on the membrane was inhibited upon binding of QR2 to PH domain of AKT whereas access to the catalytic activity site of AKT was disrupted upon binding of QR2 to kinase domain (KD) of AKT. Pre-binding by resveratrol further modulated this interaction. The findings showed both QR2 and AKT are targets of resveratrol; moreover, QR2:AKT interaction is a novel physiological regulator of AKT activation/function.

2.5. QR2 may act as a pleiotropic cellular modulator of resveratrol in chemoprevention — direct and indirect effects of QR2

In addition to aforementioned examples of QR2 in chemoprevention of resveratrol, other possible mechanisms may be proposed, as summarized below.

2.5.1. *Interaction of resveratrol with QR2 may promote the maintenance of a reductive cellular environment*

As noted, QR2 requires dihydronicotinamide riboside NRH as a cofactor, which has been suggested to be derived from NADH or NADPH.[56–58] Inhibition of QR2 by resveratrol, therefore, could reduce the need for continuous synthesis of NRH, sparing its conversion from NADH or NADPH, and favoring the preservation of these two reduced pyridine nucleotides in the cell. This cellular status might shift the prooxidant-antioxidant balance as to permit the establishment of a reduced state environment restrictive to CaP cell growth and progression.[59–61] Such a state could result in more NADH conversion to NAD, which could be used as a substrate for repair of damaged cellular DNA by poly(ADP-ribosylation) thereby decreasing the frequency and/or accumulation of genetic alterations.[61–64] Inhibition of QR2 by resveratrol also implies that less nicotinamide riboside, a reported stimulator of cell growth,[65,66] would be generated intracellularly.

2.5.2. *QR2 and chemoprevention by resveratrol — insights from the expanded role of QR1 and QR2*

QR1 and QR2 are genetically distinct but functionally homologous flavoproteins.[56,67] QR1 was discovered more than 50 years ago, as an NADPH dependent protein with an assigned role in phase 2 detoxification reactions, for defense against carcinogens and chemical toxicants.[68–74] QR1 also circumvents the one-electron redox cycling of quinones, thereby suppressing the generation of reactive oxygen species (ROS) with potentially deleterious effects.[75,76] In support of its role as a complementary deterrent against damage by free radicals, polymorphism of QR1 gene shows a correlated increased risk for diseases associated with exposure to environmental chemicals.[68,72,77,78] Animals harboring QR1 gene knockouts show myelogenous hyperplasia, suggesting that QR1 plays a role in the control of cell size and numbers.[79,80] In bone marrow of QR1-null mice, a decrease in levels of pyridine nucleotides and the tumor suppressor protein p53 is observed. Recent studies also show that QR1 affects the stability of p53.[81,82] The expression of QR1 is copiously induced by a multitude of chemicals and dietary agents including resveratrol.[83–85]

Like QR1, QR2 was first described several decades ago as an ill-defined mammalian FAD-dependent enzyme and only received much more intense scientific scrutiny in recent years.[86–88] QR2 shares a high degree of structural homology to QR1. However, it is actually a separately encoded gene product,[89] with substrate and catalytic properties distinct from those of QR1. Thus, QR2 uses dihydronicotinamide riboside as an electron donor, compared to NADPH for QR1.[56,57,67,87,88] Interestingly, however, QR2 gene knockout animals also show myelogenous hyperplasia suggesting that QR2 too plays a role in controlling cell number and size.[90] Additionally, QR2-null mice were found to have reduced p53 expression, suggesting that QR2, like QR1, may also function in the control of p53 expression and/or stability.[91]

2.5.3. *A QR2-QR1 yin-yang hypothesis in control of p53 stability and its relationship to chemoprevention by resveratrol*

Since QR1 expression is inducible by resveratrol, while QR2 binds tightly to and is potently inhibited by resveratrol, and because both QR1 and QR2 appear to be involved in controlling the stability of p53, we propose a new mechanistic framework on how this polyphenol may function in the prevention of cancer, named the QR1-QR2 yin-yang hypothesis (Figure 2.3). In this hypothesis, we suggest that QR1-QR2 — acting in yin-yang fashion with QR1 functioning as "yang" (stabilizing factor) and QR2 acting as "yin" (destabilizing factor) — serves as dual regulator of the stability of p53. In this model, the sequestration of QR2, through binding to resveratrol, in concert with upregulation of QR1 by resveratrol, results in a net increase in the level of QR1 that promotes the stabilization of the p53.

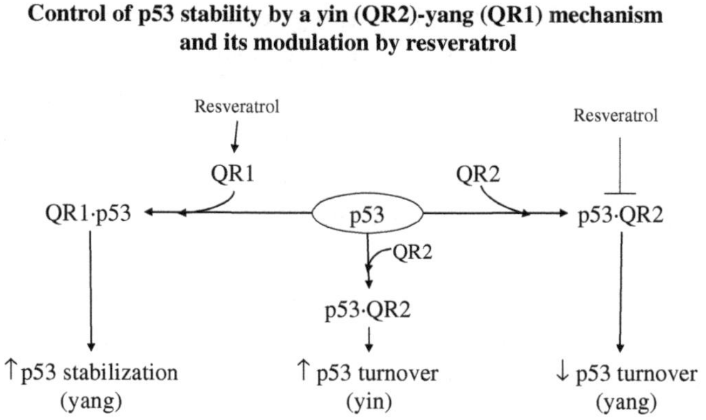

Fig. 2.3. QR1/QR2 may act by yin-yang mechanism, in coordination with resveratrol, in CaP prevention. QR2 acts as the "yin" partner to "yang" QR1 in controlling the stability of p53. By its ability to complex QR2 and induce QR1, resveratrol increases QR1, stabilizing the p53 and contributing to p53-dependent growth suppression and PSA regulation in CaP cells.

2.6. A bioinformatic approach to elucidate anti-CaP effects of resveratrol — applications and implications

A review of PubMed shows that resveratrol has had a remarkable yearly growth in citations since the late 1990s. Because of the reported diverse cellular processes and diseases resveratrol affects and benefits, the interest in resveratrol will likely continue, despite the equivocation on its mechanisms in different biological systems.

We postulate that systematic dissection and mining of publicly available data could reveal insights on disease-modulatory activities of resveratrol. Information obtained can further be analyzed to explore and connect the relationship that might exist between resveratrol and its cellular targets to provide a framework of solution on specific diseases. An effective strategy may lie in the use of the publicly available open-sourced database and gene networks. In the last section, we will review a FAN approach we established in 2016 as a new paradigm for guiding future studies of resveratrol.

Rationale. The successful launch and execution of high-throughput genomic screening studies in recent years frequently generate innumerable potential hits. The integration of the output genes can be challenging, and functional interpretation of gene sets and their extrapolation into common or unique biological themes can be equally daunting. We introduce an approach called FAN analysis that can effectively analyze and prioritize candidate genes for subsequent linkage to experimentally verifiable pathways. Our focus on resveratrol stems from the following considerations. First, based on the diversity of cellular processes and diseases resveratrol reportedly affects and benefits, a large number of clinical trials have been launched.[92–95] However, uncertainty concerning its mechanism remains. Second, the data generated on resveratrol from experiments in the same or different laboratories have been difficult to compare, interpret, and extrapolate into unifying biological themes.

Approach. As proof of principle, we deploy a Web-accessible, open portal database source, DTome (Drug-Target Interactome) to dissect and analyze the interplay between resveratrol and control of the tumor suppressor gene p53. DTome analysis identified 4 direct protein targets (DPT) and 219 DPT-associated genes for resveratrol. Among 4 DPTs identified (QR2, CSNK2A1, PTGS2 and PTGS1), the first three — QR2, CSNK2A1, PTGS2, show a direct association with p53 control via gene-protein interactions, suggesting that they underlie a main target pathway connecting resveratrol and control of p53. To elaborate further on resveratrol:p53 interplay, the DPT-associated genes were queried by WebGestalt enrichment analysis. This resulted in 10 identified KEGG pathways, 2 of which, one linked to p53 and a second one to CaP, show functional connectivity to resveratrol and its four DPTs.

Application and implication. As illustrated a resveratrol: QR2:p53 axis underlying the control of CaP proliferation and gene expression may be tested and validated experimentally using a two-prolong strategy (Figure 2.4A).

The DTome-based, functional activity network FAN analysis has BioGPS attributes capable of mapping and defining a Web-mined scientific framework that can productively and cost effectively connect resveratrol to its primary and secondary target proteins

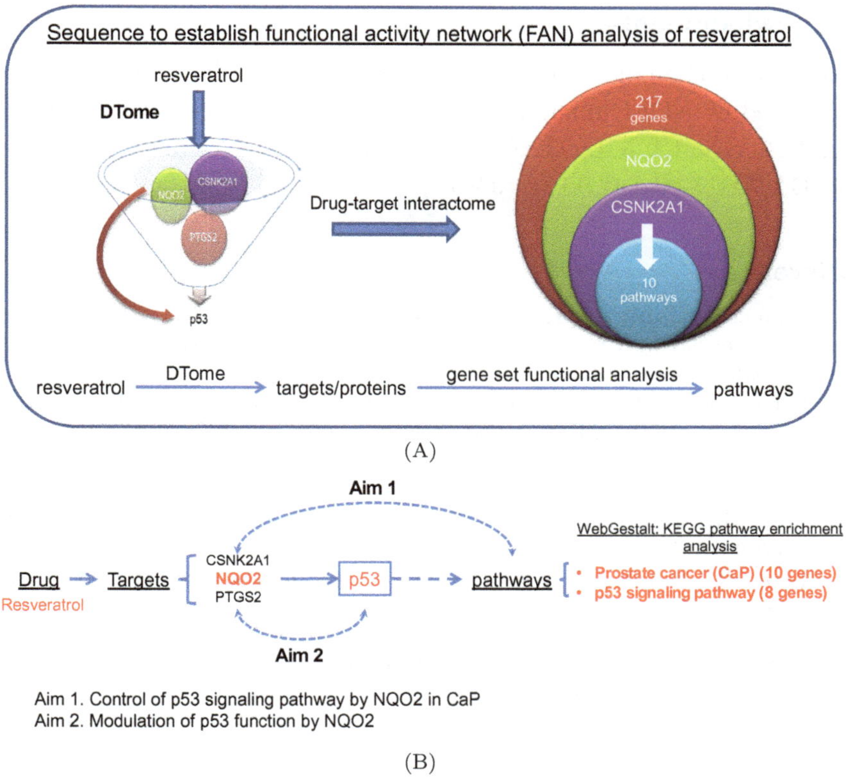

Fig. 2.4. Function and Activity Network (FAN) as related to the functional output and targets connecting the control p53 by resveratrol. **A.** The sequence of steps in performance of FAN analysis applicable to resveratrol. **B.** The use of DTome-based FAN (functional activity network) analysis as a new paradigm for analyzing resveratrol:p53 interplay in chemoprevention of CaP. **Aim 1** will utilize functional and activity assays to define role QR2 plays on control of CaP proliferation and metastasis via the p53 signaling. Focus will be on overlapping p53 and CaP genes and proteins identified by KEGG enrichment analysis. p53 downstream events — cell cycle, induction of apoptosis and metastasis will also be examined. **Aim 2** will investigate how QR2 and resveratrol modulates p53 stability and activity.

and to its biological functions, and also serve to guide studies of resveratrol using an experimentally testable and verifiable blueprint. The network-based FAN approach may help to formulate proposed studies and lay the mechanistic foundation for drug discovery in the future.

Acknowledgments

Research on chemoprevention of resveratrol in our laboratories were supported by NCI Clinical Nutrition Research Unit Grant CA 29502 and NCI Grant CA109932 (to TCH) and by NCI Grant CA104424 and DAMD17-00-1-0296 (to JMW).

References

1. Sporn MB, Dunlop NM, Newton DL, Smith JM. Prevention of chemical carcinogenesis by vitamin A and its synthetic analogs (retinoids). *Fed Proc* 1976; **35**: 1332–8.
2. O'Shaughnessy JA, Kelloff GJ, Gordon GB, Dannenberg AJ, Hong WK, Fabian CJ, *et al.* Treatment and prevention of intraepithelial neoplasia: an important target for accelerated new agent development. *Clin Cancer Res* 2002; **8**: 314–46.
3. Sporn MB, Liby KT. Cancer chemoprevention: scientific promise, clinical uncertainty. *Nat Clin Pract Oncol* 2005; **2**: 518–25.
4. Sporn MB, Suh N. Chemoprevention: an essential approach to controlling cancer. *Nat Rev Cancer* 2002; **2**: 537–43.
5. Jang M, Cai L, Udeani GO, Slowing KV, Thomas CF, Beecher CW, *et al.* Cancer chemopreventive activity of resveratrol, a natural product derived from grapes. *Science* 1997; **275**: 218–20.
6. Tsan MF, White JE, Maheshwari JG, Chikkappa G. Anti-leukemia effect of resveratrol. *Leuk Lymphoma* 2002; **43**: 983–7.
7. Morris GZ, Williams RL, Elliott MS, Beebe SJ. Resveratrol induces apoptosis in LNCaP cells and requires hydroxyl groups to decrease viability in LNCaP and DU 145 cells. *Prostate* 2002; **52**: 319–29.
8. Hsieh TC, Wu JM. Grape-derived chemopreventive agent resveratrol decreases prostate-specific antigen (PSA) expression in LNCaP cells by an androgen receptor (AR)-independent mechanism. *Anticancer Res* 2000; **20**: 225–8.
9. Hsieh TC, Wu JM. Differential effects on growth, cell cycle arrest, and induction of apoptosis by resveratrol in human prostate cancer cell lines. *Exp Cell Res* 1999; **249**: 109–15.
10. Lin HY, Shih A, Davis FB, Tang HY, Martino LJ, Bennett JA, *et al.* Resveratrol induced serine phosphorylation of p53 causes apoptosis in a mutant p53 prostate cancer cell line. *J Urol* 2002; **168**: 748–55.
11. Gao X, Xu YX, Divine G, Janakiraman N, Chapman RA, Gautam SC. Disparate in vitro and in vivo antileukemic effects of resveratrol, a natural polyphenolic compound found in grapes. *J Nutr* 2002; **132**: 2076–81.
12. Roman V, Billard C, Kern C, Ferry-Dumazet H, Izard JC, Mohammad R, *et al.* Analysis of resveratrol-induced apoptosis in human B-cell chronic leukaemia. *Br J Haematol* 2002; **117**: 842–51.

13. Joe AK, Liu H, Suzui M, Vural ME, Xiao D, Weinstein IB. Resveratrol induces growth inhibition, S-phase arrest, apoptosis, and changes in biomarker expression in several human cancer cell lines. *Clin Cancer Res* 2002; **8**: 893–903.
14. Jiang H, Zhang L, Kuo J, Kuo K, Gautam SC, Groc L, *et al.* Resveratrol-induced apoptotic death in human U251 glioma cells. *Mol Cancer Ther* 2005; **4**: 554–61.
15. Zhang W, Fei Z, Zhen HN, Zhang JN, Zhang X. Resveratrol inhibits cell growth and induces apoptosis of rat C6 glioma cells. *J Neurooncol* 2007; **81**: 231–40.
16. Park JW, Choi YJ, Suh SI, Baek WK, Suh MH, Jin IN, *et al.* Bcl-2 overexpression attenuates resveratrol-induced apoptosis in U937 cells by inhibition of caspase-3 activity. *Carcinogenesis* 2001; **22**: 1633–9.
17. Mouria M, Gukovskaya AS, Jung Y, Buechler P, Hines OJ, Reber HA, *et al.* Food-derived polyphenols inhibit pancreatic cancer growth through mitochondrial cytochrome C release and apoptosis. *Int J Cancer* 2002; **98**: 761–9.
18. Wolter F, Akoglu B, Clausnitzer A, Stein J. Downregulation of the cyclin D1/Cdk4 complex occurs during resveratrol-induced cell cycle arrest in colon cancer cell lines. *J Nutr* 2001; **131**: 2197–203.
19. Mahyar-Roemer M, Katsen A, Mestres P, Roemer K. Resveratrol induces colon tumor cell apoptosis independently of p53 and precede by epithelial differentiation, mitochondrial proliferation and membrane potential collapse. *Int J Cancer* 2001; **94**: 615–22.
20. Huang C, Ma WY, Goranson A, Dong Z. Resveratrol suppresses cell transformation and induces apoptosis through a p53-dependent pathway. *Carcinogenesis* 1999; **20**: 237–42.
21. Kuo PL, Chiang LC, Lin CC. Resveratrol-induced apoptosis is mediated by p53-dependent pathway in Hep G2 cells. *Life Sci* 2002; **72**: 23–34.
22. De Lédinghen V, Monvoisin A, Neaud V, Krisa S, Payrastre B, Bedin C, *et al.* Trans-resveratrol, a grapevine-derived polyphenol, blocks hepatocyte growth factor-induced invasion of hepatocellular carcinoma cells. *Int J Oncol* 2001; **19**: 83–8.
23. Kozuki Y, Miura Y, Yagasaki K. Resveratrol suppresses hepatoma cell invasion independently of its anti-proliferative action. *Cancer Lett* 2001; **167**: 151–6.
24. Tseng SH, Lin SM, Chen JC, Su YH, Huang HY, Chen CK, *et al.* Resveratrol suppresses the angiogenesis and tumor growth of gliomas in rats. *Clin Cancer Res* 2004; **10**: 2190–202.
25. Whitsett T, Carpenter M, Lamartiniere CA. Resveratrol, but not EGCG, in the diet suppresses DMBA-induced mammary cancer in rats. *J Carcinog* 2006; **5**: 15.
26. Martín AR, Villegas I, Sánchez-Hidalgo M, de la Lastra CA. The effects of resveratrol, a phytoalexin derived from red wines, on chronic inflammation induced in an experimentally induced colitis model. *Br J Pharmacol* 2006; **147**: 873–85.

27. Sengottuvelan M, Viswanathan P, Nalini N. Chemopreventive effect of trans-resveratrol–a phytoalexin against colonic aberrant crypt foci and cell proliferation in 1,2-dimethylhydrazine induced colon carcinogenesis. *Carcinogenesis* 2006; **27**: 1038–46.
28. Soleas GJ, Diamandis EP, Goldberg DM. Resveratrol: a molecule whose time has come? And gone? *Clin Biochem* 1997; **30**: 91–113.
29. Soleas GJ, Diamandis EP, Goldberg DM. The world of resveratrol. *Adv Exp Med Biol* 2001; **492**: 159–82.
30. Patel KR, Brown VA, Jones DJ, Britton RG, Hemingway D, Miller AS, *et al.* Clinical pharmacology of resveratrol and its metabolites in colorectal cancer patients. *Cancer Res* 2010; **70**: 7392–9.
31. Hsieh TC, Burfeind P, Laud K, Backer JM, Traganos F, Darzynkiewicz Z, *et al.* Cell cycle effects and control of gene expression by resveratrol in human breast carcinoma cell lines with different metastatic potentials. *Int J Oncol* 1999; **15**: 245–52.
32. Hsieh T, Halicka D, Lu X, Kunicki J, Guo J, Darzynkiewicz Z, *et al.* Effects of resveratrol on the G(0)-G(1) transition and cell cycle progression of mitogenically stimulated human lymphocytes. *Biochem Biophys Res Commun* 2002; **297**: 1311–7.
33. Wang Z, Hsieh TC, Zhang Z, Ma Y, Wu JM. Identification and purification of resveratrol targeting proteins using immobilized resveratrol affinity chromatography. *Biochem Biophys Res Commun* 2004; **323**: 743–9.
34. Stivala LA, Savio M, Carafoli F, Perucca P, Bianchi L, Maga G, *et al.* Specific structural determinants are responsible for the antioxidant activity and the cell cycle effects of resveratrol. *J Biol Chem* 2001; **276**: 22586–94.
35. Cho DI, Koo NY, Chung WJ, Kim TS, Ryu SY, Im SY, *et al.* Effects of resveratrol-related hydroxystilbenes on the nitric oxide production in macrophage cells: structural requirements and mechanism of action. *Life Sci* 2002; **71**: 2071–82.
36. Buryanovskyy L, Fu Y, Boyd M, Ma Y, Hsieh TC, Wu JM, *et al.* Crystal structure of quinone reductase 2 in complex with resveratrol. *Biochemistry* 2004; **43**: 11417–26.
37. Levenson AS, Gehm BD, Pearce ST, Horiguchi J, Simons LA, Ward JE, *et al.* Resveratrol acts as an estrogen receptor (ER) agonist in breast cancer cells stably transfected with ER alpha. *Int J Cancer* 2003; **104**: 587–96.
38. Bhat KP, Pezzuto JM. Resveratrol exhibits cytostatic and antiestrogenic properties with human endometrial adenocarcinoma (Ishikawa) cells. *Cancer Res* 2001; **61**: 6137–44.
39. Hsieh TC, Wang Z, Deng H, Wu JM. Identification of glutathione sulfotransferase-pi (GSTP1) as a new resveratrol targeting protein (RTP) and studies of resveratrol-responsive protein changes by resveratrol affinity chromatography. *Anticancer Res* 2008; **28**: 29–36.
40. Hsieh TC, Lin CY, Bennett DJ, Wu E, Wu JM. Biochemical and cellular evidence demonstrating AKT-1 as a binding partner for resveratrol targeting protein NQO2. *PLoS One* 2014; **9**: e101070.

41. Walle T, Hsieh F, DeLegge MH, Oatis JE, Walle UK. High absorption but very low bioavailability of oral resveratrol in humans. *Drug Metab Dispos* 2004; **32**: 1377–82.
42. Vitaglione P, Sforza S, Galaverna G, Ghidini C, Caporaso N, Vescovi PP, *et al.* Bioavailability of trans-resveratrol from red wine in humans. *Mol Nutr Food Res* 2005; **49**: 495–504.
43. Gescher AJ, Steward WP. Relationship between mechanisms, bioavailibility, and preclinical chemopreventive efficacy of resveratrol: a conundrum. *Cancer Epidemiol Biomarkers Prev* 2003; **12**: 953–7.
44. Goldberg DM, Yan J, Soleas GJ. Absorption of three wine-related polyphenols in three different matrices by healthy subjects. *Clin Biochem* 2003; **36**: 79–87.
45. Meng X, Maliakal P, Lu H, Lee MJ, Yang CS. Urinary and plasma levels of resveratrol and quercetin in humans, mice, and rats after ingestion of pure compounds and grape juice. *J Agric Food Chem* 2004; **52**: 935–42.
46. Asensi M, Medina I, Ortega A, Carretero J, Baño MC, Obrador E, *et al.* Inhibition of cancer growth by resveratrol is related to its low bioavailability. *Free Radic Biol Med* 2002; **33**: 387–98.
47. Marier JF, Vachon P, Gritsas A, Zhang J, Moreau JP, Ducharme MP. Metabolism and disposition of resveratrol in rats: extent of absorption, glucuronidation, and enterohepatic recirculation evidenced by a linked-rat model. *J Pharmacol Exp Ther* 2002; **302**: 369–73.
48. Juan ME, Planas JM. Trans-resveratrol oral administration does not affect the enzymatic activities in rat small intestine. *J Physiol Biochem* 2002; **58**: 59–60.
49. Versari A, Parpinello GP, Tornielli GB, Ferrarini R, Giulivo C. Stilbene compounds and stilbene synthase expression during ripening, wilting, and UV treatment in grape cv. *Corvina. J Agric Food Chem* 2001; **49**: 5531–6.
50. Wang Z, Chen Y, Labinskyy N, Hsieh TC, Ungvari Z, Wu JM. Regulation of proliferation and gene expression in cultured human aortic smooth muscle cells by resveratrol and standardized grape extracts. *Biochem Biophys Res Commun* 2006; **346**: 367–76.
51. Cantos E, García-Viguera C, de Pascual-Teresa S, Tomás-Barberán FA. Effect of postharvest ultraviolet irradiation on resveratrol and other phenolics of cv. Napoleon table grapes. *J Agric Food Chem* 2000; **48**: 4606–12.
52. Hsieh TC, Wang Z, Hamby CV, Wu JM. Inhibition of melanoma cell proliferation by resveratrol is correlated with upregulation of quinone reductase 2 and p53. *Biochem Biophys Res Commun* 2005; **334**: 223–30.
53. Hsieh TC. Antiproliferative effects of resveratrol and the mediating role of resveratrol targeting protein NQO2 in androgen receptor-positive, hormone-non-responsive CWR22Rv1 cells. *Anticancer Res* 2009; **29**: 3011–7.
54. Hsieh TC. Uptake of resveratrol and role of resveratrol-targeting protein, quinone reductase 2, in normally cultured human prostate cells. *Asian J Androl* 2009; **11**: 653–61.

55. Hsieh TC, Yang CJ, Lin CY, Lee YS, Wu JM. Control of stability of cyclin D1 by quinone reductase 2 in CWR22Rv1 prostate cancer cells. *Carcinogenesis* 2012; **33**: 670–7.
56. Chen S, Wu K, Knox R. Structure-function studies of DT-diaphorase (NQO1) and NRH: quinone oxidoreductase (NQO2). *Free Radic Biol Med* 2000; **29**: 276–84.
57. Wu K, Knox R, Sun XZ, Joseph P, Jaiswal AK, Zhang D, *et al.* Catalytic properties of NAD(P)H:quinone oxidoreductase-2 (NQO2), a dihydronicotinamide riboside dependent oxidoreductase. *Arch Biochem Biophys* 1997; **347**: 221–8.
58. Zhao Q, Yang XL, Holtzclaw WD, Talalay P. Unexpected genetic and structural relationships of a long-forgotten flavoenzyme to NAD(P)H:quinone reductase (DT-diaphorase). *Proc Natl Acad Sci U S A* 1997; **94**: 1669–74.
59. Ripple MO, Henry WF, Rago RP, Wilding G. Prooxidant-antioxidant shift induced by androgen treatment of human prostate carcinoma cells. *J Natl Cancer Inst* 1997; **89**: 40–8.
60. Zhong W, Oberley TD. Redox-mediated effects of selenium on apoptosis and cell cycle in the LNCaP human prostate cancer cell line. *Cancer Res* 2001; **61**: 7071–8.
61. Ouyang X, DeWeese TL, Nelson WG, Abate-Shen C. Loss-of-function of Nkx3.1 promotes increased oxidative damage in prostate carcinogenesis. *Cancer Res* 2005; **65**: 6773–9.
62. Fan R, Kumaravel TS, Jalali F, Marrano P, Squire JA, Bristow RG. Defective DNA strand break repair after DNA damage in prostate cancer cells: implications for genetic instability and prostate cancer progression. *Cancer Res* 2004; **64**: 8526–33.
63. Sharma P, Karian J, Sharma S, Liu S, Mongan PD. Pyruvate ameliorates post ischemic injury of rat astrocytes and protects them against PARP mediated cell death. *Brain Res* 2003; **992**: 104–13.
64. Nakamura J, Asakura S, Hester SD, de Murcia G, Caldecott KW, Swenberg JA. Quantitation of intracellular NAD(P)H can monitor an imbalance of DNA single strand break repair in base excision repair deficient cells in real time. *Nucleic Acids Res* 2003; **31**: e104.
65. Slade N, Storga-Tomic D, Birkmayer GD, Pavelic K, Pavelic J. Effect of extracellular NADH on human tumor cell proliferation. *Anticancer Res* 1999; **19**: 5355–60.
66. Bieganowski P, Brenner C. Discoveries of nicotinamide riboside as a nutrient and conserved NRK genes establish a Preiss-Handler independent route to NAD+ in fungi and humans. *Cell* 2004; **117**: 495–502.
67. Knox RJ, Chen S. Quinone reductase-mediated nitro-reduction: clinical applications. *Methods Enzymol* 2004; **382**: 194–221.
68. Foster CE, Bianchet MA, Talalay P, Faig M, Amzel LM. Structures of mammalian cytosolic quinone reductases. *Free Radic Biol Med* 2000; **29**: 241–5.

69. Faig M, Bianchet MA, Talalay P, Chen S, Winski S, Ross D, *et al.* Structures of recombinant human and mouse NAD(P)H:quinone oxidoreductases: species comparison and structural changes with substrate binding and release. *Proc Natl Acad Sci U S A* 2000; **97**: 3177–82.
70. Joseph P, Long DJ, Klein-Szanto AJ, Jaiswal AK. Role of NAD(P)H:quinone oxidoreductase 1 (DT diaphorase) in protection against quinone toxicity. *Biochem Pharmacol* 2000; **60**: 207–14.
71. Danson S, Ward TH, Butler J, Ranson M. DT-diaphorase: a target for new anticancer drugs. *Cancer Treat Rev* 2004; **30**: 437–49.
72. Ross D, Kepa JK, Winski SL, Beall HD, Anwar A, Siegel D. NAD(P)H: quinone oxidoreductase 1 (NQO1): chemoprotection, bioactivation, gene regulation and genetic polymorphisms. *Chem Biol Interact* 2000; **129**: 77–97.
73. Dinkova-Kostova AT, Talalay P. Persuasive evidence that quinone reductase type 1 (DT diaphorase) protects cells against the toxicity of electrophiles and reactive forms of oxygen. *Free Radic Biol Med* 2000; **29**: 231–40.
74. Ross D. Quinone reductases multitasking in the metabolic world. *Drug Metab Rev* 2004; **36**: 639–54.
75. Jaiswal AK. Jun and Fos regulation of NAD(P)H: quinone oxidoreductase gene expression. *Pharmacogenetics* 1994; **4**: 1–10.
76. Watanabe N, Forman HJ. Autoxidation of extracellular hydroquinones is a causative event for the cytotoxicity of menadione and DMNQ in A549-S cells. Arch Biochem Biophys 2003; **411**: 145–57.
77. Talalay P, Dinkova-Kostova AT. Role of nicotinamide quinone oxidoreductase 1 (NQO1) in protection against toxicity of electrophiles and reactive oxygen intermediates. *Methods Enzymol* 2004; **382**: 355–64.
78. Siegel D, Ross D. Immunodetection of NAD(P)H:quinone oxidoreductase 1 (NQO1) in human tissues. *Free Radic Biol Med* 2000; **29**: 246–53.
79. Long DJ, Gaikwad A, Multani A, Pathak S, Montgomery CA, Gonzalez FJ, *et al.* Disruption of the NAD(P)H:quinone oxidoreductase 1 (NQO1) gene in mice causes myelogenous hyperplasia. *Cancer Res* 2002; **62**: 3030–6.
80. Radjendirane V, Joseph P, Lee YH, Kimura S, Klein-Szanto AJ, Gonzalez FJ, *et al.* Disruption of the DT diaphorase (NQO1) gene in mice leads to increased menadione toxicity. *J Biol Chem* 1998; **273**: 7382–9.
81. Asher G, Lotem J, Cohen B, Sachs L, Shaul Y. Regulation of p53 stability and p53-dependent apoptosis by NADH quinone oxidoreductase 1. *Proc Natl Acad Sci U S A* 2001; **98**: 1188–93.
82. Asher G, Lotem J, Kama R, Sachs L, Shaul Y. NQO1 stabilizes p53 through a distinct pathway. *Proc Natl Acad Sci U S A* 2002; **99**: 3099–104.
83. Brooks JD, Paton VG, Vidanes G. Potent induction of phase 2 enzymes in human prostate cells by sulforaphane. *Cancer Epidemiol Biomarkers Prev* 2001; **10**: 949–54.
84. Brooks JD, Goldberg MF, Nelson LA, Wu D, Nelson WG. Identification of potential prostate cancer preventive agents through induction

of quinone reductase in vitro. *Cancer Epidemiol Biomarkers Prev* 2002; **11**: 868–75.
85. Jones SB, Brooks JD. Modest induction of phase 2 enzyme activity in the F-344 rat prostate. *BMC Cancer* 2006; **6**: 62.
86. Vella F, Ferry G, Delagrange P, Boutin JA. NRH:quinone reductase 2: an enzyme of surprises and mysteries. *Biochem Pharmacol* 2005; **71**: 1–12.
87. Jaiswal AK, Burnett P, Adesnik M, McBride OW. Nucleotide and deduced amino acid sequence of a human cDNA (NQO2) corresponding to a second member of the NAD(P)H:quinone oxidoreductase gene family. Extensive polymorphism at the NQO2 gene locus on chromosome 6. *Biochemistry* 1990; **29**: 1899–906.
88. Long DJ, Jaiswal AK. NRH:quinone oxidoreductase2 (NQO2). *Chem Biol Interact* 2000; **129**: 99–112.
89. Jaiswal AK, Bell DW, Radjendirane V, Testa JR. Localization of human NQO1 gene to chromosome 16q22 and NQO2-6p25 and associated polymorphisms. *Pharmacogenetics* 1999; **9**: 413–8.
90. Long DJ, Iskander K, Gaikwad A, Arin M, Roop DR, Knox R, *et al.* Disruption of dihydronicotinamide riboside:quinone oxidoreductase 2 (NQO2) leads to myeloid hyperplasia of bone marrow and decreased sensitivity to menadione toxicity. *J Biol Chem* 2002; **277**: 46131–9.
91. Iskander K, Paquet M, Brayton C, Jaiswal AK. Deficiency of NRH:quinone oxidoreductase 2 increases susceptibility to 7,12-dimethylbenz(a)anthracene and benzo(a)pyrene-induced skin carcinogenesis. *Cancer Res* 2004; **64**: 5925–8.
92. Smoliga JM, Baur JA, Hausenblas HA. Resveratrol and health–a comprehensive review of human clinical trials. *Mol Nutr Food Res* 2011; **55**: 1129–41.
93. Patel KR, Scott E, Brown VA, Gescher AJ, Steward WP, Brown K. Clinical trials of resveratrol. *Ann N Y Acad Sci* 2011; **1215**: 161–9.
94. Tomé-Carneiro J, Larrosa M, González-Sarrías A, Tomás-Barberán FA, García-Conesa MT, Espín JC. Resveratrol and clinical trials: the crossroad from in vitro studies to human evidence. *Curr Pharm Des* 2013; **19**: 6064–93.
95. Zordoky BN, Robertson IM, Dyck JR. Preclinical and clinical evidence for the role of resveratrol in the treatment of cardiovascular diseases. *Biochim Biophys Acta* 2015; **1852**: 1155–77.

Chapter 3

Resveratrol as an Activator or Inhibitor of Enzymes and Proteins

Avinash Kumar, Eun-Jung Park and John M. Pezzuto

Arnold and Marie Schwartz College of Pharmacy and Health Sciences, Long Island University, Brooklyn, NY 11201, USA

Resveratrol has been intensively studied over the past 20 years, primarily based on the premise of preventing or treating various disease states, or enhancing health span. In dealing with these topics, it is common to investigate molecular targets that ostensibly correlate with a physiological response. Of course correlation does not necessarily mean causation, but nonetheless, scores of targets have been reported. Here, we catalog many of these targets, explore how they can form networks, and critically evaluate how this relates to a meaningful outcome. Anti-inflammatory activity appears to be a response that is credibly mediated by resveratrol, and this represents a common denominator for modulating the myriad of ailments for which resveratrol has been studied. In addition, targets that can be modulated by topical or direct application of resveratrol, which are numerous, appear particularly meaningful since high concentrations can be achieved. Recent results further indicate that treatment of neuronal diseases warrant further investigation. It remains problematic to specifically endorse the use of resveratrol for any particular ailment, but work continues, and evidence-based recommendations should be forthcoming.

3.1. Introduction

Following the publication of our seminal paper describing the pleiotropic activities of resveratrol related to cancer chemoprevention and other disease states,[1] the compound has been the subject of intensive investigation. Thousands of manuscripts have appeared in the scientific literature, monographs have been published, patents have been issued, symposia and conferences have been conducted, derivatives and metabolites have been studied, etc. The spectrum of activities is overwhelming.

In the beginning, interest largely stemmed from the natural occurrence of resveratrol in grapes and grape products such as wine, which of course are found in the diet of humans. While dietary consumption of resveratrol from natural sources remains of interest, most animal, human, and *in vitro* studies have employed concentrations that vastly exceed even the largest quantity that could be conceivably consumed in a normal diet. For example, daily doses of 5 g have been reported, and based on a perusal of the numerous dietary supplements available on the open market, nearly all products provide high milligram levels.

A positive relationship between resveratrol and health is often touted but in fact remains moot. Nonetheless, given that resveratrol has been widely investigated for potential value in treating of preventing a broad array of disease states,[2] an interesting question relates to molecular targets. First of all, it has been clear from the outset that resveratrol does not abide by the concept of 'one drug, one target'. This dogma simply does not apply. The molecule is incredibly promiscuous, interacting with a host of targets.[3]

Obviously, however, the characteristic of promiscuity has not held back scientific inquiry, and it is only natural to beg the question, is there a key target? Although numerous treatises have been published, we attempt here to summarize many of the targets that are reported to be influenced in the presence of resveratrol. With this knowledge, we can depict a network or web of response that may exist in the biological milieu. The ultimate physiological response, if any, is less clear, but at least we can generate a somewhat more realistic picture of the mode of action.

3.2. Action of resveratrol on enzymes and proteins

3.2.1. *Introduction*

Proteins are the functional units of a cell. The majority of the proteins are enzymes, which catalyze a myriad of biochemical reactions to enable a cell to function properly and maintain homeostasis. A disease condition is a perturbance of the homeostasis due to altered levels and activity of enzymes and proteins. The premise of any intervention to treat a disease is based on the correction of this alteration and restoration of homeostasis. The health beneficiary properties of resveratrol are a consequence of its action on multiple proteins and enzymes, many of which are summarized in Table 3.1.

3.2.2. *Neuroprotective effects of resveratrol*

Neuroprotective effects of resveratrol have been reported through inhibition of AMPK, SIRT1, p38 MAP kinase;[4] AKT and the mTOR signaling pathway;[5] MUNC13-1;[6] PDE4A;[7] CRF and GR;[8] AKT, mTOR, MAPK signaling and tPA;[9] GAPDH;[10] ERK1/2,[11] and activation of SIRT1;[12] β-catenin;[13] SIRT1 and AMPK;[7] BDNF;[8] AMPK;[9] BDNF and ERK1/2.[14] Neuroprotective effects of resveratrol have also been reported as being related to anti-inflammatory activity through inhibition of TNFα and IL1β;[15] CCR3, CCR5, CCR7, CCR9, CXCR3, CXCR5;[16] IL6 and TNFα.[13] The antioxidant activity of resveratrol through activation of NRF2;[17] NRF2, NQO1, SOD[18] leading to neuroprotection has also been documented. The protective effects of resveratrol against glutamate-induced neurotoxicity through activation of Na^+K^+-ATPase and glutamine synthetase,[19] and SIRT1 and HO1[20] have also been reported. Moreover, the protective role of resveratrol in neurodegenerative diseases such as Alzheimer's,[21] Parkinson's[22,23] and Huntington's disease[24] have been studied.

Table 3.1. Action of resveratrol on proteins and enzymes.

Dose	Target	Action	Model	Disease	Reference
—	AMPK SIRT1 p38	Inhibitor Inhibitor Activator	SH-SY5Y cell line and Sprague-Dawley rats	Alcohol-induced neuron damage	Gu *et al.* 2018
—	PTEN	Inhibitor	Vascular smooth muscle cells	Atherosclerosis	Ma *et al.* 2018
300 mg/kg	SIRT1	Activator	Mouse cochlea	Noise-induced hearing loss	Xiong *et al.* 2018
30 mg/kg 10 μM	PGC1α SIRT1 FOXO3a	Activator Activator Activator	Rat pancreas and INS-1E cell line	Type 2 diabetes mellitus	Cao *et al.* 2018
15, 50, 100 μM	MDR1		CCRF-CEM cell line	Acute Lymphoblastic Leukemia	Zaidi *et al.* 2018
60 mg/kg	AKT/mTOR	Inhibitor	Rat brain	Early brain injury after subarachnoid hemorrhage	Guo *et al.* 2018
800 mg/day	NRF2 SOD	Activator Activator	Peripheral Blood Mononuclear Cells (PBMCs) from human patients	Type 2 diabetes mellitus	Seyyedebrahimi *et al.* 2018
12.5–100 μM	SIRT1 AKT1	Activator Activator	PC12 cell line	Parkinson's disease	Wang *et al.* 2018
25 μM	SIRT1 EPAC1	Activator Activator	Pancreatic islets	Normal mouse pancreas	Brouwer *et al.* 2018

(*Continued*)

Table 3.1. (*Continued*)

Dose	Target	Action	Model	Disease	Reference
40 mg/kg	SIRT1	Activator	Medial prefrontal cortex and amygdala	Mice on resveratrol supplemented high fat diet	Xu *et al.* 2018
10 mg/kg	TNFα	Inhibitor	Spinal cord tissue	Spinal cord injury	Senturk *et al.* 2017
	IL1β	Inhibitor			
—	SIRT1	Activator	Dorsal hippocampus tissue	Rats on resveratrol supplemented high fat diet	Li *et al.* 2017
	BDNF	Activator			
	ERK1/2	Inhibitor			
12 μM	Cx3	Activator	U2-OS cell line	Osteosarcoma	Xie *et al.* 2017
	E-cadherin	Activator			
	MYC	Inhibitor			
	CCND1	Inhibitor			
	MMP2	Inhibitor			
	MMP9	Inhibitor			
0.1–50 μM	NRF2	Activator	A549 cell line	Oxidative stress in lung tissue	Canella *et al.* 2017
0–100 μM	SIRT2	Activator	HepG2 cell line	Aging	Pan *et al.* 2017
100 μM	BRD2	Inhibitor	*in vitro*	BET bromodomain inhibition	Dutra *et al.* 2017
	BRD3	Inhibitor			
	BRD4	Inhibitor			
	BRDT	Inhibitor			

(*Continued*)

Table 3.1. (*Continued*)

Dose	Target	Action	Model	Disease	Reference
5 μM	HNRNPA0	Activator	HNDF cell line	Cellular senescence	Latorre *et al.* 2017
	HNRNPA1	Activator			
	HNRNPA2B1	Activator			
	HNRNPD	Activator			
	HNRNPH3	Activator			
	HNRNPK	Activator			
	HNRNPM	Activator			
	HNRNPUL2	Activator			
	AKAP17A	Activator			
	IMP3	Activator			
	LSM14A	Activator			
	LSM2	Activator			
	SF3B1	Activator			
	SRSF1	Activator			
	SRSF2	Activator			
	SRSF3	Activator			
	SRSF6	Activator			
	SRSF7	Activator			
	SRSF18	Activator			
	TRA2B	Activator			
6.25, 25 μM	MYC	Inhibitor	BCap37, MDA-MB-231, MDA-MB-468, Hs578T, SKBR3, MCF7 and HBL100 cell line	Breast cancer	Pan *et al.* 2017
	MICA	Activator			
	MICB	Activator			
100 μM	TMEM16A	Activator	Guinea pig ileum	Gastrointestinal hypomotility	Chai *et al.* 2017

(*Continued*)

Table 3.1. (*Continued*)

Dose	Target	Action	Model	Disease	Reference
10 μM	PDP1	Activator	Caco2 cell line	Colon cancer	Saunier *et al.* 2017
	PDH	Activator			
6.25–100 μM	MUNC13-1	Inhibitor	HT22 cell line	Neurodegenerative diseases	Pany *et al.* 2017
10, 20 mg/kg	TAK1	Inhibitor	Alveolar macrophages and lung fibroblasts from silica-exposed rats	Pneumoconiosis	Li *et al.* 2017
0–200 μM	TLR4	Inhibitor	Human articular chondrocytes	Osteoarthritis	Gu *et al.* 2017
	MyD88	Inhibitor			
	MMP13	Inhibitor			
	IL6	Inhibitor			
5 μM	IL1α	Inhibitor	MRC5 cell line	Cellular senescence	Menicacci *et al.* 2017
	IL1β	Inhibitor			
	IL6	Inhibitor			
	IL8	Inhibitor			
	TGFβ	Inhibitor			
	MMP2	Inhibitor			
	MMP3	Inhibitor			
	uPA	Inhibitor			
	uPAR	Inhibitor			
	IL6R	Inhibitor			
	IGF1R	Inhibitor			
	TGFR2	Inhibitor			
	CXCR4	Inhibitor			

(*Continued*)

Table 3.1. (*Continued*)

Dose	Target	Action	Model	Disease	Reference
	EGFR	Inhibitor	H460 cell line	Lung cancer	Wright *et al.* 2017
	AKT	Inhibitor			
	NF-κB	Inhibitor			
	c-FLIP	Inhibitor			
	Caspase 8	Activator			
	PARP	Activator			
10 mg/kg/day	KLF15	Activator	Myocardium of rats with chronic myocardial ischemia	Myocardial infarction	Rogers and Otis 2017
20 μM	AP1	Activator	HEK293 and HepG2 cell line	Regulation of gene transcription	Thiel and Rossler 2017
	CREB	Activator			
	EGR1	Activator			
	ELK1	Activator			
	NRF2	Activator			
75 μM	IGFBP3	Inhibitor	TALL-104 and SUP-B15 cell line	Acute lymphoblastic leukemia	Zhou *et al.* 2017
1 g/kg	IL6	Inhibitor	Mice hippocampus tissue	Aging, Neurodegeneration	Palomera-Avalos *et al.* 2017
	TNFα	Inhibitor			
	β-catenin	Activator			
20, 80 μM	ERα	Inhibitor	T47D cell line	Breast cancer	Saluzzo *et al.* 2016
	p53	Activator			
40 mg/kg/day	NF-κB	Inhibitor	Rat lung tissues	Hypoxic pulmonary hypertension	Xu *et al.* 2016
	IL1β	Inhibitor			
	IL6	Inhibitor			
	TNFα	Inhibitor			
	VEGF	Inhibitor			

(*Continued*)

Table 3.1. (*Continued*)

Dose	Target	Action	Model	Disease	Reference
	HIF1α	Inhibitor			
	MAPK/ERK1	Inhibitor			
	PI3K/AKT	Inhibitor			
	NRF2	Activator			
	TRX1	Activator			
20, 40 mg/kg	CCR3	Inhibitor	Mice CD4+ T cells	Autism, Neuroinflammatory disorders	Bakheet *et al.* 2016
	CCR5	Inhibitor			
	CCR7	Inhibitor			
	CCR9	Inhibitor			
	CXCR3	Inhibitor			
	CXCR5	Inhibitor			
0–100 μM	DDX5	Inhibitor	PC3 cell line	Prostate cancer	Taniguchi *et al.* 2016
100 μM	BRCA1	Activator	MCF7 and MDA-MB-231 cell line	Breast cancer	Medina-Aguilar *et al.* 2016
	AURKA	Inhibitor			
	PLK1	Inhibitor			
	CCND1	Inhibitor			
	CCNB1	Inhibitor			
30 mg/kg/day	SP1	Inhibitor	Rat white adipose tissue	Rats on resveratrol supplemented obesogenic diet	Gracia *et al.* 2016
	SREBP1	Inhibitor			
	FASN	Inhibitor			
50 μM	ULK1	Inhibitor	MCF7 and SW620 cell line	Cancer, inflammation	Park *et al.* 2016
	mTOR	Inhibitor			
20 mg/kg	SIRT1	Activator	Rat brain tissue	Cerebral ischemic injury (stroke)	Wan *et al.* 2016
	AMPK	Activator			
	PDE4A	Inhibitor			

(*Continued*)

Table 3.1. (*Continued*)

Dose	Target	Action	Model	Disease	Reference
30 μM	TRPA1	Inhibitor	HEK293 cell line	Nociceptive and inflammatory pain	Nalli *et al.* 2016
10, 25 μM	E-cadherin	Activator	SKOV3 and PA1 cell line	Ovarian cancer	Kim *et al.* 2016
	SRC	Inhibitor			
	HIF1α	Inhibitor			
	hTERT	Inhibitor			
1-10 μM	SIRT1	Inhibitor	HCT116 and SW480 cell line	Colon cancer	Buhrmann *et al.* 2016
	NF-κB	Inhibitor			
	MMP9	Inhibitor			
	CXCR4	Inhibitor			
0–400 μM	KITLG	Inhibitor	HT29 and HCT116 cell line	Colorectal cancer	Yang *et al.* 2015
	IL6	Inhibitor			
	TNFα	Inhibitor			
50, 100 μM	PTEN	Activator	DU145 and 22Rv1 cell line	Prostate cancer	Dhar *et al.* 2015
1, 5, 10 mg/kg	Visfatin	Inhibitor	Rat visceral adipose tissue	Type 2 diabetes mellitus	Asadi *et al.* 2015
	Vaspin	Inhibitor			
10, 20, 40 mg/kg	CRF	Inhibitor	Rat hippocampus, hypothalamus and frontal cortex tissue	Post stroke depression	Pang *et al.* 2015
	GR	Inhibitor			
	BDNF	Activator			
25 μM, 10 mg/kg	NRF2	Activator	Mice astrocyte cells and brain tissue	Cerebral ischemic injury (stroke)	Narayanan *et al.* 2015
15 μM	HER2	Inhibitor	SKBR3 and MDA-MB-231 cell line	Breast cancer	Vinod *et al.* 2015
	AKT	Inhibitor			
	ERK1/2	Inhibitor			

(*Continued*)

Table 3.1. (*Continued*)

Dose	Target	Action	Model	Disease	Reference
50 μM, 50 mg/kg	MTA1	Inhibitor	DU145 and 22Rv1 cell lines, DU145 xenograft	Prostate cancer	Dhar *et al.* 2015
	MTA2	Inhibitor			
	MTA3	Inhibitor			
	HDAC1	Inhibitor			
	HDAC2	Inhibitor			
	PTEN	Activator			
	AKT	Inhibitor			
0.25 % w/w	AMPK	Activator	4NQO-treated mice tongue tissue	Oral cancer	Shrotriya *et al.* 2015
	p62	Inhibitor			
50 μM	AMPK	Activator	HepG2 cell line	Hepatocellular carcinoma	Tameda *et al.* 2014
	Survivin	Inhibitor			
0–200 μM	TLR4	Inhibitor	Human articular chondrocytes	Osteoarthritis	Liu *et al.* 2014
	MyD88	Inhibitor			
	NF-κB	Inhibitor			
10 μM	NF-κB	Inhibitor	Human monocyte-derived dendritic cells	Inflammation	Silva *et al.* 2014
	IL-12/IL-23	Inhibitor			
	IL-10	Activator			
10–100 μM	AMPK	Activator	Rat cortical neurons	Neurological disease	Cho *et al.* 2014
	AKT	Inhibitor			
	MAPK	Inhibitor			
	mTOR	Inhibitor			
	tPA	Inhibitor			
100 μM	AKT	Activator	C2C12 cell line	Muscle atrophy/inflammation	Wang *et al.* 2014
	mTOR	Activator			
	FOXO1	Activator			
	MAFbx	Inhibitor			
	MuRF1	Inhibitor			

(*Continued*)

Table 3.1. (*Continued*)

Dose	Target	Action	Model	Disease	Reference
10, 20, 30 μM	p21	Activator	Cigarette smoke-transformed MCF10A cell line	Breast cancer	Mohapatra *et al.* 2014
	PI3K	Inhibitor			
	AKT	Inhibitor			
	NF-κB	Inhibitor			
	CDC2	Inhibitor			
	CDC6	Inhibitor			
	CCND1	Inhibitor			
	CCNB1	Inhibitor			
	CCNA	Inhibitor			
	CCNE1	Inhibitor			
	Pol-β	Inhibitor			
	Pol-δ	Inhibitor			
	Pol-ε	Inhibitor			
	Pol-η	Inhibitor			
	RPA	Inhibitor			
	Fen1	Inhibitor			
	DNA-Ligase I	Inhibitor			
	PCNA	Inhibitor			
100 μM	Na^+K^+ATPase	Activator	Glutamate-exposed rat hippocampus tissue	Neurodegenerative diseases	Quincozes-Santos *et al.* 2014
	Glutamine synthetase	Activator			
	GAPDH	Inhibitor	*in vitro*	Neurodegenerative diseases	Rodacka *et al.* 2014
174 mg/kg/day	GSK3β	Activator	Mice brain tissue	Alzheimer's disease	Varamini *et al.* 2014
	Transthyretin	Activator			
	Drebrin	Activator			

(*Continued*)

Table 3.1. (*Continued*)

Dose	Target	Action	Model	Disease	Reference
5, 25 mg/kg	DNMT3b	Inhibitor	Rat breast tumor tissue	Breast cancer	Qin *et al.* 2014
1, 25 μM	CCNA2	Inhibitor	C2C12 cell line	Muscle impairment	Montesano *et al.* 2013
	CCNB1	Inhibitor			
	CCND1	Inhibitor			
	CCNE1	Inhibitor			
	CCNF	Inhibitor			
	p21	Activator			
	Myf5	Activator			
	MyoD	Activator			
	N-cadherin	Activator			
	p120	Activator			
	α-Actinin	Activator			
	IGF1R	Activator			
	AKT	Activator			
	ERK1/2	Activator			
	AMPK	Activator			
25, 50 μM	SIRT1	Inhibitor	AGS, BGC823, SGC7901	Gastric cancer	Yang *et al.* 2013
40 mg/kg/day	CCND1	Inhibitor			
	CDK4	Inhibitor			
	CDK6	Inhibitor			
	p21	Activator			
	p16	Activator			
20, 40, 80 mg/kg	BDNF	Activator	Mice brain tissue	Depression	Wang *et al.* 2013
	ERK1/2	Activator			

(*Continued*)

Table 3.1. (*Continued*)

Dose	Target	Action	Model	Disease	Reference
0–50 μM	ERK1/2 STIM1	Inhibitor Inhibitor	HEK293 cell line	Calcium homeostasis	Casas-Rua *et al.* 2013
100 μM	BCL2 Survivin	Inhibitor Inhibitor	GH3 cell line	Pituitary adenoma	Voellger *et al.* 2013
10 mg/kg	SPHK1 MPO	Inhibitor Inhibitor	Oxozalone-treated rat colon tissue	Ulcerative colitis, Inflammatory Bowel Disorder	Abdin 2013
10, 25 μM	NF-κB NADPH oxidase	Inhibitor Inhibitor	bEnd3 cell line	Type 2 diabetes mellitus	Chen *et al.* 2013
0–100 μM	SIRT1 FOXO1 FOXO3A eNOS	Activator Activator Activator Activator	EA.hy 926 cell line	Cardiovascular disorders	Xia *et al.* 2013
50, 100 mg/kg	LDLr SR-BI	Activator Activator	Rat liver tissue	Nonalcoholic fatty liver disease	Xin *et al.* 2013
5–50 μM	DNMT1 DNMT3a DNMT3b MeCP2	Inhibitor Inhibitor Inhibitor Inhibitor	Human breast tumors	Breast cancer	Mirza *et al.* 2013
5–100 μM 0.1–0.2 % w/w	SIRT1	Activator	RWPE1, LNCaP, C42B, DU145 and PC3 cell lines and prostate-specific *Pten* knockout mice	Prostate cancer	Li *et al.* 2013
5–100 μM	HDACs	Inhibitor	HepG2, Hep3B and HuH7 cell lines	Liver cancer	Venturelli *et al.* 2013
20, 50, 100 μM	LSD1	Inhibitor	HepG2, FAO, HEK293 and C2C12 cell lines		Abdulla *et al.* 2013

(*Continued*)

Table 3.1. (*Continued*)

Dose	Target	Action	Model	Disease	Reference
0–300 μM	NF-κB	Inhibitor	EBV-positive Raji and Akata cell line	Virus infection	De Leo *et al.* 2012
	AP1	Inhibitor			
25, 50 μM	CCNA	Inhibitor	Mice preadipocytes	Obesity	Kwon *et al.* 2012
	CDK2	Inhibitor			
	IR	Inhibitor			
0.005 %, 0.02 % w/w	AMPK	Activator	Mice serum and liver, pancreas and skeletal muscle tissue	Type 2 diabetes mellitus	Do *et al.* 2012
	Gluconeogenic enzymes	Activator			
	Insulin	Activator			
	GLUT4	Activator			
	Glycolytic enzymes	Activator			
	Glycogen synthase	Activator			
	IKK	Inhibitor			
	UCP	Inhibitor			
5 μM	YAP1P	Activator	Yeast cells and SGBS cell line	Aging	Escote *et al.* 2012
	TRX2	Activator			
	TRR1	Activator			
	AHP1	Activator			
2.5–10 μM	HO1	Activator	HT22 cell line	Glutamate-induced neurotoxicity	Kim *et al.* 2012
	SIRT1	Activator			
	BCR-ABL	Inhibitor	K562 cell line	Chronic myeloid leukemia	Liao *et al.* 2012
	GLI1	Inhibitor			

(*Continued*)

Table 3.1. *(Continued)*

Dose	Target	Action	Model	Disease	Reference
100 μM	SIRT1	Activator	Rat granulosa cells	Ovary physiology	Morita *et al.* 2012
	LHr	Activator			
	StAR	Activator			
	P450 aromatase	Activator			
	Progesterone	Activator			
5, 10 μM	JNK	Inhibitor	Rat renal mesangial cell line and primary mesangial cells	Diabetic nephropathy	Zhang *et al.* 2012
	NF-κB	Inhibitor			
	NADPH oxidase	Inhibitor			
0.5–20 μM	DNMT1	Inhibitor	MCF7 cell line	Breast cancer	Papoutsis *et al.* 2012
	BRCA1	Activator			
20–120 μM	TOPOIIα	Inhibitor	U87 cell line	Glioma	Leone *et al.* 2012
	ATM	Activator			
	CHK2	Activator			
	H2AX	Activator			
5 mg/kg/day	VEGF	Inhibitor	Rat heart tissue	Myocardial Ischemia	Mukhopadhyay *et al.* 2012
	HIF1α	Inhibitor			
10, 14, 30 μM	PTEN	Activator	MCF7 cell line	Breast cancer	Stefanska *et al.* 2012
	DNMT1	Inhibitor			
	p21	Activator			
5 μM	mTOR	Inhibitor	H4-II-E cell line	Protein synthesis	Villa-Cuesta *et al.* 2011
	S6K1	Inhibitor			
	eIF4EBP1	Inhibitor			
	eIF2α	Activator			
	AMPK	Activator			
	AKT	Inhibitor			

(Continued)

Table 3.1. (*Continued*)

Dose	Target	Action	Model	Disease	Reference
0–50 μM	ASF/SF2	Activator	HEK293 cell line	RNA splicing	Markus *et al.* 2011
	hnRNPA1	Activator			
	HuR	Activator			
50 μM	AKT/PKB	Inhibitor	Sudh14, Sudh110, HBL1, OCI-LY3 and RIVA cell lines	Diffuse large B-cell lymphoma	Hussain *et al.* 2011
	FOXO1	Inhibitor			
	GSK3β	Activator			
	BAD	Inhibitor			
	BAX	Activator			
	Cytochrome C	Activator			
	Caspase 9	Activator			
	Caspase 3	Activator			
	PARP	Activator			
	DR5	Activator			
150 μM	G6PDH	Inhibitor	HT29 cell line	Colon cancer	Vanamala *et al.* 2011
	TKT	Inhibitor			
	Talin	Inhibitor			
	FAK	Inhibitor			
100 mg/kg	NRF2	Activator	Rat cerebellum tissue and cerebellar granule cells	Fetal alcohol spectrum disorders	Kumar *et al.* 2011
	NQO1	Activator			
	SOD	Activator			
30 μM	ATG5	Activator	U87MG, U251 and U138MG cell line	Glioma	Filippi-Chiela *et al.* 2011
	Beclin1	Activator			
	LC3-II	Activator			
	Caspase 3				
	BAX				
	CDC2				
	CCNA				

(*Continued*)

Table 3.1. (*Continued*)

Dose	Target	Action	Model	Disease	Reference
	CCNE	Activator			
	RB	Activator			
	CCND1	Inhibitor			
60 mg/kg/day	GLP1	Activator	Colon, liver and muscle tissues and serum from mice on resveratrol supplemented high fat diet	Type 2 diabetes mellitus	Dao *et al.* 2011
	Insulin	Activator			
	IL10	Activator			
	TGFβ	Activator			
	TNFα	Inhibitor			
100 mg	LBP	Inhibitor	Human blood samples	Obesity	Ghanim *et al.* 2011
	p47	Inhibitor			
	TLR4	Inhibitor			
	CD14	Inhibitor			
	SOCS3	Inhibitor			
	IL1β	Inhibitor			
	KEAP1	Inhibitor			
	NRF2	Activator			
	NQO1	Activator			
	GSTP1	Activator			
25, 50 μM	p53	Activator	MCF7 and MDA-MB-231 cell line	Breast cancer	Hsieh *et al.* 2011
	p21	Activator			
	p53R2	Activator			
	ERK1/2	Activator			
	p38	Activator			

(*Continued*)

Table 3.1. (*Continued*)

Dose	Target	Action	Model	Disease	Reference
1, 5, 10 μM	COX2	Activator	OVCAR3 cell line	Ovarian cancer	Lin *et al.* 2011
	ERK1/2	Activator			
	p53	Activator			
	SUMO1	Activator			
25 μM	GLUT2	Activator	INS1E cell line and islets from type 2 diabetes donor	Type 2 diabetes mellitus	Vetterli *et al.* 2011
	Glucokinase	Activator			
	PDX1	Activator			
	HNF1α	Activator			
	TFAM	Activator			
30, 50, 100 μM	TGFβ2	Inhibitor	A431 cell line and p53(+/−)/SKH1 mice	Skin cancer	Kim *et al.* 2011
	AKT/PKB	Inhibitor			
	CREB	Inhibitor			
	ERK1/2	Inhibitor			
	SMAD3	Inhibitor			
	SMA	Inhibitor			
	E-cadherin	Activator			
5 mg/kg/day	ERK1/2	Activator	Rat heart tissue	Cardiovascular disease	Mukhopadhyay *et al.* 2010
	p38	Inhibitor			
50 mg/kg	GPx	Activator	Mice brain tissue	Parkinson's disease	Anandhan *et al.* 2010
	SOD	Inhibitor			
	Catalase	Inhibitor			
25 mg/mouse/day	SIRT1	Activator	Mice brown adipose tissue	Huntington's disease	Ho *et al.* 2010
	PGC1α	Activator			
	NRF1	Activator			
	UCP1	Activator			

(*Continued*)

Table 3.1. (*Continued*)

Dose	Target	Action	Model	Disease	Reference
0–500 μM	SIRT1	Activator	*in vitro*	Pleiotropic beneficial effects	Calamini *et al.* 2010
	COX1	Inhibitor			
	COX2	Inhibitor			
	QR2	Inhibitor			
1.56, 6.25, 12.5 μM	NKG2D	Activator	NK92 cell line	NK cell cytotoxicity	Lu *et al.* 2010
	JNK1	Activator			
	ERK1/2	Activator			
	Perforin	Activator			
50 μM	CD40	Inhibitor	Human dendritic cells	Immunological function	Svajger *et al.* 2010
	CD80	Inhibitor			
	CD86	Inhibitor			
	NF-κB	Inhibitor			
	CD1a	Inhibitor			
	CCR7	Inhibitor			
	ILT3	Activator			
	ILT4	Activator			
	IL12p70	Inhibitor			
	IL10	Activator			
10–250 μM	SIRT1	Activator	HUVEC cell line	Cardiovascular disorder	Gracia-Sancho *et al.* 2010
	KLF2	Activator			
	eNOS	Activator			
	TN	Activator			
	CNP	Activator			
	MEK5	Activator			
	MEF2	Activator			

(*Continued*)

Table 3.1. (*Continued*)

Dose	Target	Action	Model	Disease	Reference
50 μM	MTA1	Inhibitor	LNCaP and DU145 cell lines	Prostate cancer	Kai *et al.* 2010
	p53	Activator			
	p21	Activator			
	BAX	Activator			
25 μM	QR2	Activator	Prostate stromal and epithelial cells	Prostate physiology	Hsieh 2009
0.1 μM	ERK1/2	Activator	$GH_3/B_6/F_{10}$ cell line	Pituitary cancer	Jeng and Watson 2009
	JNK	Activator			
	ELK1	Activator			
	ATF2	Activator			
	Caspase8	Inhibitor			
10 μM, 20 mg/kg/day	SIRT1	Activator	Human coronary artery endothelial cells and aorta from type 2 diabetes mice	Cardiovascular disease	Csiszar *et al.* 2009
	eNOS	Activator			
	PGC1α	Activator			
	NRF1	Activator			
	TFAM	Activator			
	Porin	Activator			
	Complex I	Activator			
	Complex II	Activator			
	Complex III	Activator			
	Complex IV	Activator			
	Complex V	Activator			
10 μM	NOX1	Activator	Human endothelial cells	Cardiovascular disease	Schilder *et al.* 2009
	NOX4	Activator			

(*Continued*)

Table 3.1. *(Continued)*

Dose	Target	Action	Model	Disease	Reference
5–30 μM	NOX1	Inhibitor	Macrophages	Atherosclerosis	Park *et al.* 2009
	MCP1	Inhibitor			
	AKT	Inhibitor			
	ERK1/2	Inhibitor			
	FOXO3A	Inhibitor			
20 μM	SIRT1	Activator	MCF7 cell line	Breast cancer	Bourguignon *et al.* 2009
	MDR1	Inhibitor			
	BCL-xL	Activator			
1–100 μM	pGC/KinaseG	Activator	Human coronary smooth muscle cells	Atherosclerosis	El-Mowafy *et al.* 2009
	ERK1/2	Inhibitor			
50 μM	APE/REF1	Inhibitor	Lu1205 and Wm3211 cell line	Melanoma	Yang *et al.* 2008
	JUND	Inhibitor			
	MMP1	Inhibitor			
	BCL2	Inhibitor			
	iNOS	Inhibitor			
50 μM	ERK1/2	Inhibitor	Rat hippocampus tissue	Oxidative stress-induced brain disorders	de Almeida *et al.* 2008
20 mg/kg/day	HO1	Activator	Heart tissue from rats on resveratrol supplemented high cholesterol diet	Atherosclerosis	Penumathsa *et al.* 2008
	eNOS	Activator			
	VEGF	Activator			
	CAV1	Inhibitor			
10 μM	SIRT1	Activator	HaCaT cell line	Aging	Cao *et al.* 2009
	AMPK	Activator			
	p53	Activator			
	BCL-xL	Activator			

(Continued)

Table 3.1. (*Continued*)

Dose	Target	Action	Model	Disease	Reference
25 μM	SIRT1	Activator	Rat and human mesangial cells	Mesangioproliferative glomerulonephritis	Venkatesan *et al.* 2008
	SRC	Inhibitor			
	AKT	Inhibitor			
	CCND1	Inhibitor			
	RB	Inhibitor			
	CDK2	Inhibitor			
	PDGFR	Inhibitor			
	PI3K	Inhibitor			
	ERK1/2	Inhibitor			
	PTP1B	Activator			
25, 50, 100 μM	STAT3	Inhibitor	Lu1205, Wm9, Wm35, HHMSX, LOX and SW1 cell lines	Melanoma	Johnson *et al.* 2008
	NF-κB	Inhibitor			
	cFLIP	Inhibitor			
	BCL-xL	Inhibitor			
	FOXO3A	Inhibitor			
	ERK1/2	Inhibitor			
	JNK	Activator			
	p38	Activator			
	ATM	Activator			
	CHK2	Activator			
	p53	Activator			
	p21	Activator			
	ATF2	Activator			
	TRAIL	Activator			

(*Continued*)

Table 3.1. (*Continued*)

Dose	Target	Action	Model	Disease	Reference
25 μM	GSTP1	—	LNCaP, CWR22Rv1, PC3 and DU145 cell lines	Prostate cancer	Hsieh *et al.* 2008
	ERβ	—			
	QR1	Activator			
	QR2	Activator			
	AR	Inhibitor			
	PSA	Inhibitor			
5–50 μM	COX2	Inhibitor	HT29 cell line	Colon cancer	Zykova *et al.* 2008
20 μg/mL	EGR1	Activator	MiaPaCa2, PROb and MCF7 cell lines	Cancers	Bickenbach *et al.* 2008
7.5 μg/ml	SIRT1	Activator	SIRT1 mutant mice	Cancer	Wang *et al.* 2008
0–128 μM	NOTCH	Inhibitor	MOLT4 cell line	Acute lymphoblastic leukemia	Cecchinato *et al.* 2007
	PI3K/AKT	Inhibitor			
	GSK3β	Activator			
	p53	Activator			
	p21	Activator			
	BAX	Activator			
0–150 μM	AR	Inhibitor	LNCaP and PC3 cell line	Prostate cancer	Benitez *et al.* 2007
	ERα	Inhibitor			
	PI3K	Inhibitor			
	AKT/PKB	Inhibitor			
	GSK3β	Activator			
	CCND1	Inhibitor			
0–150 μM	GPx	Activator	A549 cell line	Non-small cell lung cancer	Hu *et al.* 2007
	TRXr	Activator			
	p53	Activator			
	FAS	Activator			
	BCL2	Activator			
	Survivin	Inhibitor			

(*Continued*)

Table 3.1. (*Continued*)

Dose	Target	Action	Model	Disease	Reference
10 μM	AKT	Activator	Cardiomyocytes	Ischemic myocardial injury	Goh *et al.* 2007
	GSK3β	Inhibitor			
	p38	Activator			
0.25, 1 μmol	COX2	Inhibitor	Mouse skin tissue	Skin cancer	Kundu *et al.* 2006
	ERK1/2	Inhibitor			
	JNK	Inhibitor			
	p38	Inhibitor			
	NF-κB	Inhibitor			
	AP1	Inhibitor			
50–200 μM	PPARγ	Activator	Caco2 and HCT116 cell line	Colon cancer	Ulrich *et al.* 2006
	CK20	Activator			
	p38	Activator			
	SIRT1	Activator			
	PGC1α	Activator			
0–100 μM	ERK1/2	Activator	SJSA1 cell line	Osteosarcoma	Alkhalaf and Jaffal 2006
	p53	Activator			
	PARP	Activator			
0–100 μM	CCNA	Inhibitor	A431 cell line	Skin cancer	Kim *et al.* 2006
	CCND1	Inhibitor			
	CDK6	Inhibitor			
	p21	Activator			
	RB	Activator			
	p27	Activator			
	MEK1	Inhibitor			
	ERK1/2	Inhibitor			
	cJUN	Inhibitor			
	AP1	Inhibitor			

(*Continued*)

Table 3.1. (*Continued*)

Dose	Target	Action	Model	Disease	Reference
10 μM	FcεRI	Inhibitor	Mast cells	Immune response	Koo *et al.* 2006
	ERK1/2	Inhibitor			
	PLCγ1	Inhibitor			
10 μM	ERK1/2	Activator	Rat heart tissue	Ischemic myocardial injury	Das *et al.* 2006
	p38	Activator			
	MSK1	Activator			
	MAPKAPK2	Activator			
0.25–1 μmol	IKK	Inhibitor	Mouse skin tissue	Skin cancer	Kundu *et al.* 2006
	IκBα	Inhibitor			
	NF-κB	Inhibitor			
	ERK1/2	Inhibitor			
	p38	Inhibitor			
	COX2	Inhibitor			
50 μM	p53	Activator	WR21 cell line	Oral cancer	Young and Martin 2006
	p21	Activator			
	MDM2	Activator			
	CCNG	Activator			
30, 50 μM	TLR3	Activator	RAW264.7 cell line	Inflammatory diseases	Youn *et al.* 2005
	TLR4	Activator			
	TRIF	Activator			
	TAK1	Activator			
	RIP1	Activator			
	NF-κB	Activator			
	COX2	Activator			

(*Continued*)

Table 3.1. (*Continued*)

Dose	Target	Action	Model	Disease	Reference
0–200 μM	PI4K	Inhibitor	*in vitro*	Inflammatory diseases	Srivastava *et al.* 2005
2.5 mg/kg	iNOS	Activator	Rat heart tissue	Cardiovascular diseases	Das *et al.* 2005
	VEGF	Activator			
	KDR	Activator			
	eNOS	Activator			
0–150 μM	PI3K	Inhibitor	MCF7 cell line	Breast cancer	Pozo-Guisado *et al.* 2005
	Calpain	Inhibitor			
	NF-κB	Inhibitor			
	BCL2	Inhibitor			
	MMP9	Inhibitor			
1, 5, 25 μM	ERK1/2	Inhibitor	Mice skin tissue	Skin cancer	Kundu *et al.* 2004
	p38	Inhibitor			
	AP1	Inhibitor			
	COX2	Inhibitor			
10 μM	PGAM1B	Inhibitor	LNCaP, DU145 and PC3 cell line	Prostate cancer	Narayanan *et al.* 2004
100 μM	CK2	Inhibitor	HeLa cell line	Protein degradation	Uhle *et al.* 2003
	PKD	Inhibitor			
	cJUN	Inhibitor			
	p53	Activator			
10 μM	PSA	Inhibitor	LNCaP cell line	Prostate cancer	Narayanan *et al.* 2003
	ARA24	Inhibitor			
	NF-κB	Inhibitor			
	p53	Activator			

(*Continued*)

Table 3.1. (*Continued*)

Dose	**Target**	**Action**	**Model**	**Disease**	**Reference**
	p21	Activator			
	PIG7	Activator			
	CBP/p300	Activator			
	APAF1	Activator			
50 μM	O-acetyltransferase	Inhibitor	Human mammary epithelial cells, mammary tumor and breast reduction tissue and MCF7, ZR-75-1 cell lines	Breast cancer	Dubuisson *et al.* 2002
	Sulfotransferase	Inhibitor			
0.01 % w/v	CCND1	Inhibitor	Colon tissue from Min mice	Colon cancer	Schneider *et al.* 2001
	CCND2	Inhibitor			
	DP1	Inhibitor			
	YB1	Inhibitor			
	CTLA4	Activator			
	LIFr	Activator			
	MCP3	Activator			
	TSG101	Inhibitor			
	TGFβ	Inhibitor			
	INHBA	Inhibitor			
	DSC2	Inhibitor			
25 μM	ODC	Inhibitor	Caco2 cell line	Colon cancer	Schneider *et al.* 2000

3.2.3. *Cardiovascular protective effect of resveratrol*

Resveratrol is known to function as a cardiovascular protectant as well. Resveratrol protects against atherosclerosis through inhibition of PTEN;[25] NOX1, MCP, AKT, ERK1/2, FOXO3A;[26] CAV1,[27] and ERK1/2 signaling,[28] and activation of HO1, eNOS, VEGF[27] and pGC/Kinase G signaling.[28] Its role in the protection against ischemic myocardial injury through inhibition of VEGF, HIF1α;[29] GSK3β,[30] and activation of KLF15;[31] AKT, p38;[30] MAPK signaling[32] has also been well documented. Protection by resveratrol against other cardiovascular conditions through inhibition of p38 MAPK signaling,[33] and activation of SIRT1, FOXO1, FOXO3A, eNOS;[34] ERK1/2;[33] SIRT1, KLF2, eNOS, TN, CNP, MEK5, MEF2;[35] SIRT1, eNOS, PGC1α, NRF1, TFAM, Porin, Complex I-V;[36] NOX1, NOX4;[37] iNOS, VEGF, KDR, and eNOS[38] has been reported.

3.2.4. *Antidiabetic effect of resveratrol*

The antidiabetic effects of resveratrol are also well known. Resveratrol activation of PGC1α, SIRT1, FOXO3A;[39] NRF2, SOD;[40] AMPK, GLUT4, IKK, insulin, glycolytic and gluconeogenic enzymes, glycogen synthase;[41] GLP1, IL10, TGFβ, insulin;[42] GLUT2, glucokinase, PDX1, HNF1α, TFAM[43] and inhibition of Visfatin, Vaspin;[44] NF-κB, NOX;[45] UCP;[41] TNFα[42] leading to protective effects against type 2 diabetes mellitus have been reported. In addition, resveratrol was found to protect against diabetic nephropathy through inhibition of JNK, NF-κB and NOX.[46] In normal mouse pancreas, resveratrol was found to activate SIRT1 and EPAC1.[47]

3.2.5. *Antiobesity effect of resveratrol*

The antiobesity effects of resveratrol are known to be mediated through activation of SIRT1, BDNF;[48] SIRT1;[49] NRF2, NQO1, GSTP1,[50] and inhibition of ERK1/2 signaling;[48] SP1, SREBP1, FASN;[51] CCNA, CDK2, IR;[52] LBP, p47, TLR4, CD14, SOCS3, IL1β, KEAP1.[50]

3.2.6. *Antiaging effect of resveratrol*

Resveratrol is also reported to exhibit antiaging effects through inhibition of IL6, TNFα,[13] and activation of β-catenin;[13] SIRT2;[53] YAP1P, TRX2, TRR1, AHP1;[54] SIRT1, AMPK, p53, BCL-xL.[55]

3.2.7. *Anticancer activity of resveratrol*

The anticancer effect of resveratrol in numerous cancers through inhibition of oncogenes and activation of tumor suppressors has been extensively studied.

In leukemia, resveratrol showed inhibition of MDR1;[56] IGFBP3;[57] NOTCH and PI3K/AKT;[58] BCR-ABL and GLI1,[59] and activation of GSK3β, p53, p21 and BAX.[58]

In osteosarcoma, resveratrol showed activation of Cx3 and E-cadherin;[60] ERK1/2, p53 and PARP,[61] and inhibition of MYC, CCND1, MMP2 and MMP9.[60]

In breast cancer, resveratrol exhibited inhibition of MYC;[62] ERα;[63] AURKA, PLK1, CCND1 and CCNB1;[64] HER2, AKT and ERK1/2;[65] PI3K, NF-κB, CDC2/6, Cyclins A/B/D/E, DNA replication enzymes and PCNA;[66] DNMT3B;[67] DNMT1, DNMT3A, DNMT3B and MeCP2;[68] DNMT1;[69] DNMT1;[70] MDR1;[71] PI3K, calpain, NF-κB, BCL2 and MMP9;[72] *O*-acetyltransferase and sulfotransferase,[73] and activation of MICA and MICB;[62,63] BRCA1;[64] p21;[66] BRCA1;[69] PTEN and p21;[70] p53, p21, p53R2, p38 and ERK1/2;[74] SIRT1 and BCL-xL.[71]

In colon cancer, resveratrol showed activation of PDP1 and PDH;[75] PPARγ, PGC1α, SIRT1, p38 and CK20;[75] CTLA4, LIFr and MCP3,[76] and inhibition of KITLG, IL6 and TNFα;[77] SIRT1, NF-κB, MMP9 and CxCR4;[78] G6PDH, TKT, FAK and;[79] COX2;[80] CCND1, CCND2, DP1, YB1, TSG101, TGFβ, INHBA and DSC2;[76] ODC.[81]

In lung cancer, resveratrol exhibited inhibition of EGFR, AKT, NF-κB and cFLIP;[82] Survivin,[83] and activation of caspase 8 and PARP;[82] GPx, TRXr, p53, FAS and BCL2.[83]

In ovarian cancer, resveratrol showed activation of E-cadherin;[84] COX2, p53, SUMO1 and ERK1/2,[85] and inhibition of SRC, HIF1α and hTERT.[84]

In prostate cancer, resveratrol exhibited activation of PTEN;[86,87] SIRT1;[88] p53, p21 and BAX;[89] QR1 and QR2;[90] GSK3β;[91] p53, p21, PIG7, CBP/p300 and APAF1,[92] and inhibition of DDX5;[93] MTA1/2/3, HDAC1/2 and AKT;[87] MTA1;[89] AR and PSA;[90] AR, ERα, PI3K, AKT/PKB and CCND1;[91] PGAM1B;[94] PSA, ARA24 and NF-κB.[92]

In liver cancer, resveratrol showed activation of AMPK,[95] and inhibition of survivin;[95] HDACs.[96]

In gastric cancer, resveratrol exhibited inhibition of SIRT1, CCND1 and CDK4/6, and activation of p21 and p16.[97]

In oral cancer, resveratrol showed activation of AMPK;[98] p53, p21, MDM2 and CCNG,[99] and inhibition of p62.[98]

In pituitary cancer, resveratrol exhibited inhibition of BCL2 and survivin;[100] caspase 8,[101] and activation of ERK1/2, JNK, ELK1 and ATF2.[101]

In glioma, resveratrol showed inhibition of TOPOIIα;[102] CCND1,[103] and activation of ATM, CHK2 and H2AX;[102] ATG5, BAX, CDC2, CCNA, CCNE, RB, LC3-II, beclin 1 and caspase 3.[103]

In lymphoma, resveratrol exhibited inhibition of AKT/PKB, FOXO1 and BAD,[104] and activation of GSK3β, BAX, caspase 3/9, cytochrome C, PARP and DR5.[104]

In skin cancer and melanoma, resveratrol showed inhibition of TGFβ2, AKT/PKB, CREB, ERK1/2, SMAD3 and SMA;[105] APE/REF1, JUND, MMP1, BCL2 and iNOS;[106] STAT3, NF-κB, cFLIP, FOXO3A, ERK1/2 and BCL-xL;[107] COX2, JNK, ERK1/2, p38, AP1 and NF-κB;[108] CCNA, CCND1, CDK6, MEK1, ERK1/2, AP1 and cJUN;[109] IKK, IκBα, NF-κB, COX2, ERK1/2 and p38;[110] AP1, COX2, ERK1/2 and p38,[111] and activation of E-cadherin;[105] ATM, CHK2, ATF2, TRAIL, p38, p53 and p21;[107] RB, p21 and p27.[109]

In multiple cancers, resveratrol also serves as an inhibitor of ULK1 and mTOR,[112] and activator of EGR1[113] and SIRT1.[114]

3.2.8. *Other beneficial effects of resveratrol*

The beneficial effects of resveratrol in other tissues such as lung;[115–117] ileum;[118] cartilage;[119,120] muscle;[121,122] colon;[123] liver;[124] ovary;[125] prostate;[126] kidney,[127] and in other processes such as epigenetic regulation;[67–70,86,87,96,128,129] cellular senescence;[130,131] RNA splicing;[130,132] transcriptional regulation;[133] pain;[134] immune response;[135–141] calcium homeostasis;[142] viral infection;[143] protein synthesis;[144] protein degradation[145] have also been studied.

3.2.9. *Quinone reductase 2 (QR2): The high affinity target of resveratrol*

In general, resveratrol activation or inhibition of proteins and enzymes involves binding to those species. Quinone reductase 2 (QR2; NQO2) was identified as the major protein bound to resveratrol *in vitro* with a dissociation constant of 35 nM, the lowest reported for binding to any protein/enzyme to date. This high affinity binding also led to potent inhibition of QR2 by resveratrol *in vitro* and may be responsible for the antioxidant activity leading to its health beneficial properties.[146] Later, a group studying the effect of resveratrol on QR2 in cellular context showed that resveratrol activated QR2 in cancer as well as normal cells of the prostate.[90,126] Further, a group again showed that resveratrol inhibited QR2 *in vitro* and may be one of the reasons for its pleiotropic beneficial properties.[136] Additional targets with relatively high affinity for resveratrol are described below.

3.2.10. *Resveratrol and Sirtuin 1 (SIRT1)*

SIRT1 is a nicotinamide adenine dinucleotide (NAD^+)-dependent, class III HDAC and represents one of the most extensively studied targets of resveratrol. Although most of the studies report activation of SIRT1 by resveratrol including SRT501,[147] a proprietary formulation of resveratrol, there are a few studies which report resveratrol as an inhibitor of SIRT1 (Table 3.1). Notably, however, a report demonstrated that resveratrol as well as other ostensible SIRT1 activators, namely SIRT1720, SIRT2183 and SIRT1460,[148] which are structurally unrelated to resveratrol, appeared to activate SIRT1 only with peptide substrate containing a covalently attached fluorophore and not with native peptide or full length protein substrates. Thereby, it was concluded that SRT1720, SRT2183,

SRT1460, and resveratrol are not direct activators of SIRT1.[149] Consistent with this, there does not appear to be X-ray diffraction data of a full length of SIRT1 interacting with resveratrol.

Not to be discouraged, however, an answer has been provided through the possibility of SIRT1 activation by resveratrol through activation of previously reported SIRT1 activators such as AMPK.[150,151] Reports show AMPK activation by resveratrol[7,9,41,55,95,98,144] (Table 3.1). A few of these reports also show SIRT1 activation but the dependence of SIRT1 activation on AMPK activation by resveratrol has not been investigated.[7,55]

In summary, SIRT1 activation as a mechanism responsible for the health beneficial effects of resveratrol remains controversial.

However, one X-ray crystallographic datum deposited in RCSB PDB (www.rcsb.org; accessed February 28, 2018)[152] demonstrates a co-crystal structure of SIRT1 in complex with resveratrol and a four-residue acetylated p53 peptide which carries 7-amino-4-methylcoumarin (AMC) as a substrate.[153] With a resolution of 3.2 Å [R-value (free) of 0.252] these results indicated that resveratrol mediates the interaction between AMC peptide and SIRT1. Therefore, the direct interaction between resveratrol and SIRT1 is still ambiguous. Nonetheless, the ChEMBL database indicates resveratrol activates SIRT1 by 239, 802, and 200%.

3.3. Use of publicly available drug-target databases to give a view of targets

3.3.1. *Introduction*

Resveratrol is known to exert pleiotropic effects by acting on multiple targets.[3] Drug promiscuity (polypharmacological bioactivity) as demonstrated by resveratrol can be considered as an unfavorable feature owing to undesirable/unexpected side effects or toxicities. In the era with the prevailing paradigm of "one drug, one target", researchers pursued drug candidates with high selectivity. However, a paradigm shift took place in recent years and emergence of the "one drug, multiple targets" concept (or multi-target strategies) in drug discovery broaden drug developers' horizon toward drug repositioning or repurposing.[154] In addition, drugs possessing multiple biological activities have a tendency to retain their activity in pre-clinical and clinical settings relative to single target drug candidates due to the complexity of disease pathways with multiple pathogenic factors.[155] Thus, researchers have explored the mechanism of action on bioactive compounds in the context of biological networks, rather than only focusing on a single target. In fact, the emergence of network pharmacology, applying network analysis using computational and experimental methods in biological systems (at the systems levels),[156] provides a more realistic understanding of mechanism of action. The negative connotation of "off-target" activity being associated with undesirable effects now bears a positive connotation in the field of drug repositioning/ repurposing.

In this section, we employ an approach of using web-based resources to provide plausible future targets based on STRING (Protein-Protein Interaction Networks, version: 10.5). One can find large amounts of data (big data) based on chemical information (chemoinformatics) once canonical simplified molecular-input line-entry system (SMILES) is created.[157] Numerous curated and integrated data on chemical structures are deposited in web-based publicly accessible databases including ChEMBL, PubChem, BindingDB, etc. ChEMBL and BindingDB contain manually extracted data from research articles focusing on compound activity, and PubChem encompasses data from 587 sources (accessed on March 1, 2018) including ChEMBL and BindingDB. Here we analyze the experimentally demonstrated bioactivities of resveratrol using databases specifically with proteins from *Homo sapiens.* Molecular docking data (*in silico* data) is not included in this chapter.

3.3.2. *UniChem: Collection of identifier codes*

Canonical SMILES is magical format like "Open Sesame" to access enormous data, yet, different resources have different ID systems. Canonical SMILES is like a DOI for searching articles, whereas ID code in each resource is like a PMID number in PubMed. For instance, canonical SMILES of resveratrol is C1=CC(=CC=C1C=CC2=CC(=CC(=C2)O)O)O whereas ChEMBL ID is CHEMBL165, which is simpler and easier to remember. Thus, UniChem is a new hub for layman style "Open Sesame" — It provides the IDs of resveratrol so researchers can easily access the database using the given ID. To find information on resveratrol, UniChem (https://www.ebi.ac.uk/), a compound identifier mapping service which is designed to optimize structure-based hyperlinks between UniChem and small molecule containing resources (~40) including ChEMBL, PubChem Compounds, DrugBank, Protein Data Bank Europe (PDBe), Kyoto Encyclopedia of Genes and Genomes (KEGG) Ligand, Chemical Entities of Biological Interest (ChEBI), ZINC, Brenda, Human Metabolome Database (HMDB), etc., is shown in Table 3.2.[158]

Table 3.2. List of resveratrol ID in various resources.

Resource	ID
ChEMBL	CHEMBL165
DrugBank	DB02709
Guide to Pharmacology	8741
KEGG Ligand	C03582
ChEBI	45713
NIH Clinical Collection	SAM001246888
ZINC	ZINC00006787
SureChEMBL	SCHEMBL19425
PharmGKB	PA165291838
HMDB	HMDB0003747
PubChem Compounds	445154
Mcule	MCULE-5678456463
Library of Integrated Network-based Cellular Signatures	LSM-42917
ACToR	31100-06-8
MolPort	MolPort-019-999-364
BindingDB	23926
LipidMaps	LMPK13090005
Metabolights	MTBLC45713
Brenda	91278
Rhea	45713

Table 3.3. Deposited data and update status of sources.

Source	Data counts	Last updated
ChEMBL	1,735,442 Live Substances 1,244,770 Live BioAssays 5,373 Annotations	2018/02/27
BindingDB	651,401 Live Substances 20 Live BioAssays	2018/02/08
KEGG	39,758 Live Substances 3,186 Annotations 48 Classifications	2018/02/28
DrugBank	10,517 Live Substances 34,574 Annotations	2018/02/27
PDB	9,324 Annotations	2018/01/03

3.3.3. *Overview of web-accessible resources*

The virtue of sources used in this section shown in Table 3.3 is relatively well-curated and integrates publicly available chemical biology resources on a large number of compounds, especially with bioactivities, with frequent updates (accessed on February 28, 2018), https://pubchem.ncbi.nlm.nih.gov/sources/.[159] Thus, three resources were used for drug target identification of resveratrol in this section. These databases can be used to predict the protein targets of a small molecule such as resveratrol.

3.3.4. *Bioactivities of resveratrol on proteins (Data collection tool: ChEMBL)*

From 2,120 bioactivities of resveratrol, in the current database (https://www.ebi.ac.uk/chembl, accessed on February 28, 2018), only *Homo sapiens* organism-based activities were selected (other organisms were filtered.). Then, those with higher confidence score values ($\geq$5) which include the bioactivities of resveratrol on homologous protein complex subunits, direct protein complex subunits, homologous single protein, direct single protein, and multiple direct protein complex subunits. The higher number implies that the assay is more closely related to target proteins. The majority of assays

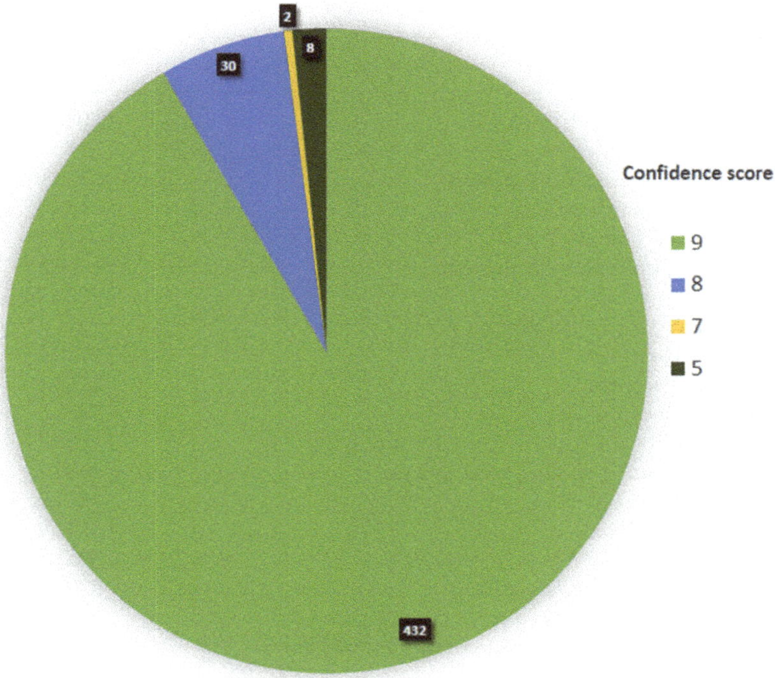

Fig. 3.1. Confidence score (≥5) of assays related to resveratrol bioactivities.

(432 out of 472) with the confidence score ≥5 falls into confidence score 9 which is highest value (Figure 3.1).

Out of 472 assays, "Not Active", "Not Determined", or "Outside typical range" were filtered. However, a datum, which has lower value of potency (higher potency) than the range, is included. The list of protein targets of resveratrol according to potency, percent inhibition, IC_{50}, Ki, and activation are summarized in Tables 3.4–3.7.[160]

3.3.5. *Bioactivities of resveratrol on proteins (Data collection tool: PubChem)*

Through PubChem, human protein targets which resveratrol exerts activities less than 1 μM are selected as shown in Table 3.8. (https://pubchem.ncbi.nlm.nih.gov, accessed on February 28, 2018).

Table 3.4. Resveratrol target proteins with greater potency ($\leq$10 μM).

Potency	**Unit, nM**
Chromobox protein homolog 1	0.04228
Prelamin-A/C	5
78 kDa glucose-regulated protein	130.9
M-phase phosphoprotein 8	158.5
Cerebroside-sulfatase	379.3
78 kDa glucose-regulated protein	1166.8
Huntingtin	2238.7
Nuclear receptor ROR-gamma	2371
Androgen Receptor	2730.6
Androgen Receptor	2755.6
Androgen Receptor	2853.2
ATPase family AAA domain-containing protein 5	2908.1
Serine/threonine-protein kinase mTOR	3689.8
Androgen Receptor	3857.1
Cytochrome P450 3A4	3981.1
Cytochrome P450 3A4	3981.1
Parathyroid hormone receptor	3981.1
Ras-related protein Rab-9A	3981.1
ATPase family AAA domain-containing protein 5	4109.5
Androgen Receptor	4403.7
Nuclear factor erythroid 2-related factor 2	4855.8
Niemann-Pick C1 protein	5623.4
Microtubule-associated protein tau	5623.4
Nuclear factor erythroid 2-related factor 2	5804.8
Nuclear factor erythroid 2-related factor 2	5955.7
Cellular tumor antigen p53	6309.6
Ataxin-2	6309.6
Androgen Receptor	6441.8
ATPase family AAA domain-containing protein 5	6510.4
Androgen Receptor	6921.8
Estrogen receptor alpha	7079.5
Microtubule-associated protein tau	7079.5
Microtubule-associated protein tau	7079.5
Androgen Receptor	7761.6
Histone-lysine N-methyltransferase, H3 lysine-9 specific 3	7943.3
Survival motor neuron protein	8912.5
Microtubule-associated protein tau	8912.5
Microtubule-associated protein tau	8912.5
Hypoxia-inducible factor 1 alpha	10000
Microtubule-associated protein tau	10000

(*Continued*)

Table 3.4. (*Continued*)

Potency	Unit, nM
Cellular tumor antigen p53	10000
Cellular tumor antigen p53	10000
15-hydroxyprostaglandin dehydrogenase [NAD+]	10000
Hypoxia-inducible factor 1 alpha	10000
Estrogen receptor alpha	10000

Table 3.5. Resveratrol target proteins with IC_{50} value under 1 μM.

Protein	IC_{50} (nM)
PI3-kinase p110-alpha subunit	250
Monoamine oxidase A	449
Quinone reductase 2	450
PI3-kinase p110-beta subunit	500
Cytochrome P450 3A4	600
Cyclooxygenase-1	625
Cyclooxygenase-2	750
Cyclooxygenase-1	830
Quinone reductase 2	913
Quinone reductase 2	960
Cyclooxygenase-2	990
Cyclooxygenase-2	996

Table 3.6. Resveratrol target proteins with % inhibition over 60%.

Protein	% Inhibition
Beta amyloid A4 protein	61.1
Beta amyloid A4 protein	61.5
Cytochrome P450 19A1	63.2
Beta amyloid A4 protein	63.4
Beta amyloid A4 protein	64.2
Beta amyloid A4 protein	64.2
Cytochrome P450 19A1	67.8
Beta amyloid A4 protein	68.51
Beta amyloid A4 protein	69.73
Cyclooxygenase-2	72.2
Beta amyloid A4 protein	78.92
Solute carrier organic anion transporter family member 1B3	116.68
Solute carrier organic anion transporter family member 1B1	157.62

Table 3.7. Resveratrol target proteins with Ki values under 10 μM.

Protein	Ki (nM)
Quinone reductase 2	88
Estrogen receptor alpha	785
Carbonic anhydrase (CA) 9	810
CA14	830
CA12	950
CA1	2210
Norepinephrine transporter	2293
CA2	2770
CA13	4090
CA7	4350
CA4	4470
CA5B	4640
CA5A	4750
CA6	8070
CA3	9090

Table 3.8. Protein targets of resveratrol from PubChem (Activity value under 1 μM).

Protein	Activity value (μM)	Assay ID name
CBX1	4.23E-05	qHTS Validation Assay for Inhibitors of HP1-beta Chromodomain Interactions with Methylated Histone Tails
LMNA	0.005	qHTS Assay for Modulators of Lamin A Splicing
ESR1	0.0088	qHTS assay to identify small molecule agonists of the estrogen receptor alpha (ER-alpha) signaling pathway
NQO2	0.054	Binding affinity to human quinone reductase 2 by ITC analysis
NQO2	0.088	Competitive inhibition of human quinone reductase 2 using menadione/N-methyldihydronicotinamide as substrate after 10 mins by double-reciprocal plot analysis
MPHOSPH8	0.1585	qHTS Validation Assay for Inhibitors for MPP8 Chromodomain Interactions with Methylated Histone Tails

(*Continued*)

Table 3.8. (*Continued*)

Protein	Activity value (μM)	Assay ID name
PIK3CA	0.25	Displacement of ATP in PI3Kalpha (unknown origin) by cell based assay
ARSA	0.3793	qHTS screen for enhancers of Arylsulfatase A (ASA1): LOPAC Validation Assay
PTGS1	0.38	Inhibition of ovine COX1
MAOA	0.449	DRUGMATRIX: Monoamine Oxidase MAO-A enzyme inhibition (substrate: Kynuramine)
NQO2	0.45	Inhibition of human recombinant NQO2 by spectrophotometry
NQO2	0.45	Inhibition of human recombinant NQO2 assessed as reduction of DCPIP by spectrophotometry
PIK3CB	0.5	Displacement of ATP in PI3Kbeta (unknown origin) by cell based assay
PTGS1	0.535	Inhibition of ovine COX1 by measuring PGE2
CYP3A4	0.6	DRUGMATRIX: CYP450, 3A4 enzyme inhibition (substrate: 7-Benzyloxy-4-(trifluoromethyl)-coumarin)
PTGS1	0.625	DRUGMATRIX: Cyclooxygenase COX-1 enzyme inhibition (substrate: Arachidonic acid)
PTGS2	0.75	Inhibition of human recombinant COX2 assessed as PGE2 production after 10 mins by ELISA
ESR1	0.785	Binding affinity for estrogen receptor (ER) alpha of MCF-7 cell cytosol incubated with 2 nM [3H]estradiol; Agonist
CA9	0.81	Inhibition of human carbonic anhydrase 9 after 15 mins by stopped flow CO2 hydration method
CA14	0.83	Inhibition of human carbonic anhydrase 14 after 15 mins by stopped flow CO2 hydration method
CA12	0.95	Inhibition of human carbonic anhydrase 12 after 15 mins by stopped flow CO2 hydration method
NQO2	0.96	Inhibition of human quinone reductase 2 expressed in Escherichia coli BL21(DE3) incubated at 23 degC by MTT assay
NQO2	0.96	Inhibition of human quinone reductase 2 expressed in Escherichia coli BL21(DE3) assessed as N-methyldihydronicotinamide oxidation per mg of protein after 1 hr by LC-MS analysis analysis
PTGS2	0.996	Inhibition of human recombinant COX2 by measuring PGE2

3.3.6. *Bioactivities of resveratrol on proteins (Data collection tool: Binding DB) (Compare with ChEMBL)*

The majority of data is from ChEMBL, so there is overlap with the previous Tables, but cytochrome P450 1B1 and estrogen receptor are from other sources. (https://www.bindingdb.org). Resveratrol target proteins with Ki values under 10 μM is listed in Table 3.9. As shown in Figure 3.2, bioactivities on cyclooxygenases (COXs) and ribosyldihydronicotinamide dehydrogenase [quinone] (NQO2) have been reported multiple times. Thus, we can have confidence in the interaction of resveratrol with COXs and NQO2.

Table 3.9. Resveratrol target proteins with Ki values under 10 μM.

Target	Curation	Ki (nM)
NQO2	ChEMBL	88
Cytochrome P450 1B1	University of Tokyo	177
Estrogen receptor	Institut Claudius Regaud	785
CA9	ChEMBL	810
CA14	ChEMBL	830
CA12	ChEMBL	950
CA2	ChEMBL	1750
CA14	ChEMBL	1950
CA2	ChEMBL	2770
CA5A	ChEMBL	2860
CA5B	ChEMBL	3240
CA4	ChEMBL	4010
CA13	ChEMBL	4090
CA7	ChEMBL	4350
CA4	ChEMBL	4470
CA5B	ChEMBL	4640
CA5A	ChEMBL	4750
CA3	ChEMBL	6750
CA6	ChEMBL	8070
CA12	ChEMBL	8130
CA13	ChEMBL	8250
CA6	ChEMBL	8630
CA7	ChEMBL	8980
CA3	ChEMBL	9090
CA9	ChEMBL	9530

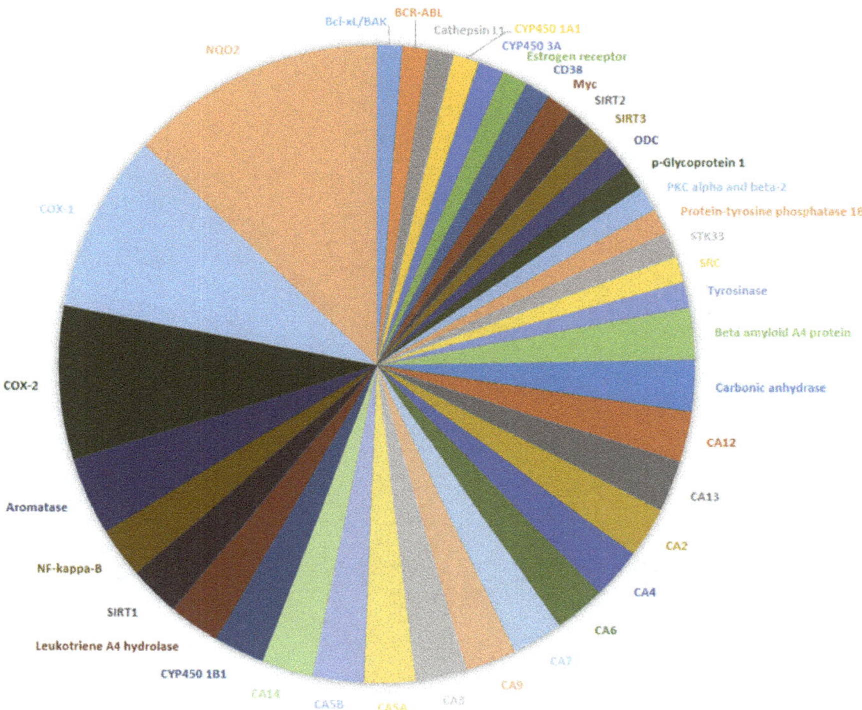

Fig. 3.2. Frequent tests on target proteins.

3.3.7. *Protein direct interaction (Protein data bank)*

There are proteins (*Homo sapiens*) which showed interaction with resveratrol using methods such as X-ray diffraction. TRANSTHYRETIN, FMN QUINONE REDUCTASE 2, tyrosyl tRNA synthetase, Estrogen receptor, PPARgamma LBD, Troponin C, Leukotriene A4 hydrolase, etc.

3.3.8. *Resveratrol-target protein map (STRING 10.5)*

Protein-protein interaction networks found as targets of resveratrol was created using STRING.[161] After combing DrugBank, ChEMBL (protein target classes), PubChem, BindingDB and manual PubMed search (ABCG2_HUMAN, AMY1_HUMAN, ANO1_HUMAN, BACE1_HUMAN, CCR2_HUMAN, NM3A_HUMAN,

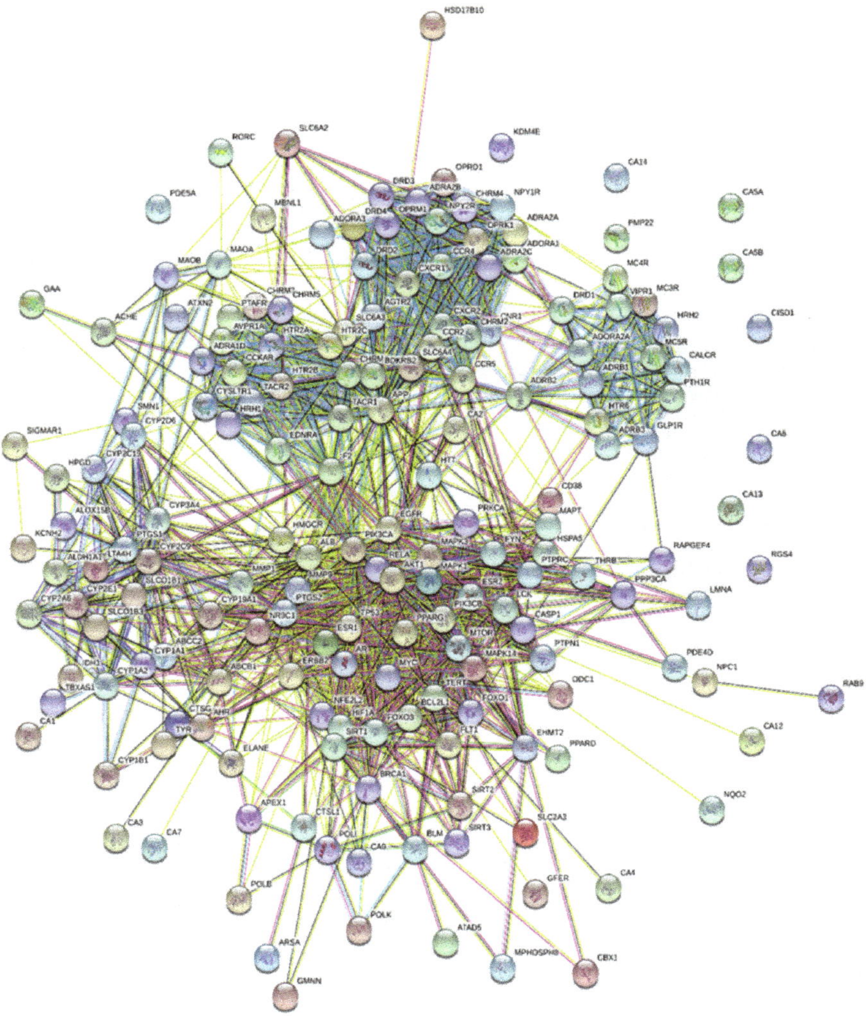

Fig. 3.3. Protein-protein interactions and plausible target pathways of resveratrol.

DNM3B_HUMAN, G3P_HUMAN, GSTP1_HUMAN, DM1A_HUMAN, TOP2A_HUMAN, VDR_HUMAN), a resveratrol-target network was constructed using STRING database (Figure 3.3). As a result, we can infer that resveratrol acts on several pathways including neuroactive ligand-receptor interaction (5.64e-47), calcium signaling pathway (1.56e-23), pathways in cancer (3.14e-19), nitrogen metabolism (1.29e-18), and serotonergic synapse (1.7e-18) with a low false discovery rate (FDR). In fact, from other curated/integrated databases, resveratrol showed potent effects on inhibition of amyloid beta, Tau proteins. Therefore, in terms of drug repositioning/repurposing, resveratrol can be a plausible candidate for the treatment of neuronal diseases.

3.4. Conclusions

First of all, it should be noted this paper follows the overall pattern of other manuscripts concerning resveratrol that have published in the scientific literature. Affirmative statements are presented. For example, we state "The antidiabetic effects of resveratrol are also well known" even though a recent report indicates no beneficial effect on metabolic syndrome in a randomized placebo-controlled clinical trial.[162] In fact, expert working groups have been unable to make specific recommendations in regard to the health benefits of resveratrol in general.[163] Nonetheless, as described herein, numerous targets have been investigated, and there is certainly a general perception some value exists.

When considering the relevance of targets, it seems rational to take into account actual blood or tissue levels that can realistically be anticipated. Thus, given that serum levels of significantly less than 1 μM are typical, high affinity targets warrant greater attention, and a few have been identified (Table 3.5). Interesting, one of these targets, cyclooxygenase, was the key target described in our seminal report.[1] Further, at the same time, we demonstrated anti-inflammatory activity using a rat model. Thus, although other targets such as SIRT have garnered great publicity even though response remains a polemic, it appears that anti-inflammation is the common denominator responsible for amelioration of a host of disease states and other beneficial effects.

Inhibition of another high affinity target, QR2, may also contribute to an anti-inflammatory effect. Further, it is particularly interesting that our network analysis described herein points toward usefulness in treating neuronal diseases, and inhibition of QR2 may play a role in this process.[164]

In addition to serum and tissue levels, of course it is important to consider resveratrol metabolism. It is well-known that resveratrol is rapidly absorbed and quickly metabolized, and the concentration of metabolites, although not exceptionally high, exceeds the concentration of resveratrol. It could be argued that these facts render the entire discourse of this manuscript moot since, aside from simple *in vitro* systems, it often is not clear if a response is mediated by resveratrol or a metabolite. However, our supposition is that resveratrol metabolites are not expected to demonstrate major differences in activity relative to resveratrol,[165] but since the concentration of metabolites are somewhat higher than resveratrol, less sensitive targets (e.g, <10 μM) may play a role in networking.

Exceptions to the discussion above include situations where high dosage levels can be achieved by direct exposure. A prime example is topical application, and another may be exposure of the gastrointestinal tract. Thus, targets affiliated with disorders such as skin or colon cancer may assume greater importance even though modulation requires high concentrations of resveratrol.

In sum, while it is interesting to muse over the milieu of protein-protein interactions and plausible pathways that are modulated by resveratrol, we still cannot discuss the physiological relevance of these effects in a definitive manner. Whereas sales of dietary supplements are prevalent, valued at about \$50 million per year,[166] evidence-based responses are sparse. Fortunately, there are no indications of chronic or acute toxicity, so as clinical trials and research continues, a clearer understanding of utility should emerge.

References

1. Jang M, Cai L, Udeani GO, Slowing KV, Thomas CF, Beecher CW, Fong HH, Farnsworth NR, Kinghorn AD, Mehta RG, Moon RC, Pezzuto JM. Cancer chemopreventive activity of resveratrol, a natural product derived from grapes. *Science* 1997; **275**: 218–20.
2. Park EJ, Pezzuto JM. The pharmacology of resveratrol in animals and humans. *Biochim Biophys Acta* 2015; **1852**: 1071–113.
3. Pezzuto JM. The phenomenon of resveratrol: redefining the virtues of promiscuity. *Ann N Y Acad Sci* 2011; **1215**: 123–30.

4. Gu X, Cai Z, Cai M, Liu K, Liu D, Zhang Q, Tan J, Ma Q. AMPK/SIRT1/p38 MAPK signaling pathway regulates alcoholinduced neurodegeneration by resveratrol. *Mol Med Rep* 2018;
5. Guo D, Xie J, Zhao J, Huang T, Guo X, Song J. Resveratrol protects early brain injury after subarachnoid hemorrhage by activating autophagy and inhibiting apoptosis mediated by the Akt/mTOR pathway. *Neuroreport* 2018; **29**: 368–79.
6. Pany S, Ghosh A, You Y, Nguyen N, Das J. Resveratrol inhibits phorbol ester-induced membrane translocation of presynaptic Munc13-1. *Biochim Biophys Acta* 2017; **1861**: 2640–51.
7. Wan D, Zhou Y, Wang K, Hou Y, Hou R, Ye X. Resveratrol provides neuroprotection by inhibiting phosphodiesterases and regulating the cAMP/AMPK/SIRT1 pathway after stroke in rats. *Brain Res Bull* 2016; **121**: 255–62.
8. Pang C, Cao L, Wu F, Wang L, Wang G, Yu Y, Zhang M, Chen L, Wang W, Lv W, Chen L, Zhu J, *et al.* The effect of trans-resveratrol on post-stroke depression via regulation of hypothalamus-pituitary-adrenal axis. *Neuropharmacology* 2015; **97**: 447–56.
9. Cho KS, Lee EJ, Kwon KJ, Gonzales EL, Kim YB, Cheong JH, Bahn GH, Lee J, Han SH, Kim YT, Shin CY. Resveratrol down-regulates a glutamate-induced tissue plasminogen activator via Erk and AMPK/mTOR pathways in rat primary cortical neurons. *Food Funct* 2014; **5**: 951–60.
10. Rodacka A, Strumillo J, Serafin E, Puchala M. Effect of resveratrol and tiron on the inactivation of glyceraldehyde-3- phosphate dehydrogenase induced by superoxide anion radical. *Curr Med Chem* 2014; **21**: 1061–9.
11. de Almeida LM, Leite MC, Thomazi AP, Battu C, Nardin P, Tortorelli LS, Zanotto C, Posser T, Wofchuk ST, Leal RB, Goncalves CA, Gottfried C. Resveratrol protects against oxidative injury induced by H2O2 in acute hippocampal slice preparations from Wistar rats. *Arch Biochem Biophys* 2008; **480**: 27–32.
12. Xiong H, Ou Y, Xu Y, Huang Q, Pang J, Lai L, Zheng Y. Resveratrol Promotes Recovery of Hearing following Intense Noise Exposure by Enhancing Cochlear SIRT1 Activity. *Audiol Neurootol* 2018; **22**: 303–10.
13. Palomera-Avalos V, Grinan-Ferre C, Puigoriol-Ilamola D, Camins A, Sanfeliu C, Canudas AM, Pallas M. Resveratrol Protects SAMP8 Brain Under Metabolic Stress: Focus on Mitochondrial Function and Wnt Pathway. *Mol Neurobiol* 2017; **54**: 1661–76.
14. Wang Z, Gu J, Wang X, Xie K, Luan Q, Wan N, Zhang Q, Jiang H, Liu D. Antidepressant-like activity of resveratrol treatment in the forced swim test and tail suspension test in mice: the HPA axis, BDNF expression and phosphorylation of ERK. *Pharmacol Biochem Behav* 2013; **112**: 104–10.
15. Senturk S, Yaman ME, Aydin HE, Guney G, Bozkurt I, Paksoy K, Abdioglu AA. Effects of Resveratrol on Inflammation and Apoptosis after Experimental Spinal Cord Injury. *Turk Neurosurg* 2017;
16. Bakheet SA, Alzahrani MZ, Nadeem A, Ansari MA, Zoheir KMA, Attia SM, Al-Ayadhi LY, Ahmad SF. Resveratrol treatment attenuates chemokine

receptor expression in the BTBR T+tf/J mouse model of autism. *Mol Cell Neurosci* 2016; **77**: 1–10.

17. Narayanan SV, Dave KR, Saul I, Perez-Pinzon MA. Resveratrol Preconditioning Protects Against Cerebral Ischemic Injury via Nuclear Erythroid 2-Related Factor 2. *Stroke* 2015; **46**: 1626–32.
18. Kumar A, Singh CK, Lavoie HA, Dipette DJ, Singh US. Resveratrol restores Nrf2 level and prevents ethanol-induced toxic effects in the cerebellum of a rodent model of fetal alcohol spectrum disorders. *Mol Pharmacol* 2011; **80**: 446–57.
19. Quincozes-Santos A, Bobermin LD, Tramontina AC, Wartchow KM, Tagliari B, Souza DO, Wyse AT, Goncalves CA. Oxidative stress mediated by NMDA, AMPA/KA channels in acute hippocampal slices: neuroprotective effect of resveratrol. *Toxicol In Vitro* 2014; **28**: 544–51.
20. Kim DW, Kim YM, Kang SD, Han YM, Pae HO. Effects of Resveratrol and trans-3,5,4'-Trimethoxystilbene on Glutamate-Induced Cytotoxicity, Heme Oxygenase-1, and Sirtuin 1 in HT22 Neuronal Cells. *Biomol Ther (Seoul)* 2012; **20**: 306–12.
21. Varamini B, Sikalidis AK, Bradford KL. Resveratrol increases cerebral glycogen synthase kinase phosphorylation as well as protein levels of drebrin and transthyretin in mice: an exploratory study. *Int J Food Sci Nutr* 2014; **65**: 89–96.
22. Wang H, Dong X, Liu Z, Zhu S, Liu H, Fan W, Hu Y, Hu T, Yu Y, Li Y, Liu T, Xie C, *et al.* Resveratrol Suppresses Rotenone-induced Neurotoxicity Through Activation of SIRT1/Akt1 Signaling Pathway. *Anat Rec (Hoboken)* 2018;
23. Anandhan A, Tamilselvam K, Vijayraja D, Ashokkumar N, Rajasankar S, Manivasagam T. Resveratrol attenuates oxidative stress and improves behaviour in 1-methyl-4-phenyl-1,2,3,6-tetrahydropyridine (MPTP) challenged mice. *Ann Neurosci* 2010; **17**: 113–9.
24. Ho DJ, Calingasan NY, Wille E, Dumont M, Beal MF. Resveratrol protects against peripheral deficits in a mouse model of Huntington's disease. *Exp Neurol* 2010; **225**: 74–84.
25. Ma SC, Zhang HP, Jiao Y, Wang YH, Zhang H, Yang XL, Yang AN, Jiang YD. Homocysteine-induced proliferation of vascular smooth muscle cells occurs via PTEN hypermethylation and is mitigated by Resveratrol. *Mol Med Rep* 2018;
26. Park DW, Baek K, Kim JR, Lee JJ, Ryu SH, Chin BR, Baek SH. Resveratrol inhibits foam cell formation via NADPH oxidase 1-mediated reactive oxygen species and monocyte chemotactic protein-1. *Exp Mol Med* 2009; **41**: 171–9.
27. Penumathsa SV, Koneru S, Samuel SM, Maulik G, Bagchi D, Yet SF, Menon VP, Maulik N. Strategic targets to induce neovascularization by resveratrol in hypercholesterolemic rat myocardium: role of caveolin-1, endothelial nitric oxide synthase, hemeoxygenase-1, and vascular endothelial growth factor. *Free Radic Biol Med* 2008; **45**: 1027–34.

28. El-Mowafy AM, Alkhalaf M, Nassar NN. Resveratrol reverses ET-1-evoked mitogenic effects in human coronary arterial cells by activating the kinase-G to inhibit ERK-enzymes. *Int J Cardiol* 2009; **136**: 263–9.
29. Mukhopadhyay P, Das S, Ahsan MK, Otani H, Das DK. Modulation of microRNA 20b with resveratrol and longevinex is linked with their potent anti-angiogenic action in the ischaemic myocardium and synergestic effects of resveratrol and gamma-tocotrienol. *J Cell Mol Med* 2012; **16**: 2504–17.
30. Goh SS, Woodman OL, Pepe S, Cao AH, Qin C, Ritchie RH. The red wine antioxidant resveratrol prevents cardiomyocyte injury following ischemia-reperfusion via multiple sites and mechanisms. *Antioxid Redox Signal* 2007; **9**: 101–13.
31. Rogers RG, Otis JS. Resveratrol-Mediated Expression of KLF15 in the Ischemic Myocardium is Associated with an Improved Cardiac Phenotype. *Cardiovasc Drugs Ther* 2017; **31**: 29–38.
32. Das S, Tosaki A, Bagchi D, Maulik N, Das DK. Potentiation of a survival signal in the ischemic heart by resveratrol through p38 mitogen-activated protein kinase/mitogen- and stress-activated protein kinase 1/cAMP response element-binding protein signaling. *J Pharmacol Exp Ther* 2006; **317**: 980–8.
33. Mukhopadhyay P, Mukherjee S, Ahsan K, Bagchi A, Pacher P, Das DK. Restoration of altered microRNA expression in the ischemic heart with resveratrol. *PLoS One* 2010; **5**: e15705.
34. Xia N, Strand S, Schlufter F, Siuda D, Reifenberg G, Kleinert H, Forstermann U, Li H. Role of SIRT1 and FOXO factors in eNOS transcriptional activation by resveratrol. *Nitric Oxide* 2013; **32**: 29–35.
35. Gracia-Sancho J, Villarreal G, Jr., Zhang Y, Garcia-Cardena G. Activation of SIRT1 by resveratrol induces KLF2 expression conferring an endothelial vasoprotective phenotype. *Cardiovasc Res* 2010; **85**: 514–9.
36. Csiszar A, Labinskyy N, Pinto JT, Ballabh P, Zhang H, Losonczy G, Pearson K, de Cabo R, Pacher P, Zhang C, Ungvari Z. Resveratrol induces mitochondrial biogenesis in endothelial cells. *Am J Physiol Heart Circ Physiol* 2009; **297**: H13–20.
37. Schilder YD, Heiss EH, Schachner D, Ziegler J, Reznicek G, Sorescu D, Dirsch VM. NADPH oxidases 1 and 4 mediate cellular senescence induced by resveratrol in human endothelial cells. *Free Radic Biol Med* 2009; **46**: 1598–606.
38. Das S, Alagappan VK, Bagchi D, Sharma HS, Maulik N, Das DK. Coordinated induction of iNOS-VEGF-KDR-eNOS after resveratrol consumption: a potential mechanism for resveratrol preconditioning of the heart. *Vascul Pharmacol* 2005; **42**: 281–9.
39. Cao MM, Lu X, Liu GD, Su Y, Li YB, Zhou J. Resveratrol attenuates type 2 diabetes mellitus by mediating mitochondrial biogenesis and lipid metabolism via Sirtuin type 1. *Exp Ther Med* 2018; **15**: 576–84.
40. Seyyedebrahimi S, Khodabandehloo H, Nasli Esfahani E, Meshkani R. The effects of resveratrol on markers of oxidative stress in patients with type 2

diabetes: a randomized, double-blind, placebo-controlled clinical trial. *Acta Diabetol* 2018;

41. Do GM, Jung UJ, Park HJ, Kwon EY, Jeon SM, McGregor RA, Choi MS. Resveratrol ameliorates diabetes-related metabolic changes via activation of AMP-activated protein kinase and its downstream targets in db/db mice. *Mol Nutr Food Res* 2012; **56**: 1282–91.
42. Dao TM, Waget A, Klopp P, Serino M, Vachoux C, Pechere L, Drucker DJ, Champion S, Barthelemy S, Barra Y, Burcelin R, Seree E. Resveratrol increases glucose induced GLP-1 secretion in mice: a mechanism which contributes to the glycemic control. *PLoS One* 2011; **6**: e20700.
43. Vetterli L, Brun T, Giovannoni L, Bosco D, Maechler P. Resveratrol potentiates glucose-stimulated insulin secretion in INS-1E beta-cells and human islets through a SIRT1-dependent mechanism. *J Biol Chem* 2011; **286**: 6049–60.
44. Asadi S, Goodarzi MT, Saidijam M, Karimi J, Azari RY, Farimani AR, Salehi I. Resveratrol attenuates visfatin and vaspin genes expression in adipose tissue of rats with type 2 diabetes. *Iran J Basic Med Sci* 2015; **18**: 537–43.
45. Chen F, Qian LH, Deng B, Liu ZM, Zhao Y, Le YY. Resveratrol protects vascular endothelial cells from high glucose-induced apoptosis through inhibition of NADPH oxidase activation-driven oxidative stress. *CNS Neurosci Ther* 2013; **19**: 675–81.
46. Zhang L, Pang S, Deng B, Qian L, Chen J, Zou J, Zheng J, Yang L, Zhang C, Chen X, Liu Z, Le Y. High glucose induces renal mesangial cell proliferation and fibronectin expression through JNK/NF-kappaB/NADPH oxidase/ROS pathway, which is inhibited by resveratrol. *Int J Biochem Cell Biol* 2012; **44**: 629–38.
47. Brouwer S, Hoffmeister T, Gresch A, Schonhoff L, Dufer M. Resveratrol Influences Pancreatic Islets by Opposing Effects on Electrical Activity and Insulin Release. *Mol Nutr Food Res* 2018;
48. Li SW, Yu HR, Sheen JM, Tiao MM, Tain YL, Lin IC, Lin YJ, Chang KA, Tsai CC, Huang LT. A maternal high-fat diet during pregnancy and lactation, in addition to a postnatal high-fat diet, leads to metabolic syndrome with spatial learning and memory deficits: beneficial effects of resveratrol. *Oncotarget* 2017; **8**: 111998–2013.
49. Xu L, Xu S, Lin L, Gu X, Fu C, Fang Y, Li X, Wang X. High-fat Diet Mediates Anxiolytic-like Behaviors in a Time-dependent Manner Through the Regulation of SIRT1 in the Brain. *Neuroscience* 2018; **372**: 237–45.
50. Ghanim H, Sia CL, Korzeniewski K, Lohano T, Abuaysheh S, Marumganti A, Chaudhuri A, Dandona P. A resveratrol and polyphenol preparation suppresses oxidative and inflammatory stress response to a high-fat, high-carbohydrate meal. *J Clin Endocrinol Metab* 2011; **96**: 1409–14.
51. Gracia A, Miranda J, Fernandez-Quintela A, Eseberri I, Garcia-Lacarte M, Milagro FI, Martinez JA, Aguirre L, Portillo MP. Involvement

of miR-539-5p in the inhibition of de novo lipogenesis induced by resveratrol in white adipose tissue. *Food Funct* 2016; **7**: 1680–8.
52. Kwon JY, Seo SG, Yue S, Cheng JX, Lee KW, Kim KH. An inhibitory effect of resveratrol in the mitotic clonal expansion and insulin signaling pathway in the early phase of adipogenesis. *Nutr Res* 2012; **32**: 607–16.
53. Pan Y, Zhang H, Zheng Y, Zhou J, Yuan J, Yu Y, Wang J. Resveratrol Exerts Antioxidant Effects by Activating SIRT2 To Deacetylate Prx1. *Biochemistry* 2017; **56**: 6325–28.
54. Escote X, Miranda M, Menoyo S, Rodriguez-Porrata B, Carmona-Gutierrez D, Jungwirth H, Madeo F, Cordero RR, Mas A, Tinahones F, Clotet J, Vendrell J. Resveratrol induces antioxidant defence via transcription factor Yap1p. *Yeast* 2012; **29**: 251–63.
55. Cao C, Lu S, Kivlin R, Wallin B, Card E, Bagdasarian A, Tamakloe T, Wang WJ, Song X, Chu WM, Kouttab N, Xu A, *et al.* SIRT1 confers protection against UVB- and H2O2-induced cell death via modulation of p53 and JNK in cultured skin keratinocytes. *J Cell Mol Med* 2009; **13**: 3632–43.
56. Zadi Heydarabad M, Nikasa M, Vatanmakanian M, Azimi A, Farshdousti Hagh M. Regulatory effect of resveratrol and prednisolone on MDR1 gene expression in acute lymphoblastic leukemia cell line (CCRF-CEM): An epigenetic perspective. *J Cell Biochem* 2018;
57. Zhou W, Wang S, Ying Y, Zhou R, Mao P. miR-196b/miR-1290 participate in the antitumor effect of resveratrol via regulation of IGFBP3 expression in acute lymphoblastic leukemia. *Oncol Rep* 2017; **37**: 1075–83.
58. Cecchinato V, Chiaramonte R, Nizzardo M, Cristofaro B, Basile A, Sherbet GV, Comi P. Resveratrol-induced apoptosis in human T-cell acute lymphoblastic leukaemia MOLT-4 cells. *Biochem Pharmacol* 2007; **74**: 1568–74.
59. Liao HF, Su YC, Zheng ZY, Jhih Cai C, Hou MH, Chao KS, Chen YJ. Sonic hedgehog signaling regulates Bcr-Abl expression in human chronic myeloid leukemia cells. *Biomed Pharmacother* 2012; **66**: 378–83.
60. Xie D, Zheng GZ, Xie P, Zhang QH, Lin FX, Chang B, Hu QX, Du SX, Li XD. Antitumor activity of resveratrol against human osteosarcoma cells: a key role of Cx43 and Wnt/beta-catenin signaling pathway. *Oncotarget* 2017; **8**: 111419–32.
61. Alkhalaf M, Jaffal S. Potent antiproliferative effects of resveratrol on human osteosarcoma SJSA1 cells: Novel cellular mechanisms involving the ERKs/p53 cascade. *Free Radic Biol Med* 2006; **41**: 318–25.
62. Pan J, Shen J, Si W, Du C, Chen D, Xu L, Yao M, Fu P, Fan W. Resveratrol promotes MICA/B expression and natural killer cell lysis of breast cancer cells by suppressing c-Myc/miR-17 pathway. *Oncotarget* 2017; **8**: 65743–58.
63. Saluzzo J, Hallman KM, Aleck K, Dwyer B, Quigley M, Mladenovik V, Siebert AE, Dinda S. The regulation of tumor suppressor protein, p53, and estrogen receptor (ERalpha) by resveratrol in breast cancer cells. *Genes Cancer* 2016; **7**: 414–25.
64. Medina-Aguilar R, Marchat LA, Arechaga Ocampo E, Gariglio P, Garcia Mena J, Villegas Sepulveda N, Martinez Castillo M, Lopez-Camarillo C.

Resveratrol inhibits cell cycle progression by targeting Aurora kinase A and Polo-like kinase 1 in breast cancer cells. *Oncol Rep* 2016; **35**: 3696–704.

65. Vinod BS, Nair HH, Vijayakurup V, Shabna A, Shah S, Krishna A, Pillai KS, Thankachan S, Anto RJ. Resveratrol chemosensitizes HER-2-overexpressing breast cancer cells to docetaxel chemoresistance by inhibiting docetaxel-mediated activation of HER-2-Akt axis. *Cell Death Discov* 2015; **1**: 15061.
66. Mohapatra P, Satapathy SR, Das D, Siddharth S, Choudhuri T, Kundu CN. Resveratrol mediated cell death in cigarette smoke transformed breast epithelial cells is through induction of p21Waf1/Cip1 and inhibition of long patch base excision repair pathway. *Toxicol Appl Pharmacol* 2014; **275**: 221–31.
67. Qin W, Zhang K, Clarke K, Weiland T, Sauter ER. Methylation and miRNA effects of resveratrol on mammary tumors vs. normal tissue. *Nutr Cancer* 2014; **66**: 270–7.
68. Mirza S, Sharma G, Parshad R, Gupta SD, Pandya P, Ralhan R. Expression of DNA methyltransferases in breast cancer patients and to analyze the effect of natural compounds on DNA methyltransferases and associated proteins. *J Breast Cancer* 2013; **16**: 23–31.
69. Papoutsis AJ, Borg JL, Selmin OI, Romagnolo DF. BRCA-1 promoter hypermethylation and silencing induced by the aromatic hydrocarbon receptor-ligand TCDD are prevented by resveratrol in MCF-7 cells. *J Nutr Biochem* 2012; **23**: 1324–32.
70. Stefanska B, Salame P, Bednarek A, Fabianowska-Majewska K. Comparative effects of retinoic acid, vitamin D and resveratrol alone and in combination with adenosine analogues on methylation and expression of phosphatase and tensin homologue tumour suppressor gene in breast cancer cells. *Br J Nutr* 2012; **107**: 781–90.
71. Bourguignon LY, Xia W, Wong G. Hyaluronan-mediated CD44 interaction with p300 and SIRT1 regulates beta-catenin signaling and NFkappaB-specific transcription activity leading to MDR1 and Bcl-xL gene expression and chemoresistance in breast tumor cells. *J Biol Chem* 2009; **284**: 2657–71.
72. Pozo-Guisado E, Merino JM, Mulero-Navarro S, Lorenzo-Benayas MJ, Centeno F, Alvarez-Barrientos A, Fernandez-Salguero PM. Resveratrol-induced apoptosis in MCF-7 human breast cancer cells involves a caspase-independent mechanism with downregulation of Bcl-2 and NF-kappaB. *Int J Cancer* 2005; **115**: 74–84.
73. Dubuisson JG, Dyess DL, Gaubatz JW. Resveratrol modulates human mammary epithelial cell O-acetyltransferase, sulfotransferase, and kinase activation of the heterocyclic amine carcinogen N-hydroxy-PhIP. *Cancer Lett* 2002; **182**: 27–32.
74. Hsieh TC, Wong C, John Bennett D, Wu JM. Regulation of p53 and cell proliferation by resveratrol and its derivatives in breast cancer cells: an in silico and biochemical approach targeting integrin alphavbeta3. *Int J Cancer* 2011; **129**: 2732–43.

75. Saunier E, Antonio S, Regazzetti A, Auzeil N, Laprevote O, Shay JW, Coumoul X, Barouki R, Benelli C, Huc L, Bortoli S. Resveratrol reverses the Warburg effect by targeting the pyruvate dehydrogenase complex in colon cancer cells. *Sci Rep* 2017; **7**: 6945.
76. Schneider Y, Duranton B, Gosse F, Schleiffer R, Seiler N, Raul F. Resveratrol inhibits intestinal tumorigenesis and modulates host-defense-related gene expression in an animal model of human familial adenomatous polyposis. *Nutr Cancer* 2001; **39**: 102–7.
77. Yang S, Li W, Sun H, Wu B, Ji F, Sun T, Chang H, Shen P, Wang Y, Zhou D. Resveratrol elicits anti-colorectal cancer effect by activating miR-34c-KITLG in vitro and in vivo. *BMC Cancer* 2015; **15**: 969.
78. Buhrmann C, Shayan P, Popper B, Goel A, Shakibaei M. Sirt1 Is Required for Resveratrol-Mediated Chemopreventive Effects in Colorectal Cancer Cells. *Nutrients* 2016; **8**: 145.
79. Vanamala J, Radhakrishnan S, Reddivari L, Bhat VB, Ptitsyn A. Resveratrol suppresses human colon cancer cell proliferation and induces apoptosis via targeting the pentose phosphate and the talin-FAK signaling pathways-A proteomic approach. *Proteome Sci* 2011; **9**: 49.
80. Zykova TA, Zhu F, Zhai X, Ma WY, Ermakova SP, Lee KW, Bode AM, Dong Z. Resveratrol directly targets COX-2 to inhibit carcinogenesis. *Mol Carcinog* 2008; **47**: 797–805.
81. Schneider Y, Vincent F, Duranton B, Badolo L, Gosse F, Bergmann C, Seiler N, Raul F. Anti-proliferative effect of resveratrol, a natural component of grapes and wine, on human colonic cancer cells. *Cancer Lett* 2000; **158**: 85–91.
82. Wright C, Iyer AKV, Yakisich JS, Azad N. Anti-Tumorigenic Effects of Resveratrol in Lung Cancer Cells Through Modulation of c-FLIP. *Curr Cancer Drug Targets* 2017; **17**: 669–80.
83. Hu Y, Rahlfs S, Mersch-Sundermann V, Becker K. Resveratrol modulates mRNA transcripts of genes related to redox metabolism and cell proliferation in non-small-cell lung carcinoma cells. *Biol Chem* 2007; **388**: 207–19.
84. Kim SH, Cho KH, Kim YN, Jeong BY, Park CG, Hur GM, Lee HY. Resveratrol attenuates norepinephrine-induced ovarian cancer invasiveness through downregulating hTERT expression. *Arch Pharm Res* 2016; **39**: 240–8.
85. Lin C, Crawford DR, Lin S, Hwang J, Sebuyira A, Meng R, Westfall JE, Tang HY, Lin S, Yu PY, Davis PJ, Lin HY. Inducible COX-2-dependent apoptosis in human ovarian cancer cells. *Carcinogenesis* 2011; **32**: 19–26.
86. Dhar S, Kumar A, Rimando AM, Zhang X, Levenson AS. Resveratrol and pterostilbene epigenetically restore PTEN expression by targeting oncomiRs of the miR-17 family in prostate cancer. *Oncotarget* 2015; **6**: 27214–26.
87. Dhar S, Kumar A, Li K, Tzivion G, Levenson AS. Resveratrol regulates PTEN/Akt pathway through inhibition of MTA1/HDAC unit of the NuRD complex in prostate cancer. *Biochim Biophys Acta* 2015; **1853**: 265–75.

88. Li G, Rivas P, Bedolla R, Thapa D, Reddick RL, Ghosh R, Kumar AP. Dietary resveratrol prevents development of high-grade prostatic intraepithelial neoplastic lesions: involvement of SIRT1/S6K axis. *Cancer Prev Res (Phila)* 2013; **6**: 27–39.
89. Kai L, Samuel SK, Levenson AS. Resveratrol enhances p53 acetylation and apoptosis in prostate cancer by inhibiting MTA1/NuRD complex. *Int J Cancer* 2010; **126**: 1538–48.
90. Hsieh TC, Wang Z, Deng H, Wu JM. Identification of glutathione sulfotransferase-pi (GSTP1) as a new resveratrol targeting protein (RTP) and studies of resveratrol-responsive protein changes by resveratrol affinity chromatography. *Anticancer Res* 2008; **28**: 29–36.
91. Benitez DA, Pozo-Guisado E, Clementi M, Castellon E, Fernandez-Salguero PM. Non-genomic action of resveratrol on androgen and oestrogen receptors in prostate cancer: modulation of the phosphoinositide 3-kinase pathway. *Br J Cancer* 2007; **96**: 1595–604.
92. Narayanan BA, Narayanan NK, Re GG, Nixon DW. Differential expression of genes induced by resveratrol in LNCaP cells: P53-mediated molecular targets. *Int J Cancer* 2003; **104**: 204–12.
93. Taniguchi T, Iizumi Y, Watanabe M, Masuda M, Morita M, Aono Y, Toriyama S, Oishi M, Goi W, Sakai T. Resveratrol directly targets DDX5 resulting in suppression of the mTORC1 pathway in prostate cancer. *Cell Death Dis* 2016; **7**: e2211.
94. Narayanan NK, Narayanan BA, Nixon DW. Resveratrol-induced cell growth inhibition and apoptosis is associated with modulation of phosphoglycerate mutase B in human prostate cancer cells: two-dimensional sodium dodecyl sulfate-polyacrylamide gel electrophoresis and mass spectrometry evaluation. *Cancer Detect Prev* 2004; **28**: 443–52.
95. Tameda M, Sugimoto K, Shiraki K, Inagaki Y, Ogura S, Kasai C, Yoneda M, Okamoto R, Yamamoto N, Takei Y, Ito M, Nobori T. Resveratrol sensitizes HepG2 cells to TRAIL-induced apoptosis. *Anticancer Drugs* 2014; **25**: 1028–34.
96. Venturelli S, Berger A, Bocker A, Busch C, Weiland T, Noor S, Leischner C, Schleicher S, Mayer M, Weiss TS, Bischoff SC, Lauer UM, *et al.* Resveratrol as a pan-HDAC inhibitor alters the acetylation status of histone [corrected] proteins in human-derived hepatoblastoma cells. *PLoS One* 2013; **8**: e73097.
97. Yang Q, Wang B, Zang W, Wang X, Liu Z, Li W, Jia J. Resveratrol inhibits the growth of gastric cancer by inducing G1 phase arrest and senescence in a Sirt1-dependent manner. *PLoS One* 2013; **8**: e70627.
98. Shrotriya S, Tyagi A, Deep G, Orlicky DJ, Wisell J, Wang XJ, Sclafani RA, Agarwal R, Agarwal C. Grape seed extract and resveratrol prevent 4-nitroquinoline 1-oxide induced oral tumorigenesis in mice by modulating AMPK activation and associated biological responses. *Mol Carcinog* 2015; **54**: 291–300.
99. Young LF, Martin KR. Time-dependent resveratrol-mediated mRNA and protein expression associated with cell cycle in WR-21 cells containing mutated human c-Ha-Ras. *Mol Nutr Food Res* 2006; **50**: 70–7.

100. Voellger B, Kirches E, Wilisch-Neumann A, Weise A, Tapia-Perez JH, Rupa R, Mawrin C, Firsching R. Resveratrol decreases B-cell lymphoma-2 expression and viability in GH3 pituitary adenoma cells of the rat. *Onco Targets Ther* 2013; **9**: 1269–76.
101. Jeng YJ, Watson CS. Proliferative and anti-proliferative effects of dietary levels of phytoestrogens in rat pituitary GH3/B6/F10 cells — the involvement of rapidly activated kinases and caspases. *BMC Cancer* 2009; **9**: 334.
102. Leone S, Basso E, Polticelli F, Cozzi R. Resveratrol acts as a topoisomerase II poison in human glioma cells. *Int J Cancer* 2012; **131**: E173–8.
103. Filippi-Chiela EC, Villodre ES, Zamin LL, Lenz G. Autophagy interplay with apoptosis and cell cycle regulation in the growth inhibiting effect of resveratrol in glioma cells. *PLoS One* 2011; **6**: e20849.
104. Hussain AR, Uddin S, Bu R, Khan OS, Ahmed SO, Ahmed M, Al-Kuraya KS. Resveratrol suppresses constitutive activation of AKT via generation of ROS and induces apoptosis in diffuse large B cell lymphoma cell lines. *PLoS One* 2011; **6**: e24703.
105. Kim KH, Back JH, Zhu Y, Arbesman J, Athar M, Kopelovich L, Kim AL, Bickers DR. Resveratrol targets transforming growth factor-beta2 signaling to block UV-induced tumor progression. *J Invest Dermatol* 2011; **131**: 195–202.
106. Yang Z, Yang S, Misner BJ, Chiu R, Liu F, Meyskens FL, Jr. Nitric oxide initiates progression of human melanoma via a feedback loop mediated by apurinic/apyrimidinic endonuclease-1/redox factor-1, which is inhibited by resveratrol. *Mol Cancer Ther* 2008; **7**: 3751–60.
107. Johnson GE, Ivanov VN, Hei TK. Radiosensitization of melanoma cells through combined inhibition of protein regulators of cell survival. *Apoptosis* 2008; **13**: 790–802.
108. Kundu JK, Shin YK, Kim SH, Surh YJ. Resveratrol inhibits phorbol ester-induced expression of COX-2 and activation of NF-kappaB in mouse skin by blocking IkappaB kinase activity. *Carcinogenesis* 2006; **27**: 1465–74.
109. Kim AL, Zhu Y, Zhu H, Han L, Kopelovich L, Bickers DR, Athar M. Resveratrol inhibits proliferation of human epidermoid carcinoma A431 cells by modulating MEK1 and AP-1 signalling pathways. *Exp Dermatol* 2006; **15**: 538–46.
110. Kundu JK, Shin YK, Surh YJ. Resveratrol modulates phorbol ester-induced pro-inflammatory signal transduction pathways in mouse skin in vivo: NF-kappaB and AP-1 as prime targets. *Biochem Pharmacol* 2006; **72**: 1506–15.
111. Kundu JK, Chun KS, Kim SO, Surh YJ. Resveratrol inhibits phorbol ester-induced cyclooxygenase-2 expression in mouse skin: MAPKs and AP-1 as potential molecular targets. *Biofactors* 2004; **21**: 33–9.
112. Park D, Jeong H, Lee MN, Koh A, Kwon O, Yang YR, Noh J, Suh PG, Park H, Ryu SH. Resveratrol induces autophagy by directly inhibiting mTOR through ATP competition. *Sci Rep* 2016; **6**: 21772.
113. Bickenbach KA, Veerapong J, Shao MY, Mauceri HJ, Posner MC, Kron SJ, Weichselbaum RR. Resveratrol is an effective inducer of CArG-driven TNF-alpha gene therapy. *Cancer Gene Ther* 2008; **15**: 133–9.

114. Wang RH, Sengupta K, Li C, Kim HS, Cao L, Xiao C, Kim S, Xu X, Zheng Y, Chilton B, Jia R, Zheng ZM, *et al.* Impaired DNA damage response, genome instability, and tumorigenesis in SIRT1 mutant mice. *Cancer Cell* 2008; **14**: 312–23.
115. Canella R, Benedusi M, Martini M, Cervellati F, Cavicchio C, Valacchi G. Role of Nrf2 in Preventing Oxidative Stress Induced Chloride Current Alteration in Human Lung Cells. *J Cell Physiol* 2017;
116. Li J, Liang C, Zhang ZK, Pan X, Peng S, Lee WS, Lu A, Lin Z, Zhang G, Leung WN, Zhang BT. TAK1 inhibition attenuates both inflammation and fibrosis in experimental pneumoconiosis. *Cell Discov* 2017; **3**: 17023.
117. Xu D, Li Y, Zhang B, Wang Y, Liu Y, Luo Y, Niu W, Dong M, Liu M, Dong H, Zhao P, Li Z. Resveratrol alleviate hypoxic pulmonary hypertension via anti-inflammation and anti-oxidant pathways in rats. *Int J Med Sci* 2016; **13**: 942–54.
118. Chai R, Chen Y, Yuan H, Wang X, Guo S, Qi J, Zhang H, Zhan Y, An H. Identification of resveratrol, an herbal compound, as an activator of the calcium-activated chloride channel, TMEM16A. *J Membr Biol* 2017; **250**: 483–92.
119. Gu H, Jiao Y, Yu X, Li X, Wang W, Ding L, Liu L. Resveratrol inhibits the IL-1beta-induced expression of MMP-13 and IL-6 in human articular chondrocytes via TLR4/MyD88-dependent and — independent signaling cascades. *Int J Mol Med* 2017;
120. Liu L, Gu H, Liu H, Jiao Y, Li K, Zhao Y, An L, Yang J. Protective effect of resveratrol against IL-1beta-induced inflammatory response on human osteoarthritic chondrocytes partly via the TLR4/MyD88/NF-kappaB signaling pathway: an "in vitro study". *Int J Mol Sci* 2014; **15**: 6925–40.
121. Wang DT, Yin Y, Yang YJ, Lv PJ, Shi Y, Lu L, Wei LB. Resveratrol prevents TNF-alpha-induced muscle atrophy via regulation of Akt/mTOR/FoxO1 signaling in C2C12 myotubes. *Int Immunopharmacol* 2014; **19**: 206–13.
122. Montesano A, Luzi L, Senesi P, Mazzocchi N, Terruzzi I. Resveratrol promotes myogenesis and hypertrophy in murine myoblasts. *J Transl Med* 2013; **11**: 310.
123. Abdin AA. Targeting sphingosine kinase 1 (SphK1) and apoptosis by colon-specific delivery formula of resveratrol in treatment of experimental ulcerative colitis in rats. *Eur J Pharmacol* 2013; **718**: 145–53.
124. Xin P, Han H, Gao D, Cui W, Yang X, Ying C, Sun X, Hao L. Alleviative effects of resveratrol on nonalcoholic fatty liver disease are associated with up regulation of hepatic low density lipoprotein receptor and scavenger receptor class B type I gene expressions in rats. *Food Chem Toxicol* 2013; **52**: 12–8.
125. Morita Y, Wada-Hiraike O, Yano T, Shirane A, Hirano M, Hiraike H, Koyama S, Oishi H, Yoshino O, Miyamoto Y, Sone K, Oda K, *et al.* Resveratrol promotes expression of SIRT1 and StAR in rat ovarian granulosa cells: an implicative role of SIRT1 in the ovary. *Reprod Biol Endocrinol* 2012; **10**: 14.

126. Hsieh TC. Uptake of resveratrol and role of resveratrol-targeting protein, quinone reductase 2, in normally cultured human prostate cells. *Asian J Androl* 2009; **11**: 653–61.
127. Venkatesan B, Ghosh-Choudhury N, Das F, Mahimainathan L, Kamat A, Kasinath BS, Abboud HE, Choudhury GG. Resveratrol inhibits PDGF receptor mitogenic signaling in mesangial cells: role of PTP1B. *FASEB J* 2008; **22**: 3469–82.
128. Dutra LA, Heidenreich D, Silva G, Man Chin C, Knapp S, Santos JLD. Dietary Compound Resveratrol Is a Pan-BET Bromodomain Inhibitor. *Nutrients* 2017; **9**:
129. Abdulla A, Zhao X, Yang F. Natural polyphenols inhibit lysine-specific demethylase-1 in vitro. *J Biochem Pharmacol Res* 2013; **1**: 56–63.
130. Latorre E, Birar VC, Sheerin AN, Jeynes JCC, Hooper A, Dawe HR, Melzer D, Cox LS, Faragher RGA, Ostler EL, Harries LW. Small molecule modulation of splicing factor expression is associated with rescue from cellular senescence. *BMC Cell Biol* 2017; **18**: 31.
131. Menicacci B, Laurenzana A, Chilla A, Margheri F, Peppicelli S, Tanganelli E, Fibbi G, Giovannelli L, Del Rosso M, Mocali A. Chronic Resveratrol Treatment Inhibits MRC5 Fibroblast SASP-Related Protumoral Effects on Melanoma Cells. *J Gerontol A Biol Sci Med Sci* 2017; **72**: 1187–95.
132. Markus MA, Marques FZ, Morris BJ. Resveratrol, by modulating RNA processing factor levels, can influence the alternative splicing of pre-mRNAs. *PLoS One* 2011; **6**: e28926.
133. Thiel G, Rossler OG. Resveratrol regulates gene transcription via activation of stimulus-responsive transcription factors. *Pharmacol Res* 2017; **117**: 166–76.
134. Nalli M, Ortar G, Moriello AS, Morera E, Di Marzo V, De Petrocellis L. TRPA1 channels as targets for resveratrol and related stilbenoids. *Bioorg Med Chem Lett* 2016; **26**: 899–902.
135. Silva AM, Oliveira MI, Sette L, Almeida CR, Oliveira MJ, Barbosa MA, Santos SG. Resveratrol as a natural anti-tumor necrosis factor-alpha molecule: implications to dendritic cells and their crosstalk with mesenchymal stromal cells. *PLoS One* 2014; **9**: e91406.
136. Calamini B, Ratia K, Malkowski MG, Cuendet M, Pezzuto JM, Santarsiero BD, Mesecar AD. Pleiotropic mechanisms facilitated by resveratrol and its metabolites. *Biochem J* 2010; **429**: 273–82.
137. Lu CC, Chen JK. Resveratrol enhances perforin expression and NK cell cytotoxicity through NKG2D-dependent pathways. *J Cell Physiol* 2010; **223**: 343–51.
138. Svajger U, Obermajer N, Jeras M. Dendritic cells treated with resveratrol during differentiation from monocytes gain substantial tolerogenic properties upon activation. *Immunology* 2010; **129**: 525–35.
139. Koo N, Cho D, Kim Y, Choi HJ, Kim KM. Effects of resveratrol on mast cell degranulation and tyrosine phosphorylation of the signaling components of the IgE receptor. *Planta Med* 2006; **72**: 659–61.

140. Youn HS, Lee JY, Fitzgerald KA, Young HA, Akira S, Hwang DH. Specific inhibition of MyD88-independent signaling pathways of TLR3 and TLR4 by resveratrol: molecular targets are TBK1 and RIP1 in TRIF complex. *J Immunol* 2005; **175**: 3339–46.
141. Srivastava R, Ratheesh A, Gude RK, Rao KV, Panda D, Subrahmanyam G. Resveratrol inhibits type II phosphatidylinositol 4-kinase: a key component in pathways of phosphoinositide turn over. *Biochem Pharmacol* 2005; **70**: 1048–55.
142. Casas-Rua V, Alvarez IS, Pozo-Guisado E, Martin-Romero FJ. Inhibition of STIM1 phosphorylation underlies resveratrol-induced inhibition of store-operated calcium entry. *Biochem Pharmacol* 2013; **86**: 1555–63.
143. De Leo A, Arena G, Lacanna E, Oliviero G, Colavita F, Mattia E. Resveratrol inhibits Epstein Barr Virus lytic cycle in Burkitt's lymphoma cells by affecting multiple molecular targets. *Antiviral Res* 2012; **96**: 196–202.
144. Villa-Cuesta E, Boylan JM, Tatar M, Gruppuso PA. Resveratrol inhibits protein translation in hepatic cells. *PLoS One* 2011; **6**: e29513.
145. Uhle S, Medalia O, Waldron R, Dumdey R, Henklein P, Bech-Otschir D, Huang X, Berse M, Sperling J, Schade R, Dubiel W. Protein kinase CK2 and protein kinase D are associated with the COP9 signalosome. *EMBO J* 2003; **22**: 1302–12.
146. Buryanovskyy L, Fu Y, Boyd M, Ma Y, Hsieh TC, Wu JM, Zhang Z. Crystal structure of quinone reductase 2 in complex with resveratrol. *Biochemistry* 2004; **43**: 11417–26.
147. Scuto A, Kirschbaum M, Buettner R, Kujawski M, Cermak JM, Atadja P, Jove R. SIRT1 activation enhances HDAC inhibition-mediated upregulation of GADD45G by repressing the binding of NF-kappaB/STAT3 complex to its promoter in malignant lymphoid cells. *Cell Death Dis* 2013; **4**: e635.
148. Milne JC, Lambert PD, Schenk S, Carney DP, Smith JJ, Gagne DJ, Jin L, Boss O, Perni RB, Vu CB, Bemis JE, Xie R, *et al.* Small molecule activators of SIRT1 as therapeutics for the treatment of type 2 diabetes. *Nature* 2007; **450**: 712–6.
149. Pacholec M, Bleasdale JE, Chrunyk B, Cunningham D, Flynn D, Garofalo RS, Griffith D, Griffor M, Loulakis P, Pabst B, Qiu X, Stockman B, *et al.* SRT1720, SRT2183, SRT1460, and resveratrol are not direct activators of SIRT1. *J Biol Chem* 2010; **285**: 8340–51.
150. Canto C, Gerhart-Hines Z, Feige JN, Lagouge M, Noriega L, Milne JC, Elliott PJ, Puigserver P, Auwerx J. AMPK regulates energy expenditure by modulating NAD+ metabolism and SIRT1 activity. *Nature* 2009; **458**: 1056–60.
151. Ruderman NB, Xu XJ, Nelson L, Cacicedo JM, Saha AK, Lan F, Ido Y. AMPK and SIRT1: a long-standing partnership? *Am J Physiol Endocrinol Metab* 2010; **298**: E751–60.
152. Berman HM, Westbrook J, Feng Z, Gilliland G, Bhat TN, Weissig H, Shindyalov IN, Bourne PE. The Protein Data Bank. *Nucleic Acids Res* 2000; **28**: 235–42.

153. Cao D, Wang M, Qiu X, Liu D, Jiang H, Yang N, Xu RM. Structural basis for allosteric, substrate-dependent stimulation of SIRT1 activity by resveratrol. *Genes Dev* 2015; **29**: 1316–25.
154. Reddy AS, Zhang S. Polypharmacology: drug discovery for the future. *Expert Rev Clin Pharmacol* 2013; **6**: 41–7.
155. Chen C, He Y, Wu J, Zhou J. Creation of a free, Internet-accessible database: the Multiple Target Ligand Database. *J Cheminform* 2015; **7**: 14.
156. Hopkins AL. Network pharmacology: the next paradigm in drug discovery. *Nat Chem Biol* 2008; **4**: 682–90.
157. Weininger D. SMILES, a chemical language and information system. 1. Introduction to methodology and encoding rules. *J. Chem. Inf. Comput. Sci.* 1988; **28**: 31–36.
158. Chambers J, Davies M, Gaulton A, Hersey A, Velankar S, Petryszak R, Hastings J, Bellis L, McGlinchey S, Overington JP. UniChem: a unified chemical structure cross-referencing and identifier tracking system. *J Cheminform* 2013; **5**: 3.
159. Kim S, Thiessen PA, Bolton EE, Chen J, Fu G, Gindulyte A, Han L, He J, He S, Shoemaker BA, Wang J, Yu B, *et al.* PubChem Substance and Compound databases. *Nucleic Acids Res* 2016; **44**: D1202–13.
160. Bento AP, Gaulton A, Hersey A, Bellis LJ, Chambers J, Davies M, Kruger FA, Light Y, Mak L, McGlinchey S, Nowotka M, Papadatos G, *et al.* The ChEMBL bioactivity database: an update. *Nucleic Acids Res* 2014; **42**: D1083–90.
161. Szklarczyk D, Morris JH, Cook H, Kuhn M, Wyder S, Simonovic M, Santos A, Doncheva NT, Roth A, Bork P, Jensen LJ, von Mering C. The STRING database in 2017: quality-controlled protein-protein association networks, made broadly accessible. *Nucleic Acids Res* 2017; **45**: D362–D68.
162. Kjaer TN, Ornstrup MJ, Poulsen MM, Stodkilde-Jorgensen H, Jessen N, Jorgensen JOL, Richelsen B, Pedersen SB. No beneficial effects of resveratrol on the metabolic syndrome: a randomized placebo-controlled clinical trial. *J Clin Endocrinol Metab* 2017; **102**: 1642–51.
163. Vang O, Ahmad N, Baile CA, Baur JA, Brown K, Csiszar A, Das DK, Delmas D, Gottfried C, Lin HY, Ma QY, Mukhopadhyay P, *et al.* What is new for an old molecule? Systematic review and recommendations on the use of resveratrol. *PLoS One* 2011; **6**: e19881.
164. Benoit CE, Bastianetto S, Brouillette J, Tse Y, Boutin JA, Delagrange P, Wong T, Sarret P, Quirion R. Loss of quinone reductase 2 function selectively facilitates learning behaviors. *J Neurosci* 2010; **30**: 12690–700.
165. Hoshino J, Park EJ, Kondratyuk TP, Marler L, Pezzuto JM, van Breemen RB, Mo S, Li Y, Cushman M. Selective synthesis and biological evaluation of sulfate-conjugated resveratrol metabolites. *J Med Chem* 2010; **53**: 5033–43.
166. Bomgardner MM. Reviving resveratrol. *Chemical & Engineering News* 2017; **95**: 38–39.

Chapter 4

Roles of Resveratrol as Signaling Sensor and Gatekeeper

Ya-Jung Shih[*,†,**], Yi-Ru Chen[*,†,**], Kuan Wang[*,‡], Jacqueline Whang-Peng[*,#], Heng-Yuan Tang[§], Hung-Yun Lin[*,†,¶,§,‡‡] and Paul J. Davis[§,||]

[*] *Taipei Cancer Center, 250 Wu-Xing Street, 11031 Taipei, Taiwan*

[†] *PhD Program for Cancer Molecular Biology and Drug Discovery, College of Medical Science and Technology, 250 Wu-Xing Street, 11031 Taipei, Taiwan*

[‡] *Graduate Institute of Nanomedicine and Medical Engineering, College of Medical Engineering, Taipei Medical University, 250 Wu-Xing Street, 11031 Taipei, Taiwan*

[#] *Cancer Center, Wan Fang Hospital, Taipei Medical University, Taipei, Taiwan*

[¶] *Traditional Herbal Medicine Research Center of Taipei Medical University Hospital, 252 Wu-xing Street, Xin-yi District, Taipei 110, Taiwan*

[‡‡] *TMU Research Center of Cancer Translational Medicine, 250 Wu-Xing Street, 11031 Taipei, Taiwan*

[§] *Pharmaceutical Research Institute, Albany College of Pharmacy and Health Sciences, One Discovery Drive, Rensselaer, NY 12144, USA*

[||] *Albany Medical College, 43 New Scotland Avenue Albany, NY 12208, USA*

[**]These authors contributed equally to this book chapter.

Resveratrol (trans-3,4′,5-trihydroxystilbene) is a natural phytoalexin that has pharmacological effects on several human diseases: cancer, coronary artery disease and neurodegenerative diseases. Many studies have reported the application of resveratrol as a preventive drug for these important conditions. In addition, recent reports indicate that resveratrol may also increase the lifespan of yeast and fruit flies. The positive effects of resveratrol have largely been demonstrated *in vitro* or in animal studies, with only limited and inconsistent effects reported in clinical studies. In addition, roles of resveratrol as signaling sensor and gatekeeper are still not clear. In this chapter, we will summarize a group of potentially important roles for resveratrol in diverse biological systems, particularly in metabolism and in cancer.

4.1. Introduction

Resveratrol is a natural antioxidant. It has been described as a chemically prophylactic nutrient that has a significant health benefit, for example, in the setting of excessive reactive oxygen species (ROS) generation. ROS produced by cellular respiration and inflammation produce cytotoxic effects that may contribute to carcinogenesis and to cardiovascular disease and neurodegenerative diseases such as Alzheimer's disease and Parkinson's disease. Resveratrol has been shown to possess ROS-scavenging activity like other antioxidants.[1]

The 4′-hydroxyl group of resveratrol is more reactive than those of the 3- and 5-positions because of resonance effects.[2] After oxidation, the dominant structure of resveratrol radicals is a semiquinone in which the unpaired electron mainly is distributed on the para position of the O-atom.[2,3] The hydroxyl groups and the double bond of resveratrol hinder the autooxidation of ROS produced *in vivo*,[2] endowing resveratrol with excellent antioxidant properties. However, concentrations of resveratrol required for successful *in vitro* studies are not obtained *in vivo* in tumor tissue.[4] The lack of congruence in preclinical and clinical studies of resveratrol reflects, at least in part, the low bioavailability of resveratrol compounds, a limited effectiveness of resveratrol metabolites and the systemic toxicity that has served to limit dosing.[5] It is important to point out that meta-analyses of clinical studies that have supported existence of anticancer activity of polyphenols have not been based on prospective studies, but on a case–control study design, as noted by recent review articles.[6–8] Thus, much of the *in vitro* experience with resveratrol is not relevant to clinical practice. Incomplete understanding of the mechanisms of the actions of resveratrol in cancer settings, in the intact animal or patient, may also be limiting our ability to translate preclinical results into clinical cancer management.[5–8]

4.2. Resveratrol activates integrin $\alpha v\beta 3$ signaling

Resveratrol has been observed to act intracellularly and it is known to be rapidly metabolized within cells. We propose that the first step in defining the mechanisms of resveratrol action is to determine whether it initiates certain of its actions on the cell surface or whether it is exclusively dependent on intracellular targets. There are multiple resveratrol-induced effects on cancer cells and normal cells. Some of these effects may be cell-specific, tissue-specific or species-dependent. Cellular and extracellular resveratrol targets have been proposed and, as discussed below, a cell surface receptor for the agent has been defined. A requirement that resveratrol must achieve cell entry and bind to one or more discrete intracellular targets is limiting because of its rapid metabolism once it is in the cell.

Integrins are transmembrane proteins that critically support the interaction of cells with the extracellular environment — matrix proteins and other cells — and are critically important to the progression of cancers.[9] Intracellular transcriptional factors are the principal mediators of integrin action. Among the transcriptional factors regulated by resveratrol are NFκB, activator protein AP-1, AP-2, and cyclic AMP-responsive element-binding protein (CREB).[10] A study of integrin $\alpha v\beta 3$ relied on competition assays using a known ligand and binding studies using membrane fractions.[11] Dysregulation of the $\beta 3$ integrins has been implicated in cancer pathogenesis.[12] Tumor growth and associated angiogenesis, particularly as mediated by vascular endothelial growth factor, are enhanced in $\beta 3$-null mice.[13] However, overexpression of integrin $\beta 3$ can suppress tumor growth of a human glioma model in rats.[14] Such results suggest that promotion of integrin $\beta 3$ expression in cancer cells may be a therapeutic goal in the setting of cancer. Integrins contain an on/off switch that regulates their ligand binding affinity and this is referred to as integrin activation. Reducing agents, such as dithiothreitol (DTT), can activate integrins. DTT reduces two disulfide bonds within the integrin's cysteine-rich domain in the β3 chain of integrin αIIbβ3, leading to global conformational changes within integrin αIIbβ3 and to the opening of RGD and fibrinogen binding sites.[15]

The importance of the foregoing to the current review is that integrin αvβ3 contains a cell surface receptor for resveratrol[16] and because of its properties, resveratrol is likely to associate with cell membranes readily, making it difficult to assess the specificity of extracellular binding studies. However, using radiolabeled resveratrol, we recently determined that the stilbene binds dissociably to integrin αvβ3, leading to activation of mitogen-activated protein kinase (MAPK; ERK1/2) signal transduction cascade. Blocking integrin αvβ3 binding site with either RGD peptide or anti-integrin αvβ3 antibody, we were able to inhibit resveratrol-induced apoptosis in different cancer cell lines.[17–20] The RGD peptide binding site contains a cysteine-rich domain and it is possible that resveratrol interacts with cysteinyl thiol in this domain. This interaction will activate downstream signal transduction pathways. We have demonstrated a functional relationship between the resveratrol receptor and p53-dependent apoptosis in cancer cells.[17]

The cytoplasmic domains of β subunits do not contain intrinsic tyrosine kinase activity, therefore signaling occurs primarily via recruitment of intracellular signaling molecules.[21] Predominantly integrin signal is through the recruitment and activation of Src-family kinases (SFKs).[22] Integrins recruit focal adhesion kinase (FAK) via β subunits. FAK functions as a phosphorylation-regulated scaffold to recruit Src to focal adhesions. Resveratrol interacts with integrin αvβ3 to recruit signaling molecules that include the FAK.[23] Resveratrol is able to activate ERK1/2 in an estrogen receptor (ER)-, mitogen-activated protein kinase (MEK)-, matrix metalloproteinase (MMP)-, Src-, and endothelial growth factor-receptor (EGF-R)-dependent manner in endothelial cells at nanomolar level.[24] However, a specific inhibitor of the estrogen receptor, ICI 182,780, is unable to block resveratrol-induced ERK1/2 activation in human breast cancer MCF-7 cells.[25] This indicates the distinctive nature of the resveratrol receptor.

FAK is also able to activate signaling from phosphatidylinositol 3-kinase (PI3K) to AKT/protein kinase B (PKB) through phosphatidylinositol-3,4,5-trisphosphate (PtdIns(3,4,5)P3).[22] Although Pozo-Guisado *et al.* have shown that resveratrol increases ERα-dependent PI3K activity with a maximum stimulatory effect at concentrations of about 10 μM; concentrations $> 50\,\mu$M decreased PI3K activity.[26] There are no studies of a possible connection between integrin αvβ3 and ERα, therefore resveratrol-bound integrin αvβ3 may not play a role in PI3K activities. Since the binding of resveratrol to integrin αvβ3 does not activate PI3K, the uptake of resveratrol should be due to endocytosis. But resveratrol does inhibit PI3K signal transduction pathways to induce apoptosis.[27] The stilbene also inhibits high glucose-induced interleukin-17 (IL-17) expression in primary mouse cardiac fibroblasts and Ang II-induced vascular smooth muscle cell hypertrophy in the rat.[28] This may be due to inhibition of the PI3K/Akt/ERK signaling pathway. There may be other cell surface binding sites for resveratrol. For example, Colin *et al.* have shown that radiolabeled resveratrol accumulates in sphingomyelin- and cholesterol-enriched colon carcinoma and leukemia cell fractions. ERK, c-Jun NH2-terminal kinases (JNK), and Akt also accumulate in lipid rafts in response to resveratrol exposure for 48 h at 30 μM via a monensin (5–20 μM)-sensitive process thought to reflect clathrin-independent endocytosis.[23] Insulin growth factor receptor (IGFR) is another extracellular potential receptor that has been proposed for resveratrol.[29] However, no direct binding was demonstrated in this study.

4.3. Resveratrol activates signaling sensors

Beneficial effects of resveratrol on age-related metabolic deterioration have been reported. The protective role of resveratrol in metabolic diseases, such as type 2 diabetes mellitus (T2D) and obesity has been demonstrated in animal models. Resveratrol action mimics caloric restriction by extending the lifespan of different microorganisms, including *S. cerevisiae* and *C. elegans*, and by delaying several aging phenotypes in mice. Resveratrol has also been protective against the deregulation of energy homeostasis in mouse models for metabolic syndromes via activation of key metabolic sensor proteins. The concentrations of resveratrol used in such studies are listed in Table 4.1.

4.3.1. *Fasting and diabetes*

The complex metabolic effects of resveratrol involve the fuel-sensing AMP-activated kinase (AMPK), an $\alpha\beta\gamma$ heterotrimer, which is a nutrient and energy sensor that maintains energy homeostasis (Fig. 4.1).[30] AMPK is activated by metabolic stresses that decrease ATP levels and lead to an increase in cellular AMP, an allosteric activator of AMPK. AMPK acts as a master metabolic regulator to maintain cellular energy balance following increased energy demand and the described increases in the AMP/ATP ratio. AMPK is also activated under the physiological conditions of energy stress such as hypoxia, fasting and exercise.[31–33] Upon activation, AMPK diminishes the rates of most energy consuming metabolic programs — such as lipid and protein biosynthesis — and triggers energy producing biochemical processes. The latter include lipid oxidation and glycolysis to counter energy depletion.[31,32] In addition, AMPK exerts long term changes in energy homeostasis by triggering transcriptional programs involving mitochondrial biogenesis and the use of oxidative metabolism to meet cellular energy demands.[34] Given the conserved ability of AMPK to sense energy stress and trigger metabolic adaptations, it has also been speculated that AMPK could act as mediator of the beneficial effects of calorie restricted diet.[35] AMPK activation mimics certain aspects of calorie-restriction and associated promotion of longevity. It is not clear whether AMPK is an effector promoting longevity and healthy aging during calorie restriction.[35] The SIRT1/AMPK signaling pathway is a ubiquitously expressed cellular energy sensor.[36–38] Among the seven sirtuin enzymes (SIRT1-7) in mammals, SIRT1 has been the most extensively studied[41] and has a wide range of substrates, among which are key regulators of mitochondrial respiration, lipid

Table 4.1. Effects of resveratrol on life span, homeostasis, and signal transductions.

Types of studies	Concentration of resveratrol	References
To extend lifespans of *S. cerevisiae* and *C. elegans*.		Developmental Cell. 2005 Nov; 9(5): 605–15.
To delay aging phenotypes in mice.		Cell. 2012 Feb 3; 148(3): 421–33.
To protect deregulation of energy homeostasis in mice.		Proc Natl Acad Sci USA. 2008 Jul 1; 105(28): 9793–798.
To attenuate LPS-induced expression of NO, PGE2, iNOS, TNFα and IL-1β in RAW 264.7 cells.	1 to 10 μM	PLoS One. 2012; 7(8): e44107.
To activate PI3K, AKT and mTOR. It has been shown in other studies to increase AKT phosphorylation in a time-dependent manner.	5 and 10 μM	PLoS One. 2012; 7(8): e44107.
To reverse LPS-induced activation of macrophage model in RAW 264.7 cells is PI3K-dependent.	5 and 10 μM	PLoS One. 2012; 7(8): e44107.
To inhibit strongly the production of iNOS and NO in cytokine/LPS-treated L6 myocytes and 3T3-L1 adipocytes. No effect of resveratrol was observed in iNOS induction in FAO hepatocytes.	10 to 80 μM	Am J Physiol Endocrinol Metab. 2011 No; 301(5): E922–30.
To reduce iNOS and NO production induced by LPS in skeletal muscle and white adipose tissue, but not in liver *in vivo*.	40 mg/kg ip	Am J Physiol Endocrinol Metab. 2011 No; 301(5): E922–30.
To reverse retinal AMPK dephosphorylation and SIRT1 deactivation in mice with inducible diabetes. In addition, to reduce diabetes-induced retinal leukocyte adhesion and expression of ICAM-1 and VEGF in the retina.	C57BL/6 mice 50 mg/kg BW by gastric intubation daily for 7 days	Invest Ophthalmol Vis Sci. 2011 Nov 25; 52(12): 9142–8.

(*Continued*)

Table 4.1. (*Continued*)

Types of studies	**Concentration of resveratrol**	**References**
To disable the activation of SIRT1 directl.	30 & 100 μM	J Biol Chem. 2010 Mar 12; 285(11): 8340–51.
To disable combined benefit in leucine-activated Sirt1 by lowering its Km for NAD+ and -potentiated sirtuin/ AMPK-activators, which improves insulin sensitivity.		Metabolism. 2016 Nov; 65(11): 1679–91.
To stimulate mitochondrial biogenesis in an AMPK-dependent mechanism via activated LKB1 without Sirt1 in neurons and to contribute to the neuroprotective effects.	10 μM	Proc Natl Acad Sci U S A. 2007 Apr 2; 104(17): 7217–22.
To activate PI3K to phosphorylate antioxidant transcription factor Nrf2.		Mol Cell Biol. 2016 Jun 29; 36(14): 1931–42.
To inhibit class IA PI3K and its downstream signaling in cultured cells.	100 μM, which activates SIRT1 in cultured cells.	Biochem J. 2007 Sep 15; 406(3): 511–8.

metabolism and the aging process, such as the ForkheadObox (FOXO) family of transcription factors.[42]

Against this extensive metabolic background, we can review the relevant metabolic actions of resveratrol. Resveratrol activates AMPK activation within minutes. This is considerably faster than the time usually required for SIRT1 activation in most cell lines tested, including hepatic and cultured skeletal muscle cell models.[43–46] Resveratrol significantly attenuates lipopolysaccharide (LPS)-induced expression of nitric oxide (NO), prostaglandin E2 (PGE2), inducible nitric oxide synthase (iNOS), tumor necrosis factor-α (TNF-α) and IL-1β in RAW 264.7 cells.[47] Other studies indicate that resveratrol (10 to 80 μM) strongly inhibits the production of iNOS protein and NO in cytokine/LPS-treated L6 myocytes and 3T3-L1 adipocytes.[48] However, no effect of resveratrol was observed on iNOS induction in FAO hepatocytes.[48] *In vivo* studies also demonstrate that when LPS-challenged mice are treated with resveratrol (40 mg/kg ip), iNOS and NO production are reduced in skeletal muscle and white adipose tissue, but not in liver.[48]

Induction of diabetes in mice has led to retinal AMPK dephosphorylation and SIRT1 deactivation that are reversed by resveratrol.[49] Resveratrol also reduces diabetes-induced retinal leukocyte adhesion and expression of ICAM-1 and VEGF in the retina.[49] These findings are consistent with the concept that diabetes-induced retinal inflammation stems from down-regulation of the AMPK pathway, subsequent Sirt1 deactivation and NFκB activation. It also raises the possibility of usefulness of an AMPK activator, such as resveratrol, as a therapeutic agent for diabetic retinopathy.

The possible role of SIRT1 in resveratrol-induced AMPK activation is controversial. The largely overlapping outcomes of AMPK activation and SIRT1 activation, together with the strong links of resveratrol to activation of both metabolic sensors, has raised the possibility that their actions could be intermingled.[50] Since resveratrol requires AMPK and SIRT1 activities for triggering mitochondrial biogenesis *in vivo*,[50, 51] it has been suggested that AMPK and SIRT1 may be part of a common signaling pathway or might act coordinately. Studies using inhibitors of SIRT1 reveal that the deacetylase enzyme is not involved in resveratrol action on iNOS. However, the inhibitory effect of resveratrol on NO production depends upon activation of AMPK, and can be blunted in L6 myocytes with Compound C or RNA. These results indicated that resveratrol specifically inhibits iNOS induction in muscle cells through a mechanism involving AMPK and not via SIRT1 activation.[48] Activation of AMPK induced by resveratrol in neurons also does not require NAD-dependent protein deacetylase, SIRT1. Studies by Pacholec *et al.* point out that resveratrol is not a direct activator of SIRT1.[52] In other studies, leucine activates SIRT1 by lowering its Km for NAD+ and potentiates other sirtuin/AMPK-activators; this improves insulin sensitivity. Since metformin (Met) converges in this pathway, leucine would amplify its gluco-regulatory effects. Addition of resveratrol does not confer further benefit.[53]

4.3.2. *Neurodegeneration*

Resveratrol promotes anti-aging pathways via activation of several metabolic sensors, including AMPK. Alzheimer's disease is an age-related neurodegenerative disorder characterized by amyloid-β (Abeta) peptide deposition into cerebral plaques.[54] Resveratrol stimulates mitochondrial biogenesis in an AMPK-dependent mechanism. In neurons, the stilbene stimulates AMPK activity via activated LKB1, but does not require the NAD-dependent protein deacetylase, SIRT1. These findings suggest that neuronal activation of AMPK by resveratrol could positively affect neuronal energy homeostasis and contribute to the neuroprotective effects of resveratrol.[55]

Activation of PI3K for phosphorylation of antioxidant transcription factor Nrf2 has been shown directly by resveratrol.[56] However, resveratrol inhibits class IA PI3K and its downstream signaling in the same concentration range at which it activates SIRTs in cultured cells.[57] While resveratrol has been shown to inhibit phosphorylation/activation of PI3K, AKT and mTOR, it has been shown in other studies to increase Akt phosphorylation in a time-dependent manner.[47] A specific PI3K inhibitor, wortmannin, blocks the effects of resveratrol on LPS-induced activation of macrophage model RAW 264.7 cells.[47] PI3K is essential for resveratrol-dependent phosphorylation of AMPK and expression of SIRT1. Among the effects of the latter is inhibition of NFκB, a facilitator of low-grade inflammation that may contribute to aging. Thus, the observations briefly summarized here are consistent with a role for resveratrol in anti-aging.

4.4. Resveratrol acts as a gatekeeper

Gatekeeper genes such as *p53*, *PTEN* and retinoblastoma encode gene products that prevent growth of potential cancer cells and accumulation of mutated cells in which proliferation may proceed unchecked. In contrast to gatekeeper genes, caretaker genes are a group of genes that, when inactivated, do not directly promote tumors, but their inactivation results in genetic instabilities and an increased mutation rate affecting many genes — including gatekeeper genes — which directly regulate tumor growth. *BRCA1* and *BRCA2* are examples of caretaker genes.[58] We will discuss the role of several gatekeeper genes in resveratrol-induced anti-proliferation in the following sections.

4.4.1. *p53 pathway*

Resveratrol is a regulator of gatekeeper genes that directly regulate tumor growth by inhibiting growth and promoting cell death. The cancer suppressor gene, *p53*, is a primary example. Resveratrol induces expression of the *p53* gene in different cancer cells.[59] We have shown that resveratrol inhibits cancer growth via an ERK1/2-dependent mechanism that culminates in cell nucleus accumulation of cyclooxygenase-2 (COX-2) and phosphorylation/activation of p53. This was demonstrated in human breast cancer cells, glioma cells and ovarian cancer cells.[59] The process is initiated by resveratrol at its receptor site on integrin $\alpha v\beta 3$. Downstream, the ERK1/2 signal transduction pathway is activated, resulting in nuclear uptake of COX-2/activated *ERK1/2* complexes committed to specific phosphorylation of p53.[20,60] Activated p53 promotes apoptosis of cancer cells. *NS-398*, a *COX-2* inhibitor, is able to inhibit the formation of complexes of nuclear *COX-2* and activated *ERK1/2*. The nuclear *COX-2* and *ERK1/2* associate with activated *p53* to form the complex that functions as a co-activator for *p53*-responsive genes. Interestingly, resveratrol is able to normalize the pro-apoptotic function mutated *p53*.

Resveratrol has been shown to sensitize tumor cells to chemotherapeutic agents through modulation of p53. It improves the chemotherapeutic potential of doxorubicin in chemoresistant B16 melanoma cells through up-regulation of p53.[61] Interestingly, resveratrol enhances the cancer chemotherapeutic activity of other anti-cancer drugs by one or more mechanisms that are p53-independent.[62] Studies in *in vitro* and *in vivo* indicate that resveratrol can diminish chemoresistance of tumor cells by modifying apoptotic pathways, reducing the expression of plasma membrane drug transporters, and inhibiting the expression of proteins involved in tumor cell proliferation. In C57BL/6 mice, cisplatin treatment induced p53 phosphorylation and protein accumulation, which were accompanied by the development of acute kidney injury.[63] SIRT1 activation by resveratrol decreased cisplatin-induced apoptosis while improving glomerular filtration rate. Resveratrol also blocked the cisplatin-induced decrease of expression of anti-apoptotic Bcl-xL.[64] In addition, resveratrol-induced SIRT1 has been shown to ameliorate cisplatin-induced renal injury through deacetylation of p53.

4.4.2. *PTEN pathway*

Phosphatase and tensin homolog deleted on chromosome 10 (PTEN) is a gatekeeper tumor suppressor gene responsible for the control of the Ras-PI3K pathway linked to Cowden syndrome, an inherited disorder that results in many noncancerous tumors throughout the body.[65] Individuals suffering from Cowden's syndrome are at an increased risk for developing certain forms of cancer, such as those of the breast and thyroid gland. Treatment with low nutritional concentrations (1 nM) of resveratrol for 48 h reduces levels of peroxide in human breast cancer MCF-7 cells. This inhibitory effect is mediated by activation of PTEN/AKT pathway and results in upregulated expression of catalase and MnSOD.[66] Decreased expression of PTEN has been shown to be correlated with the progression of prostate cancer.[67] PTEN negatively regulates the cytoplasmic PI3K/AKT anti-apoptotic pathway. In addition, nuclear PTEN affects the cell cycle by also negatively regulating ERK1/2 pathway via Cyclin D.[68] Resveratrol has been shown to inhibit androgen receptor (AR) transcriptional activity in both androgen-dependent and -independent prostate cancer cells.[69] On the other hand, the stilbene also induces expression of PTEN through AR inhibition.[67] Acting as an antioxidant, it induces expression of antioxidant enzymes at nutritionally relevant concentrations through a mechanism involving the PTEN/AKT signaling pathway.[66]

Resveratrol reduces cancer cell migration and invasion *in vitro* through transcriptional and epigenetic regulation of *MMP-2* gene in osteosarcoma cells.[70] These results were further validated by *in vivo* experiments showing a decreased incidence of lung metastasis.[70]

Reduction of *MMP-2* activity may be achieved via inhibition of CREB–DNA-binding activity and up-regulation of *miR-328*.[70] The *JNK* and *p38 MAPK* signaling pathways are also involved in the regulation of *MMP-2*.[70]

4.4.3. *Retinoblastoma pathway*

The retinoblastoma (*RB1*; encoding *Rb*) gene was the first tumor suppressor gene to be described and is often cited as a gatekeeper. Its inactivation is a rate-limiting step for tumor initiation. This analysis led to the hypothesis that disease initiation requires two steps involving loss of function of both copies of the affected gene.[71] Thus, Rb was recognized to have a tumor suppressor function before the gene was demonstrated to be inactivated by mutation of one copy and loss or silencing of the second copy.[72] The Rb protein (pRb) also plays a key role in integrating diverse signals from intra- and extracellular sources and thus driving cell cycle progression. Rb represses the transcriptional activity of activator class of E2F transcription factors (E2F1–E2F3).[73, 74] Progression through the G1 phase of the cell cycle is mediated by phosphorylation of pRb, resulting in the release of essential transcription factors such as E2F-1. The phosphorylation of pRb is regulated positively by Cyclin D1/CDK4 and negatively by CDK inhibitors, such as p16 (CDKN2/MTS-1/INK4A). The p16/Cyclin D1/Rb pathway plays a critical role in tumorigenesis, and a number of tumor types display a high frequency of inactivation of at least one component of this pathway.[75]

Although evidence indicated that *E2Fs* regulate the expression of an increasing number of genes, the transcriptional regulation of genes required for cell cycle regulation is best defined to include genes involved in DNA replication, nucleotide synthesis, and checkpoint control.[76] When cells are in G0/G1 phase, hypophosphorylated Rbs bind to the transcription factor E2F family to suppress E2F-mediated gene transcription, which encodes a variety of genes involved in cell cycle progression, DNA replication, DNA damage repair, cell cycle checkpoint, and apoptosis.[77] Therefore, inhibition of the E2F family by pRb results in the suppression of cell cycle progression. Hsieh *et al.* discovered that quinone reductase 2 (NQO2) played a role in resveratrol-induced anti-proliferation in prostate cancer cells by targeting GSK-3β/Cyclin D1-mediated cell cycle control via a posttranslational regulatory mechanism in prostate cancer CWR22Rv1 cells.[78] Knockdown of NQO2 results in growth suppression and G1→S arrest, as well as down-regulation of Rb and Cyclin D1 phosphorylation. Additionally, resveratrol has been shown to block platelet-derived growth factor (PDGF)-stimulated c-Src and AKT kinase activation, resulting in reduced Cyclin D1 expression and attenuated pRb phosphorylation and cyclin-dependent kinase-2 (CDK2) activity.[79] These studies suggest that although Rb has been reported to be a gatekeeper protein, it may play a role of caretaker gene due to its inactivation. Loss or mutation of Rb is uncommon but inactivation by loss, mutation and/or methylation of p16 (CDKN2/MTS-1/INK4A) is regularly observed. Loss-of-function mutations in p16INK4a, a cell cycle regulatory protein involved in tumor suppression in the pRb pathway, occur frequently in human cancers.[80,81] In addition, an inherited germline mutation in the Rb gene was the first identified tumor suppressor gene implicated in several types of neoplasia, including familial retinoblastoma, small-cell lung cancer, and osteosarcoma.[82,83] Therefore, loss or mutation of these tumor suppressor genes may contribute to evading growth-inhibitory and/or cell death signals in tumors that succeed in sustaining proliferation and progressing to states of high-grade malignancy. In addition to controlling cell cycle progression, pRb is also involved in the regulation of DNA replication, differentiation, and apoptosis.[84] However, the Cyclin D/CDK4 complex phosphorylates pRb and allows E2Fs to bind their target genes by disrupting the formation of the pRb-E2F complex.[84]

Recently, it has been reported that the activation of the MAPK kinase pathway leads to pRb inactivation by sustaining cyclin levels and consequently activating CDKs.[85–87] However, the signals linking receptor kinase activity to neutralization of pRb family members have not yet been fully characterized. Constitutive cell surface kinase receptors and persistent phosphorylation/inactivation of the pRb family proteins (pRb, p107, and pRb2/p130) have been implicated in the uncontrolled growth of melanoma cells.[88] Moreover, overexpression of Cyclin D1 has been found in most human colon and breast cancers.[89,90] In addition, there is an absolute requirement for Cyclin D1 overexpression in breast malignancy trasformation that cannot be complemented by the other related Cyclins D2 and D3, supporting the concept of anti-Cyclin D1 therapy for breast cancer.[91]

4.4.4. *AMPK/SIRT pathway*

In response to resveratrol, AMPK activation is an early event leading to downstream activation of SIRT1. The activation of AMPK occurs via the phosphorylation of T172 of the α subunit, either by tumor suppressor liver kinase B1 (LKB1) or by Ca^{2+} calmodulin kinase kinase B (CamKKB) supported by an increase in intracellular Ca^{2+} levels.[32] Numerous drugs and xenobiotics, including resveratrol, indirectly activate AMPK by inhibiting ATP synthesis and causing increased cellular AMP.[92] Several different mechanisms have been proposed to link AMPK and SIRT1 activities. AMPK activation increases NAD+ availability, both in cultured myocytes and skeletal muscle.[93,94] NAD+ can be rate-limiting for SIRT1 activity. Consequently, an increase in NAD+ availability would favor SIRT1 action. This increase in NAD+ availability might derive from at least two complementary actions: on one hand, the increase in fat oxidation rates induced by AMPK would be permissive for an increase in NAD+ within hours after treatment.[93] On the other hand, AMPK triggers nicotinamide phosphoribosyltransferase (Nampt) expression, which will increase NAD+ salvage from nicotinamide (NAM).[94] In addition, AMPK will directly phosphorylate SIRT1.[95] This phosphorylation event will disrupt the interaction of SIRT1 with its endogenous negative regulator, deleted in Breast Cancer 1 (DBC1),[95] rendering SIRT1 active. AMPK is at the apex of the energy-sensing machinery and translates the message to SIRT1, which then initiates cellular transcriptional responses intended to optimize energy production from non-carbohydrate sources. LKB1, the main upstream kinase for AMPK activation,[96,97] is a target for SIRT1 deacetylation in cultured cells and rodent tissues.[98] Acetylated LKB1 fails to shuttle from the nucleus to the cytoplasm.[98]

Upon deacetylation by SIRT1, however, LKB1 shuttles efficiently to the cytoplasm, allowing the binding of STRAD and MO25 and the formation of the final active kinase complex.[98] Consequently, SIRT1 could potentially modulate AMPK activity by influencing the activity of LKB1.

Resveratrol has been described as a potential SIRT activator. Treatment of gastric cancer cells with resveratrol has been shown to induce a G1 phase arrest, to decrease the levels of Cyclin D1, CDK4 and CDK6 and to increase the levels of p21. In knockout cells that can express SIRT1, resveratrol is not capable of inhibiting cell proliferation[99] and thus it inhibits cell proliferation in a SIRT1 dependent way.[99] In a study using breast cancer cells, resveratrol similarly inhibited cell proliferation by stimulating SIRT1.[100] Activation of the AMPK pathway leads to mTOR activation and, consequently, to stimulation of cell proliferation. Resveratrol can block AMPK phosphorylation by SIRT1 activity overexpressed in tumor cells.[101]

However, several pieces of evidence suggest that SIRT1 may also act as an upstream modulator of AMPK. A SIRT1 inhibitor, Splitomicin, is able to prevent resveratrol-induced AMPK phosphorylation in HepG2 cells.[102] In a parallel study, when HepG2 cells were incubated with NAM, an end-product inhibitor of SIRT1 activity also led to decreased basal AMPK activity.[46] Specific shRNAs or generation of dominant negative models decreased SIRT1 activity and inhibited both basal and resveratrol-triggered AMPK phosphorylation, in HEK293 cells.[98,102] But overexpression of SIRT1 has been consistently shown to increase basal AMPK activity in HepG2 cells, liver and skeletal muscle.[102,103] SIRT1 can also influence AMPK activity. A recent study has helped to clarify these observations by demonstrating that the molecular pathway through which resveratrol leads to AMPK activation is dose-dependent. For example, low resveratrol doses (~25 mg/kg/day in mice; 25 μM concentrations in cultured myotubes) led to SIRT1-dependent AMPK activation, while high doses (~220 mg/kg/day in mice; 50 μM in cultured myotubes) did so in a SIRT1-independent fashion.[50] Thus, resveratrol dosing has to be carefully examined when exploring mechanistic insights on its action.

SIRT1 and AMPK are required to obtain the metabolic actions of resveratrol and both effectors are very likely required to achieve any health benefits that the stilbene possesses. For example, blocking either AMPK or SIRT1 is sufficient to prevent resveratrol-induced mitochondrial biogenesis.[50,51] The common need for both actors might also be explained by a "double hit" hypothesis in which critical mediators require inputs from both enzymes in order for them to become fully active or inactive. In addition, it could explain why models of modest SIRT1 overexpression do not necessarily show higher mitochondrial biogenesis,[104] as the signal from AMPK might be missing. In line with this observation, AMPK and SIRT1 share a number of downstream targets, including p53 and the FOXO family of transcription factors, among others.[34] This suggests further interactions of both signaling systems. Finally, the AMPK/SIRT1/PGC-1α pathway has proved to be a highly conserved axis of action through which mitochondrial biogenesis is regulated by pharmacological and nutritional cues, as well as by diverse key metabolic hormones, including adiponectin,[105] leptin[106] and fibroblast growth factor 21 (FGF21).[107] In conclusion, the above sections demonstrate that AMPK and SIRT1 are core effectors of resveratrol action and that their activities are very much intertwined.

4.4.5. *Resveratrol induces chemosensitization in tumor cells*

Multiple mechanisms have been proposed for the sensitization of cancer cells to anticancer drugs. Resveratrol has been shown to chemosensitize tumor cells to doxorubicin, cytarabine (AraC), actinomycin D, taxol, and methotrexate by modulating cell survival protein abundance and increasing apoptosis in a variety of cells.[62] Resveratrol has potentiated apoptosis induced by chemotherapeutic agents such as cisplatin, gefitinib, and paclitaxel in multidrug-resistant non-small cell lung cancer cells by a decrease in survivin expression.[108] On the other hand, inhibition of Bcl-xL expression by resveratrol is critical to its sensitization of myeloma cells and non-Hodgkin's lymphoma cells to paclitaxel-induced apoptosis.[108]

Interestingly, resveratrol at high concentrations synergistically promotes 5-FU-mediated apoptosis, while at lower concentrations it inhibits 5-FU-triggered apoptosis.[109] In addition, resveratrol overcomes chemoresistance by inhibiting NFκB and signal transducer and activator of transcription 3 (STAT-3) signaling pathways. Resveratrol has been shown to sensitize tumor cells by inducing caspase activation. The stilbene also exerted synergistic effects with 5-FU in a caspase-6 dependent manner in human colon cancer HCT116 cells. Interestingly, the potentiating effect of resveratrol was concentration-dependent.

In a few cases, resveratrol has been shown to sensitize tumor cells to chemotherapeutic agents through modulation of the tumor suppressor gene *p53*. For example, resveratrol was shown to enhance the chemotherapeutic potential of doxorubicin in chemoresistant B16 melanoma cells through up-regulation of *p53*.[61] In another study, enhancement of the sensitization effect of resveratrol on apoptosis induced by various drugs in cancer cell lines was found to be p53 independent. In some tumor cells, resveratrol has been shown to eliminate chemoresistance by decreasing the activity of membrane transporters such as P-gp and multi-drug resistance-associated protein (MRP).

One study has shown that resveratrol in a concentration-dependent manner caused increased accumulation of daunorubicin in human multidrug-resistant carcinoma KB-C2 cells. Although the authors did not examine the effect of resveratrol on chemosensitivity, they observed that the increase in daunorubicin accumulation was associated with a decrease in P-gp level.[110] In a recent study, Quan *et al.* examined the potential of resveratrol to overcome chemoresistance of oral epidermoid carcinoma KBv200 cells to vincristine, adriamycin, and paclitaxel.[111] Resveratrol produced a synergistic effect when combined with the chemotherapeutic agents and reversed the multidrug-resistant phenotype of KBv200 cells. The reversal of chemoresistance phenotype was associated with a decrease in the expression of P-gp. Using three doxorubicin-resistant acute myeloid leukemia cell lines (AML-2/DX30, AML-2/DX100, AML-2/DX300), Kweon *et al.* recently reported the potential of resveratrol to reverse the resistance phenotype of tumor cells.[112] They found an enhanced expression of the MRP1 gene in chemoresistant cells. When resveratrol was administered, the expression of MRP1 was down-regulated, while cellular uptake of doxorubicin into the resistant cells was increased. The group concluded that this resveratrol effect on MRP1 expression should be tested clinically, for example in sensitizing doxorubicin-resistant AML cells to anti-leukemic agents.

Resveratrol has also been shown to enhance chemosensitivity of tumor cells by arresting cells at different stages of the cell cycle and by down-regulating the genes involved in cell proliferation. Resveratrol enhanced doxorubicin-induced cytotoxicity in chemoresistant B16 melanoma by a down-regulation of Cyclin D1 with an increase in cell cycle arrest at the G(1)-S phase.[61] Resveratrol potentiates paclitaxel-induced anti-proliferation in lung cancer cells (A549, EBC-1, Lu65) by the induction of the expression of cell cycle regulator p21/waf1 and reduction of the paclitaxel threshold required for killing tumor cells.[113] Resveratrol has been shown to overcome chemoresistance through down-regulation of NFκB activation via inhibition of IκBα phosphorylation and IKK activation. These observations are further supported by a report of inhibition of NFκB in multiple myeloma patients treated with resveratrol in whom there was enhanced apoptotic and anti-proliferative potential of velcade and thalidomide.[114] In addition, resveratrol has been shown to overcome chemoresistance through down-regulation of STAT-3 pathway. STAT-3 signaling is critical to molecular machinery that regulates cancer cell survival pathways. STAT-3 is an abundant latent cytoplasmic transcription factor. Similar to FOXO proteins, the biological activity of STAT-3 is regulated by post-transcriptional mechanisms that regulate subcellular localization of the protein.

Treatment with resveratrol is linked to the accumulation of sub-G1 cell population, an increase of Bax release, and activation of caspase-3. It is also associated with down-regulation of gene products of various proliferation control and anti-apoptosis genes which include *Cyclin D1*, *cIAP-2*, *XIAP*, *BIRC5*, *Bcl-2*, *Bcl-xL*, *Bfl-1/A1*, and *TRAF2*.

4.5. Summary statement

In addition to modulating activation of AMPK and SIRT1 to control diverse transcriptional regulators against metabolic and age-related complications, resveratrol may also have preventive and therapeutic effects on cancers as shown in Figure 4.1. However, anti-growth signaling, mainly mediated by tumor suppressor proteins, and immune surveillance are the two major hurdles that must be overcome for cancer cells to acquire limitless proliferation. Despite many beneficial actions of resveratrol, one major concern is its poor solubility and absorption when given orally. Poor bioavailability of resveratrol is attributed to its extensive hepatic gluconuridation and sulfation. A recent study[115] of ApcMin mice — which model colorectal carcinogenesis — that were fed a high-fat diet, a low resveratrol dose suppressed intestinal adenoma development more potently than did the higher dose. The efficacy of resveratrol is correlated with activation of AMPK. Surprisingly, the effectiveness of low dose resveratrol in protecting against colon cancer in mouse colon cancer cells and perhaps in human colorectal tissues was attributable to enhanced AMPK phosphorylation, autophagy and expression of the cytoprotective NAD(P)H dehydrogenase. This suggests that lower doses of diet-derived agents can paradoxically be more effective than higher doses to prevent cancer. These results emphasize the need to perform dose-response studies and develop better methods to deliver diet-derived chemopreventive molecules such as resveratrol to the target tissues by using modern technologies such as microencapsulation or nanoparticles[116,117] to derive their beneficial actions. If the cancer-relevant actions of resveratrol are expressed by the cell surface receptor for the stilbene, then the effectiveness of lower doses is more understandable and the rapid cellular uptake and metabolism of the stilbene are less important. Cancer is a multi-factorial and heterogeneous process that involves the accumulation of multiple genetic and epigenetic changes. Resistance arises mainly due to the development of alternative pathways for cell survival and thus multi-targeted approaches with effects on multiple hallmarks of cancer are critical. Therefore, future directions should explore the development of natural compounds, not only for prevention but also for therapy.

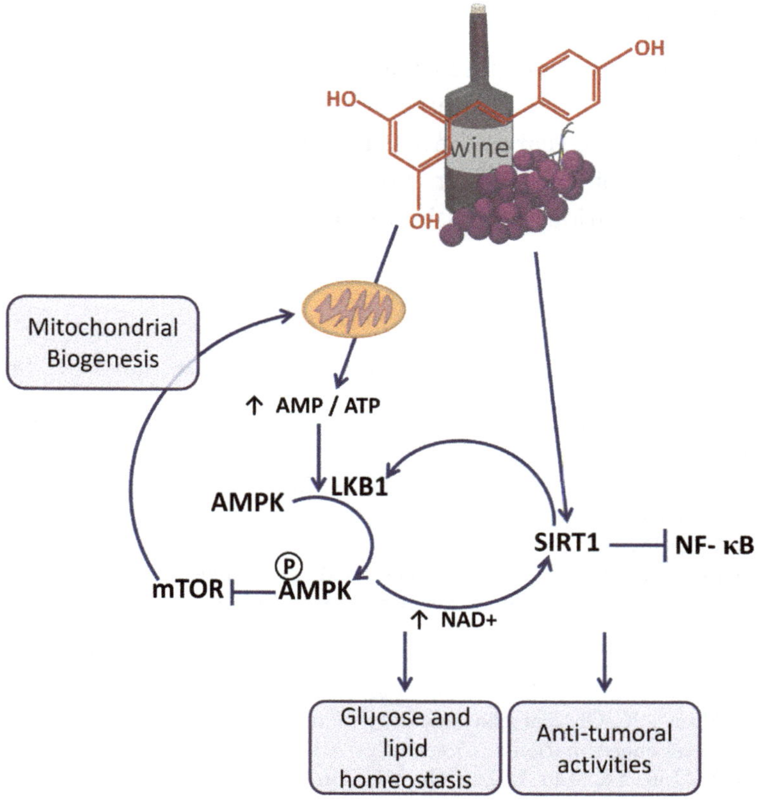

Fig. 4.1. Roles of resveratrol as signaling sensor and gatekeeper. As a mild mitochondrial poison, resveratrol inhibits mitochondrial ATP production, leading to a higher AMP/ATP ratio and an LKB1-dependent activation of AMPK. The activated AMPK enhances NAD^+ availability, which would overcome the rate-limitation that this cofactor exerts on SIRT1 enzymatic activity. Consecutively, SIRT1, a possible direct target of resveratrol, can deacetylate LKB1 to facilitate the formation of an active kinase complex through the association of LKB1. This way, SIRT1 could also positively control AMPK activity. AMPK and SIRT1 coordinately create a positive feed-forward loop to amplify the adaptation response to nutrient scarcity. Through the modulation of diverse transcriptional regulators, the actions of AMPK and SIRT1 can explain many of the beneficial effects of resveratrol against metabolic and age-related complications. In addition, resveratrol-activated AMPK would lead to the inhibition of mTOR signaling whereas SIRT1 activity would repress NFκB activity, which has been shown to involve in resveratrol-modulated cellular growth, autophagy, and immune responses.

Acknowledgments

Work from our laboratory described in this chapter was supported in part by the "TMU Research Center of Cancer Translational Medicine" from The Featured Areas Research Center Program within the framework of the Higher Education Sprout Project by the Ministry of Education (MOE) in Taiwan.

References

1. Leonard SS, Xia C, Jiang BH, Stinefelt B, Klandorf H, Harris GK, Shi X. Resveratrol scavenges reactive oxygen species and effects radical-induced cellular responses. *Biochem Biophys Res Commun* 2003 Oct 3; **309**(4): 1017–26.
2. Cao H, Pan X, Li C, Zhou C, Deng F, Li T. Density functional theory calculations for resveratrol. *Bioorganic & Medicinal Chemistry Letters* 2003; **13**: 1869–1871.
3. Queiroz AN, Gomes BA, Moraes WM Jr, Borges RS. A theoretical antioxidant pharmacophore for resveratrol. *Eur J Med Chem* 2009 Apr; **44**(4): 1644–9.
4. Patel KR, Brown VA, Jones DJ, Britton RG, Hemingway D, Miller AS, West KP, Booth TD, Perloff M, Crowell JA, Brenner DE, Steward WP, Gescher AJ, Brown K. Clinical pharmacology of resveratrol and its metabolites in colorectal cancer patients. *Cancer Res* 2010 Oct 1; **70**(19): 7392–9.
5. Ho Y, Lin YS, Liu HL, Shih YJ, Lin SY, Shih A, Chin YT, Chen YR, Lin HY, Davis PJ. Biological mechanisms by which antiproliferative actions of resveratrol are minimized. *Nutrients* 2017 Sep 21; **9**(10): pii: E1046.
6. Singh CK, Ndiaye MA, Ahmad N. Resveratrol and cancer: challenges for clinical translation. *Biochim Biophys Acta* 2015 Jun; **1852**(6): 1178–85.
7. Erdogan CS, Vang O. Challenges in analyzing the biological effects of resveratrol. *Nutrients* 2016 Jun 9; **8**(6): pii: E353.
8. Berman AY, Motechin RA, Wiesenfeld MY, Holz MK. The therapeutic potential of resveratrol: a review of clinical trials. *NPJ Precis Oncol* 2017; **1**: pii: 35.
9. Watt FM, Fujiwara H. Cell-extracellular matrix interactions in normal and diseased skin. *Cold Spring Harb Perspect Biol* 2011 Apr; **3**(4): a005124.
10. Varoni EM, Lo Faro AF, Sharifi-Rad J, Iriti M. Anticancer molecular mechanisms of resveratrol. *Front Nutr* 2016 Apr 12; **3**: 8.
11. Subramanian L, Youssef S, Bhattacharya S, Kenealey J, Polans AS, van Ginkel PR. Resveratrol: challenges in translation to the clinic–a critical discussion. *Clin Cancer Res* 2010 Dec 15; **16**(24): 5942–5948.
12. Switala-Jelen K, Dabrowska K, Opolski A, Lipinska L, Nowaczyk M, Gorski A. The biological functions of beta3 integrins. *Folia Biologica (Praha)* 2004; **50**: 143–152.

13. Niu G, Chen X. Vascular endothelial growth factor as an anti-angiogenic target for cancer therapy. *Curr Drug Targets* 2010 Aug; **11**(8): 1000–17.
14. Kanamori M, Vanden Berg SR, Bergers G, Berger MS, Pieper RO. Integrin beta3 overexpression suppresses tumor growth in a human model of gliomagenesis: implications for the role of beta3 overexpression in glioblastoma multiforme. *Cancer Res* 2004 Apr 15; **64**(8): 2751–8.
15. Yan B, Smith JW. Mechanism of integrin activation by disulfide bond reduction. *Biochemistry* 2001 Jul 31; **40**(30): 8861–7.
16. Lin HY, Chin YT, Yang YC, Lai HY, Wang-Peng J, Liu LF, Tang HY, Davis PJ. Thyroid Hormone, Cancer, and Apoptosis. *Compr Physiol* 2016 Jun 13; **6**(3): 1221–37.
17. Lin HY, Lansing L, Merillon JM, Davis FB, Tang HY, Shih A, Vitrac X, Krisa S, Keating T, Cao HJ, Bergh J, Quackenbush S, Davis PJ. Integrin alphaVbeta3 contains a receptor site for resveratrol. *FASEB J.* 2006 Aug; **20**(10): 1742–4.
18. Lin HY, Tang HY, Keating T, Wu YH, Shih A, Hammond D, Sun M, Hercbergs A, Davis FB, Davis PJ. Resveratrol is pro-apoptotic and thyroid hormone is anti-apoptotic in glioma cells: both actions are integrin and ERK mediated. *Carcinogenesis.* 2008 Jan; **29**(1): 62–9.
19. Lin HY, Sun M, Tang HY, Simone TM, Wu YH, Grandis JR, Cao HJ, Davis PJ, Davis FB. Resveratrol causes COX-2- and p53-dependent apoptosis in head and neck squamous cell cancer cells. *J Cell Biochem* 2008 Aug 15; **104**(6): 2131–42.
20. Lin HY, Delmas D, Vang O, Hsieh TC, Lin S, Cheng GY, Chiang HL, Chen CE, Tang HY, Crawford DR, Whang-Peng J, Hwang J, Liu LF, Wu JM. Mechanisms of ceramide-induced COX-2-dependent apoptosis in human ovarian cancer OVCAR-3 cells partially overlapped with resveratrol. *J Cell Biochem* 2013 Aug; **114**(8): 1940–54.
21. Zhao R, Pathak AS, Stouffer GA. beta(3)-Integrin cytoplasmic binding proteins. *Arch Immunol Ther Exp (Warsz)* 2004 Sep-Oct; **52**(5): 348–55.
22. Guo W, Giancotti FG. Integrin signalling during tumour progression. *Nat Rev Mol Cell Biol* 2004 Oct; **5**(10): 816–26.
23. Colin D, Limagne E, Jeanningros S, Jacquel A, Lizard G, Athias A, Gambert P, Hichami A, Latruffe N, Solary E, Delmas D. Endocytosis of resveratrol via lipid rafts and activation of downstream signaling pathways in cancer cells. *Cancer Prev Res (Phila)* 2011 Jul; **4**(7): 1095–106.
24. Klinge CM, Blankenship KA, Risinger KE, Bhatnagar S, Noisin EL, Sumanasekera WK, Zhao L, Brey DM, Keynton RS. Resveratrol and estradiol rapidly activate MAPK signaling through estrogen receptors alpha and beta in endothelial cells. *J Biol Chem* 2005 Mar 4; **280**(9): 7460–8.
25. Zhang S, Cao HJ, Davis FB, Tang HY, Davis PJ, Lin HY. Oestrogen inhibits resveratrol-induced post-translational modification of p53 and apoptosis in breast cancer cells. *Br J Cancer* 2004 Jul 5; **91**(1): 178–85.
26. Pozo-Guisado E, Lorenzo-Benayas MJ, Fernández-Salguero PM. Resveratrol modulates the phosphoinositide 3-kinase pathway through an estrogen receptor alpha-dependent mechanism: relevance in cell proliferation. *Int J Cancer* 2004 Mar 20; **109**(2): 167–73.

27. Chen Q, Ganapathy S, Singh KP, Shankar S, Srivastava RK. Resveratrol induces growth arrest and apoptosis through activation of FOXO transcription factors in prostate cancer cells. *PLoS One* 2010 Dec 14; **5**(12): e15288.
28. Haider UG, Roos TU, Kontaridis MI, Neel BG, Sorescu D, Griendling KK, Vollmar AM, Dirsch VM. Resveratrol inhibits angiotensin II- and epidermal growth factor-mediated Akt activation: role of Gab1 and Shp2. *Mol Pharmacol* 2005 Jul; **68**(1): 41–8.
29. Singh SK, Moretta D, Almaguel F, Wall NR, De León M, De León D. Differential effect of proIGF-II and IGF-II on resveratrol induced cell death by regulating survivin cellular localization and mitochondrial depolarization in breast cancer cells. *Growth Factors* 2007 Dec; **25**(6): 363–72.
30. Hardie DG. AMP-activated protein kinase: an energy sensor that regulates all aspects of cell function. *Genes Dev* 2011 Sep 15; **25**(18): 1895–908.
31. Long YC, Zierath JR. AMP-activated protein kinase signaling in metabolic regulation. *J Clin Invest* 2006 Jul; **116**(7): 1776–83.
32. Hardie DG, Ross FA, Hawley SA. AMPK: a nutrient and energy sensor that maintains energy homeostasis. *Nat Rev Mol Cell Biol* 2012 Mar 22; **13**(4): 251–62.
33. Zhang BB, Zhou G, Li C. AMPK: an emerging drug target for diabetes and the metabolic syndrome. *Cell Metab* 2009 May; **9**(5): 407–16.
34. Cantó C, Auwerx J. AMP-activated protein kinase and its downstream transcriptional pathways. *Cell Mol Life Sci* 2010 Oct; **67**(20): 3407–23.
35. Crane JM, Magee LA, Lee T, Synnes A, von Dadelszen P, Dahlgren L, De Silva DA, Liston R, *et al.* Maternal and perinatal outcomes of pregnancies delivered at 23 weeks' gestation. *J Obstet Gynaecol Can* 2015 Mar; **37**(3): 214–224.
36. Wang CZ, Wang Y, Di A, Magnuson MA, Ye H, Roe MW, Nelson DJ, Bell GI, Philipson LH. 5-amino-imidazole carboxamide riboside acutely potentiates glucose-stimulated insulin secretion from mouse pancreatic islets by KATP channel-dependent and -independent pathways. *Biochem Biophys Res Commun* 2005 May 20; **330**(4): 1073–9.
37. Bergeron R, Ren JM, Cadman KS, Moore IK, Perret P, Pypaert M, Young LH, Semenkovich CF, Shulman GI. Chronic activation of AMP kinase results in NRF-1 activation and mitochondrial biogenesis. *Am J Physiol Endocrinol Metab* 2001 Dec; **281**(6): E1340–6.
38. Kukidome D, Nishikawa T, Sonoda K, Imoto K, Fujisawa K, Yano M, Motoshima H, Taguchi T, Matsumura T, Araki E. Activation of AMP-activated protein kinase reduces hyperglycemia-induced mitochondrial reactive oxygen species production and promotes mitochondrial biogenesis in human umbilical vein endothelial cells. *Diabetes* 2006 Jan; **55**(1): 120–7.
39. Fu L, Bruckbauer A, Li F, Cao Q, Cui X, Wu R, Shi H, Zemel MB, Xue B. Leucine amplifies the effects of metformin on insulin sensitivity and glycemic control in diet-induced obese mice. *Metabolism* 2015 Jul; **64**(7): 845–56.

40. Daskalopoulos EP, Dufeys C, Bertrand L, Beauloye C, Horman S. AMPK in cardiac fibrosis and repair: Actions beyond metabolic regulation. *J Mol Cell Cardiol* 2016 Feb; **91**: 188–200.
41. Lin Z, Fang D. The Roles of SIRT1 in Cancer. *Genes Cancer* 2013 Mar; **4**(3–4): 97–104.
42. Finkel T. The metabolic regulation of aging. *Nat Med* 2015 Dec; **21**(12): 1416–23.
43. Baur JA, Pearson KJ, Price NL, Jamieson HA, Lerin C, Kalra A, Prabhu VV, Allard JS, Lopez-Lluch G, Lewis K, Pistell PJ, Poosala S, Becker KG, Boss O, Gwinn D, Wang M, Ramaswamy S, Fishbein KW, Spencer RG, Lakatta EG, Le Couteur D, Shaw RJ, Navas P, Puigserver P, Ingram DK, de Cabo R, Sinclair DA. Resveratrol improves health and survival of mice on a high-calorie diet. *Nature* 2006 Nov 16; **444**(7117): 337–42.
44. Breen DM, Sanli T, Giacca A, Tsiani E. Stimulation of muscle cell glucose uptake by resveratrol through sirtuins and AMPK. *Biochem Biophys Res Commun* 2008 Sep 12; **374**(1): 117–22.
45. Zang M, Xu S, Maitland-Toolan KA, Zuccollo A, Hou X, Jiang B, Wierzbicki M, Verbeuren TJ, Cohen RA. Polyphenols stimulate AMP-activated protein kinase, lower lipids, and inhibit accelerated atherosclerosis in diabetic LDL receptor-deficient mice. *Diabetes* 2006 Aug; **55**(8): 2180–91.
46. Suchankova G1, Nelson LE, Gerhart-Hines Z, Kelly M, Gauthier MS, Saha AK, Ido Y, Puigserver P, Ruderman NB. Concurrent regulation of AMP-activated protein kinase and SIRT1 in mammalian cells. *Biochem Biophys Res Commun* 2009 Jan 23; **378**(4): 836–41.
47. Zong Y, Sun L, Liu B, Deng YS, Zhan D, Chen YL, He Y, Liu J, Zhang ZJ, Sun J, Lu D. Resveratrol inhibits LPS-induced MAPKs activation via activation of the phosphatidylinositol 3-kinase pathway in murine RAW 264.7 macrophage cells. *PLoS One* 2012; **7**(8): e44107.
48. Centeno-Baez C, Dallaire P, Marette A. Resveratrol inhibition of inducible nitric oxide synthase in skeletal muscle involves AMPK but not SIRT1. *Am J Physiol Endocrinol Metab* 2011 Nov; **301**(5): E922–30.
49. Kubota S, Ozawa Y, Kurihara T, Sasaki M, Yuki K, Miyake S, Noda K, Ishida S, Tsubota K. Roles of AMP-activated protein kinase in diabetes-induced retinal inflammation. *Invest Ophthalmol Vis Sci* 2011 Nov 25; **52**(12): 9142–8.
50. Price NL, Gomes AP, Ling AJ, Duarte FV, Martin-Montalvo A, North BJ, Agarwal B, Ye L, Ramadori G, Teodoro JS, Hubbard BP, Varela AT, Davis JG, Varamini B, Hafner A, Moaddel R, Rolo AP, Coppari R, Palmeira CM, de Cabo R, Baur JA, Sinclair DA. SIRT1 is required for AMPK activation and the beneficial effects of resveratrol on mitochondrial function. *Cell Metab* 2012 May 2; **15**(5): 675–90.
51. Um JH, Park SJ, Kang H, Yang S, Foretz M, McBurney MW, Kim MK, Viollet B, Chung JH. AMP-activated protein kinase-deficient mice are resistant to the metabolic effects of resveratrol. *Diabetes.* 2010 Mar; **59**(3): 554–63.

52. Pacholec M, Bleasdale JE, Chrunyk B, Cunningham D, Flynn D, Garofalo RS, Griffith D, Griffor M, Loulakis P, Pabst B, Qiu X, Stockman B, Thanabal V, Varghese A, Ward J, Withka J, Ahn K. Metabolism. SRT1720, SRT2183, SRT1460, and resveratrol are not direct activators of SIRT1. *J Biol Chem* 2010 Mar 12; **285**(11): 8340–51.
53. Banerjee J, Bruckbauer A, Zemel MB. Activation of the AMPK/Sirt1 pathway by a leucine-metformin combination increases insulin sensitivity in skeletal muscle, and stimulates glucose and lipid metabolism and increases life span in Caenorhabditis elegans. *Metabolism* 2016 Nov; **65**(11): 1679–1691.
54. Vingtdeux V, Giliberto L, Zhao H, Chandakkar P, Wu Q, Simon JE, Janle EM, Lobo J, Ferruzzi MG, Davies P, Marambaud P. AMP-activated protein kinase signaling activation by resveratrol modulates amyloid-beta peptide metabolism. *J Biol Chem* 2010 Mar 19; **285**(12): 9100–13.
55. Dasgupta B1, Milbrandt J. Resveratrol stimulates AMP kinase activity in neurons. *Proc Natl Acad Sci U S A* 2007 Apr 24; **104**(17): 7217–22.
56. Joo MS, Kim WD, Lee KY, Kim JH, Koo JH, Kim SG. AMPK facilitates nuclear accumulation of Nrf2 by phosphorylating at serine 550. *Mol Cell Biol* 2016 Jun 29; **36**(14): 1931–42.
57. Fröjdö S, Cozzone D, Vidal H, Pirola L. Resveratrol is a class IA phosphoinositide 3-kinase inhibitor. *Biochem J* 2007 Sep 15; **406**(3): 511–8.
58. Frank SA. Somatic mutation: early cancer steps depend on tissue architecture. *Curr Biol* 2003 Apr 1; **13**(7): R261-3.
59. Lin HY, Hsieh MT, Cheng GY, Lai HY, Chin YT, Shih YJ, Nana AW, Lin SY, Yang YSH, Tang HY, Chiang IJ, Wang K. Mechanisms of action of nonpeptide hormones on resveratrol-induced antiproliferation of cancer cells. *Ann N Y Acad Sci* 2017 Sep; **1403**(1): 92–100.
60. Chin YT, Yang SH, Chang TC, Changou CA, Lai HY, Fu E, HuangFu WC, Davis PJ, Lin HY, Liu LF. Mechanisms of dihydrotestosterone action on resveratrol-induced anti-proliferation in breast cancer cells with different ERα status. *Oncotarget* 2015 Nov 3; **6**(34): 35866–79.
61. Gatouillat G, Balasse E, Joseph-Pietras D, Morjani H, Madoulet C. Resveratrol induces cell-cycle disruption and apoptosis in chemoresistant B16 melanoma. *J Cell Biochem* 2010 Jul 1; **110**(4): 893–902.
62. Fulda S, Debatin KM. Sensitization for anticancer drug-induced apoptosis by the chemopreventive agent resveratrol. *Oncogene* 2004 Sep 2; **23**(40): 6702–11.
63. Wei Q, Dong G, Yang T, Megyesi J, Price PM, Dong Z. Activation and involvement of p53 in cisplatin-induced nephrotoxicity. *Am J Physiol Renal Physiol* 2007 Oct; **293**(4): F1282–91.
64. Kim DH, Jung YJ, Lee JE, Lee AS, Kang KP, Lee S, Park SK, Han MK, Lee SY, Ramkumar KM, Sung MJ, Kim W. SIRT1 activation by resveratrol ameliorates cisplatin-induced renal injury through deacetylation of p53. *Am J Physiol Renal Physiol* 2011 Aug; **301**(2): F427–35.

65. Wang X, Jiang X. PTEN: a default gate-keeping tumor suppressor with a versatile tail. *Cell Res* 2008 Aug; **18**(8): 807–16.
66. Inglés M, Gambini J, Miguel MG, Bonet-Costa V, Abdelaziz KM, El Alami M, Viña J, Borrás C. PTEN mediates the antioxidant effect of resveratrol at nutritionally relevant concentrations. *Biomed Res Int* 2014; **2014**: 580852.
67. Wang Y, Romigh T, He X, Orloff MS, Silverman RH, Heston WD, Eng C. Resveratrol regulates the PTEN/AKT pathway through androgen receptor-dependent and -independent mechanisms in prostate cancer cell lines. *Hum Mol Genet* 2010 Nov 15; **19**(22): 4319–4329.
68. Chung JH, Ostrowski MC, Romigh T, Minaguchi T, Waite KA, Eng C. The ERK1/2 pathway modulates nuclear PTEN-mediated cell cycle arrest by Cyclin D1 transcriptional regulation. *Hum Mol Genet* 2006 Sep 1; **15**(17): 2553–9.
69. Harada N, Murata Y, Yamaji R, Miura T, Inui H, Nakano Y. Resveratrol down-regulates the androgen receptor at the post-translational level in prostate cancer cells. *J Nutr Sci Vitaminol (Tokyo)* 2007 Dec; **53**(6): 556–60.
70. Yang T, Wang L, Zhu M, Zhang L, Yan L. Properties and molecular mechanisms of resveratrol: a review. *Pharmazie* 2015 Aug; **70**(8): 501–6.
71. Amin ARMR, Karpowicz PA, Carey TE, Arbiser J, Nahta R, Chen ZG, Dong JT, Kucuk O, Khan GN, Huang GS, Mi S, Lee HY, Reichrath J, Honoki K, Georgakilas AG, Amedei A, Amin A, Helferich B, Boosani CS, Ciriolo MR, Chen S, Mohammed SI, Azmi AS, Keith WN, Bhakta D, Halicka D, Niccolai E1, Fujii H, Aquilano K, Ashraf SS, Nowsheen S, Yang X, Bilsland A, Shin DM. Evasion of anti-growth signaling: a key step in tumorigenesis and potential target for treatment and prophylaxis by natural compounds. *Semin Cancer Biol* 2015 Dec; **35** Suppl: S55–S77.
72. Knudson AG Jr. Mutation and cancer: statistical study of retinoblastoma. *Proc Natl Acad Sci U S A* 1971 Apr; **68**(4): 820–3.
73. Burkhart DL, Sage J. Cellular mechanisms of tumour suppression by the retinoblastoma gene. *Nat Rev Cancer* 2008 Sep; **8**(9): 671–82.
74. Chen HZ, Tsai SY, Leone G. Emerging roles of E2Fs in cancer: an exit from cell cycle control. *Nat Rev Cancer* 2009 Nov; **9**(11): 785–97.
75. Okami K, Reed AL, Cairns P, Koch WM, Westra WH, Wehage S, Jen J, Sidransky D. Cyclin D1 amplification is independent of p16 inactivation in head and neck squamous cell carcinoma. *Oncogene* 1999 Jun 10; **18**(23): 3541–5.
76. Polager S, Kalma Y, Berkovich E, Ginsberg D. E2Fs up-regulate expression of genes involved in DNA replication, DNA repair and mitosis. *Oncogene* 2002 Jan 17; **21**(3): 437–46.
77. Bracken AP, Ciro M, Cocito A, Helin K. E2F target genes: unraveling the biology. *Trends Biochem Sci* 2004 Aug; **29**(8): 409–17.
78. Hsieh TC, Yang CJ, Lin CY, Lee YS, Wu JM. Control of stability of Cyclin D1 by quinone reductase 2 in CWR22Rv1 prostate cancer cells. *Carcinogenesis.* 2012 Mar; **33**(3): 670–7.

79. Venkatesan B, Ghosh-Choudhury N, Das F, Mahimainathan L, Kamat A, Kasinath BS, Abboud HE, Choudhury GG. Resveratrol inhibits PDGF receptor mitogenic signaling in mesangial cells: role of PTP1B. *FASEB J* 2008 Oct; **22**(10): 3469–82.
80. Ruas M, Peters G. The p16INK4a/CDKN2A tumor suppressor and its relatives. *Biochim Biophys Acta* 1998 Oct 14; **1378**(2): F115–77.
81. Sherr CJ. Cell cycle control and cancer. *Harvey Lect* 2000–2001; **96**: 73–92.
82. Giacinti C, Giordano A. RB and cell cycle progression. *Oncogene* 2006 Aug 28; **25**(38): 5220–7.
83. Chinnam M, Goodrich DW. RB1, development, and cancer. *Curr Top Dev Biol* 2011; **94**: 129–69.
84. Classon M, Harlow E. The retinoblastoma tumour suppressor in development and cancer. *Nat Rev Cancer* 2002 Dec; **2**(12): 910–7.
85. Kaelin WG Jr. Functions of the retinoblastoma protein. *Bioessays* 1999 Nov; **21**(11): 950–8.
86. Zheng L, Chen Y, Riley DJ, Chen PL, Lee WH. Retinoblastoma protein enhances the fidelity of chromosome segregation mediated by hsHec1p. *Mol Cell Biol* 2000 May; **20**(10): 3529–37.
87. Adams PD. Regulation of the retinoblastoma tumor suppressor protein by cyclin/cdks. *Biochim Biophys Acta* 2001 Mar 21; **1471**(3): M123–33.
88. von Willebrand M, Zacksenhaus E, Cheng E, Glazer P, Halaban R. The tyrphostin AG1024 accelerates the degradation of phosphorylated forms of retinoblastoma protein (pRb) and restores pRb tumor suppressive function in melanoma cells. *Cancer Res* 2003 Mar 15; **63**(6): 1420–9.
89. Steeg PS, Zhou Q. Cyclins and breast cancer. *Breast Cancer Res Treat* 1998; **52**(1–3): 17–28.
90. Tetsu O, McCormick F. Beta-catenin regulates expression of Cyclin D1 in colon carcinoma cells. *Nature* 1999 Apr 1; **398**(6726): 422–6.
91. Yu Q, Geng Y, Sicinski P. Specific protection against breast cancers by Cyclin D1 ablation. *Nature* 2001 Jun 28; **411**(6841): 1017–21.
92. Hawley SA, Ross FA, Chevtzoff C, Green KA, Evans A, Fogarty S, Towler MC, Brown LJ, Ogunbayo OA, Evans AM, Hardie DG. Use of cells expressing gamma subunit variants to identify diverse mechanisms of AMPK activation. *Cell Metab* 2010 Jun 9; **11**(6): 554–65.
93. Cantó C, Gerhart-Hines Z, Feige JN, Lagouge M, Noriega L, Milne JC, Elliott PJ, Puigserver P, Auwerx J. AMPK regulates energy expenditure by modulating NAD^+ metabolism and SIRT1 activity. *Nature* 2009 Apr 23; **458**(7241): 1056–60.
94. Fulco M, Cen Y, Zhao P, Hoffman EP, McBurney MW, Sauve AA, Sartorelli V. Glucose restriction inhibits skeletal myoblast differentiation by activating SIRT1 through AMPK-mediated regulation of Nampt. *Dev Cell* 2008 May; **14**(5): 661–73.
95. Nin V, Escande C, Chini CC, Giri S, Camacho-Pereira J, Matalonga J, Lou Z, Chini EN. Role of deleted in breast cancer 1 (DBC1) protein in SIRT1 deacetylase activation induced by protein kinase A

and AMP-activated protein kinase. *J Biol Chem* 2012 Jul 6; **287**(28): 23489–501.

96. Lizcano JM, Göransson O, Toth R, Deak M, Morrice NA, Boudeau J, Hawley SA, Udd L, Mäkelä TP, Hardie DG, Alessi DR. LKB1 is a master kinase that activates 13 kinases of the AMPK subfamily, including MARK/PAR-1. *EMBO J* 2004 Feb 25; **23**(4): 833–43.
97. Sakamoto K, McCarthy A, Smith D, Green KA, Grahame Hardie D, Ashworth A, Alessi DR. Deficiency of LKB1 in skeletal muscle prevents AMPK activation and glucose uptake during contraction. *EMBO J* 2005 May 18; **24**(10): 1810–20.
98. Lan F, Cacicedo JM, Ruderman N, Ido Y. SIRT1 modulation of the acetylation status, cytosolic localization, and activity of LKB1. Possible role in AMP-activated protein kinase activation. *J Biol Chem* 2008 Oct 10; **283**(41): 27628–35.
99. Yang Q, Wang B, Zang W, Wang X, Liu Z, Li W, Jia J. Resveratrol inhibits the growth of gastric cancer by inducing G1 phase arrest and senescence in a Sirt1-dependent manner. *PLoS One* 2013 Nov 21; **8**(11): e70627.
100. Buhrmann C, Shayan P, Popper B, Goel A, Shakibaei M. Sirt1 is required for resveratrol-mediated chemopreventive effects in colorectal cancer cells. *Nutrients* 2016 Mar 5; **8**(3): 145.
101. Lin JN, Lin VC, Rau KM, Shieh PC, Kuo DH, Shieh JC, Chen WJ, Tsai SC, Way TD. Resveratrol modulates tumor cell proliferation and protein translation via SIRT1-dependent AMPK activation. *J Agric Food Chem* 2010 Feb 10; **58**(3): 1584–92.
102. Hou X, Xu S, Maitland-Toolan KA, Sato K, Jiang B, Ido Y, Lan F, Walsh K, Wierzbicki M, Verbeuren TJ, Cohen RA, Zang M. SIRT1 regulates hepatocyte lipid metabolism through activating AMP-activated protein kinase. *J Biol Chem* 2008 Jul 18; **283**(29): 20015–26.
103. Boutant M, Cantó C. SIRT1 metabolic actions: integrating recent advances from mouse models. *Mol Metab* 2013 Oct 23; **3**(1): 5–18.
104. Pfluger PT, Herranz D, Velasco-Miguel S, Serrano M, Tschöp MH. Sirt1 protects against high-fat diet-induced metabolic damage. *Proc Natl Acad Sci U S A* 2008 Jul 15; **105**(28): 9793–8.
105. Iwabu M, Yamauchi T, Okada-Iwabu M, Sato K, Nakagawa T, Funata M, Yamaguchi M, Namiki S, Nakayama R, Tabata M, Ogata H, Kubota N, Takamoto I, Hayashi YK, Yamauchi N, Waki H, Fukayama M, Nishino I, Tokuyama K, Ueki K, Oike Y, Ishii S, Hirose K, Shimizu T, Touhara K, Kadowaki T. Adiponectin and AdipoR1 regulate PGC-1alpha and mitochondria by Ca(2+) and AMPK/SIRT1. *Nature* 2010 Apr 29; **464**(7293): 1313–9.
106. Li L, Pan R, Li R, Niemann B, Aurich AC, Chen Y, Rohrbach S. Mitochondrial biogenesis and peroxisome proliferator-activated receptor-γ coactivator-1α (PGC-1α) deacetylation by physical activity: intact adipocytokine signaling is required. *Diabetes* 2011 Jan; **60**(1): 157–67.

107. Chau MD, Gao J, Yang Q, Wu Z, Gromada J. Fibroblast growth factor 21 regulates energy metabolism by activating the AMPK-SIRT1-PGC-1alpha pathway. *Proc Natl Acad Sci U S A* 2010 Jul 13; **107**(28): 12553–8.
108. Zhao W, Bao P, Qi H, You H. Resveratrol down-regulates survivin and induces apoptosis in human multidrug-resistant SPC-A-1/CDDP cells. *Oncol Rep* 2010 Jan; **23**(1): 279–86.
109. Gupta SC, Kannappan R, Reuter S, Kim JH, Aggarwal BB. Chemosensitization of tumors by resveratrol. *Ann N Y Acad Sci* 2011 Jan; **1215**: 150–60.
110. Nabekura T, Kamiyama S, Kitagawa S. Effects of dietary chemopreventive phytochemicals on P-glycoprotein function. *Biochem Biophys Res Commun* 2005 Feb 18; **327**(3): 866–70.
111. Quan F, Pan C, Ma Q, Zhang S, Yan L. Reversal effect of resveratrol on multidrug resistance in KBv200 cell line. *Biomed Pharmacother* 2008 Nov; **62**(9): 622–9.
112. Kweon SH, Song JH, Kim TS. Resveratrol-mediated reversal of doxorubicin resistance in acute myeloid leukemia cells via downregulation of MRP1 expression. *Biochem Biophys Res Commun* 2010 Apr 23; **395**(1): 104–10.
113. Kubota T, Uemura Y, Kobayashi M, Taguchi H. Combined effects of resveratrol and paclitaxel on lung cancer cells. *Anticancer Res* 2003 Sep–Oct; **23**(5A): 4039–46.
114. Bhardwaj A, Sethi G, Vadhan-Raj S, Bueso-Ramos C, Takada Y, Gaur U, Nair AS, Shishodia S, Aggarwal BB. Resveratrol inhibits proliferation, induces apoptosis, and overcomes chemoresistance through down-regulation of STAT3 and nuclear factor-kappaB-regulated antiapoptotic and cell survival gene products in human multiple myeloma cells. *Blood* 2007 Mar 15; **109**(6): 2293–302.
115. Cai H, Scott E, Kholghi A, Andreadi C, Rufini A, Karmokar A, Britton RG, Horner-Glister E, Greaves P, Jawad D, James M, Howells L, Ognibene T, Malfatti M, Goldring C, Kitteringham N, Walsh J, Viskaduraki M, West K, Miller A, Hemingway D, Steward WP, Gescher AJ, Brown K. Cancer chemoprevention: evidence of a nonlinear dose response for the protective effects of resveratrol in humans and mice. *Sci Transl Med* 2015 Jul 29; **7**(298): 298ra117.
116. Penalva R, Esparza I, Larraneta E, González-Navarro CJ, Gamazo C, Irache JM. Zein-Based Nanoparticles Improve the Oral Bioavailability of Resveratrol and Its Anti-inflammatory Effects in a Mouse Model of Endotoxic Shock. *J Agric Food Chem* 2015 Jun 17; **63**(23): 5603–11.
117. da Rocha Lindner G1, Bonfanti Santos D, Colle D, Gasnhar Moreira EL, Daniel Prediger R, Farina M, Khalil NM, Mara Mainardes R. Improved neuroprotective effects of resveratrol-loaded polysorbate 80-coated poly(lactide) nanoparticles in MPTP-induced Parkinsonism. *Nanomedicine (Lond)* 2015; **10**(7): 1127–38.

Chapter 5

Control of AKT Survival Kinase Activation/Function by Resveratrol and Its Target Protein NQO2

Tze-chen Hsieh and Joseph M. Wu

Biochemistry and Molecular Biology,
New York Medical College,
Valhalla, NY 10595, USA

The PI3K/AKT pathway signals events impinge on a cascade of responses involved in the control of apoptosis, cell growth and survival, and motility as an underpinning driver of tumor progression. Because of the frequent deregulation of oncoprotein AKT found in various cancers, it is regarded as the therapeutic target of choice. This has provided the impetus to discover anti-cancer small molecules and/or identify proteins that can modulate AKT signaling pathway activities as potential drug candidates.

Resveratrol is a grape-derived polyphenol with cancer preventative activity and also has been reported to inhibit AKT. However molecular details of the underlying mechanisms have not been fully elucidated. Rodents fed resveratrol-fortified diets are forestalled in prostate cancer (CaP) progression while *in vitro* studies show that resveratrol inhibits proliferation, induces apoptosis, and down-regulates prostate specific antigen (PSA) expression in androgen-dependent (AD) LNCaP cells, and suppresses proliferation of androgen independent AI (also known as hormone refractory state, HRPC) cells. How resveratrol elicits its anti-CaP activity has only been partially elucidated and is the primary research focus of this laboratory.

To elucidate the mechanism of CaP prevention by resveratrol, we focus on the identification of its target protein and discovered quinone reductase 2 (NQO2) as an intracellular target of resveratrol. Knockdown NQO2 in CWR22Rv1 CaP cells slowed cell proliferation, accumulated cells in G1 phase, and decreased state of phosphorylation of retinoblastoma Rb and cyclin D1. In addition, an increase in AKT activity was observed in NQO2 knockdown cells. These observations suggest that NQO2 contributes to the intracellular control of AKT. We hypothesize that NQO2 interacts with AKT altering its activity and/or function, with or without NQO2 inhibitor/ligand resveratrol. This hypothesis was examined using resveratrol affinity column chromatography; AKT was shown to co-elute with NQO2. NQO2 may act as a *bona fide* physiological partner and modulator of the frequently activated oncoprotein AKT, and in turn, cyclin D1 expression and growth control.

The aforementioned results showing control of AKT signaling by NQO2 as relevant to cancer prevention by resveratrol are summarized. Implications of these findings pertaining to CaP management are presented, specifically on the aspects of (i) PSA regulation mediated by androgen receptor (AR) under control by AKT and (ii) growth control resulting from interaction between AKT, GSK3β and cyclin D1.

5.1. Studies on resveratrol in CaP cells

Population-based studies have revealed a highly variable incidence rate of early and advanced CaP in men residing in different geographic localities. This has been attributed to the diet and in particular dietary ingredients with chemopreventive potential. Among dietary factors investigated intensely is resveratrol, a polyphenol present in abundance in a number of commonly consumed food and drinks.

We tested the effects of resveratrol using CaP cells stemming from the seminal publication of Pezzuto and collaborators in 1997 showing the chemopreventive activities of resveratrol on skin and breast carcinomas in animals.[1] Our laboratory first reported that resveratrol is active both on AD and AI CaP cells, exerting inhibition on proliferation, inducing apoptosis, disrupting cell cycle phase transition, and blocking the expression of CaP-specific genes PSA and AR.[2,3] Similar findings were shown by other laboratories.[4–6] Although these results raise the possibility that resveratrol may be considered for CaP prevention in humans, its low potency, poor bioavailability and delivery, and only partially elucidated mechanism as a whole diminish the potential of resveratrol in chemoprevention.

To circumvent these lingering issues resveratrol faces, we embarked on studies to identify the cellular target proteins of resveratrol. We developed a resveratrol-affinity proteomic strategy using resveratrol as a biospecific bait linked to a solid matrix and combined with MALDI-TOF mass spectrometry we purified and identified NQO2[7] as a high affinity target protein of resveratrol with a K_D of 35 ± 15 nM.[8] X-ray crystal analysis of NQO2 in complex with resveratrol reveals that resveratrol fits snuggly in an active site defined by the interaction between the side-chains of NQO2 with the isoalloxazine ring of the bound co-factor flavin adenine dinucleotide (FAD), forming hydrogen bond with all three hydroxy groups in resveratrol.[8] Since following ingestion the measured resveratrol is in the range of 0.5–4.0 μM in humans, it is conceivable that its bioactive level in responsive cells/tissues can reach a sufficiently high level to not only bind NQO2 but also inhibit its enzymatic activity, thereby providing support for the bioefficacy of resveratrol *in vivo*.

5.2. Emerging roles of NQO2

NQO2 is a multi-decade old, phase II detoxification enzyme.[9] It is an atypical reductase since it utilizes N-ribosyl dihydronicotinamide (NRH) as electron donor. Since cellular availability and *de novo* biosynthesis of NRH have not been elucidated,[10–12] the biological role for NQO2 remains ill-defined. In CaP, we found that NQO2 level is low in normal prostate epithelial cells, peaks as CaP transits from AD to AI, and declines thereafter suggesting that it is involved in prostate carcinogenesis.[7] Cells harboring partially silenced NQO2 was used to further understand role NQO2 plays in CaP. We found that ~50% depletion of cellular NQO2 is accompanied by ~50% increase in AKT activity, concomitant with increased Ser219 phosphorylation and deactivation of GSK-3β and the accompanying decrease in cyclin D1 phosphorylation at threonine 286 (Thr286) and resulting in stabilization of cyclin D1.[13] These results reveal hitherto unreported role of NQO2 in the control of AKT/GSK-3β/cyclin D1 and point to NQO2 involved in degradation of cyclin D1 and growth control, pertinent to chemoprevention by resveratrol.

The non-classic role of NQO2 also surfaced from other studies, such as (i) participation in melatonin function,[14,15] (ii) control of cell number/size,[16,17] (iii) coupling of signaling pathways,[18–20] and (iv) modulation of the tumor suppressor protein p53.[21,22]

Studies using keratinocytes from NQO2-null mice have shown that deletion of NQO2 potentiated apoptosis by abrogating TNF-induced survival kinases, respectively, JNK, AKT, p38, and p44/p42 MAPK.[19,20] These results are insightful as they implicate a non-enzymatic, kinase-modulatory role of NQO2 relevant to the control of carcinogenesis. This possibility was also supported by several biochemical studies where designer kinase inhibitors are used as biospecific affinity purification bait and capturing NQO2 as a non-kinase target; examples include bisindolylmaleimide, an inhibitor of PKC,[23,24] and imatinib, an inhibitor of Bcr-Abl kinase.[25,26] Do these results merely show that NQO2 is an "accidental" off target of kinases? Or, is it possible that NQO2 is a novel "designer" modulator of kinases or by some hitherto unreported mechanisms?

5.3. NQO2 acts as a novel modulator of AKT through protein-protein interaction

Recently we found that NQO2-knockdown CaP cells showed an increase AKT activity suggesting that NQO2 is a negative modulator of AKT.[13] We hypothesize NQO2 may alter AKT via direct binding (Figure 5.1). Both NQO2 and AKT activity is inhibited by resveratrol,[8,27–29] raising the possibility that both NQO2 and AKT can form a complex with resveratrol, potentially serving as an "on/off" switch to fine-tune binding of NQO2 with AKT.

To test this hypothesis K562 cell extracts with copious expression of NQO2 were fractionated on resveratrol affinity or control columns and analyzed for target proteins. Authenticity and specific binding of prospective resveratrol target proteins to the affinity column was confirmed by competition with resveratrol, i.e., extracts were incubated with excess resveratrol prior to fractionation on the affinity column. We checked for co-elution of kinases with NQO2 by immunoblotting, systematically screening for potential candidate kinases in various fractions eluted from resveratrol affinity column. We found that AKT

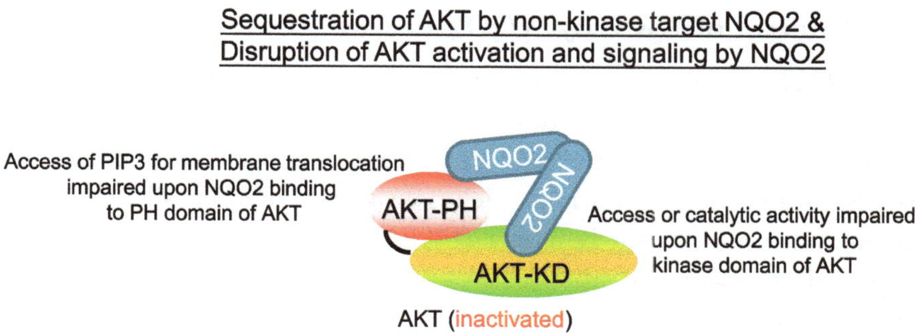

Fig. 5.1. NQO2 is a physiologic partner of AKT and modulates AKT activation as well as AKT-mediated signaling events via protein-protein interaction (adopted from[30]).

co-eluted with NQO2, appearing only in the resveratrol-eluted and not ATP-eluted fractions.[30] Importantly, binding of both NQO2 and AKT to the affinity column were abrogated by prior competition with excess resveratrol.[30] These results are consistent with the interpretation that (i) AKT binds directly with resveratrol as a second novel resveratrol target protein, (ii) NQO2:AKT may exist as a complex intracellularly whose avidity of binding may be further modulated by exposure to resveratrol. Complex formation and co-elution of NQO2 with AKT was also evident by gel filtration using Superdex 200 columns with K562 cell extracts.[30] These findings constitute the basis for our novel hypothesis that (i) NQO2 plays a hitherto unidentified role in diverse physiological and disease process[31] by functioning as physiological partner of AKT and modulator of its activity and (ii) interaction between NQO2 and AKT may be fine-tuned by prior or concurrent exposure to resveratrol. As mentioned, AKT is considered critical for cancer cell survival; thus, its ability to complex NQO2 not only provides support for a novel role NQO2 plays as a cancer risk modulator, but by extension, as a physiologically relevant regulator of AKT signaling. These observations may be translationally applied to CaP prevention by resveratrol, i.e., binding and sequestration of NQO2 by resveratrol could consequencially inhibit AKT, significantly and coordinately, imposing restriction on CaP cell proliferation, survival, and gene expression.

5.4. Control of AKT by NQO2 in CaP

PI3K/AKT pathway has been found to be frequently dysregulated in cancer[32] and is aberrantly activated either by somatic mutation in the gene encoding the p110α catalytic subunit of PI3K, or by loss-of-function mutations, deletions, or promoter silencing of phosphatase and tensin homolog (PTEN). Tumor cell survival and increased resistance to apoptosis is associated with over activated AKT kinases.[33] The impetus for investigation of control AKT pathway in CaP is prompted by that even more since ~80% of CaP tumors show mutation or deletion of PTEN.[34–37] Moreover, genomic alterations of PI3K in metastatic castration-resistant prostate cancer (mCRPC) patients has also been revealed in recent multi-institutional integrative clinical genomics study.[38] Loss of PTEN increases the phosphorylation of the PI3K/AKT/mTOR kinases, enhancing cell survival and eliciting radiotherapy/drug resistance.[39] Conversely, down regulation of AKT lowers the apoptotic threshold and sensitizes cells to cytotoxic drugs.[40,41] Especially in castration-resistant prostate cancer (CRPC) patients, AKT could activate AR via phosphorylation, creating an androgen-independent 'outlaw' receptor in control cell survival and PSA expression.

Our recent studies show that AKT interacts with NQO2 and that both proteins are the target of resveratrol.[8,30] Thus, we hypothesis is that NQO2 plays a critical role on resveratrol mediated CaP growth and gene expression by acting as a *bona fide* physiologic partner of frequently activated oncoprotein AKT in modulating AR-cyclin D1 regulatory axis (Figure 5.2). This hypothesis is particularly relevant to CaP patients diagnosed with CRPC, specifically in men who expresses high AKT despite their PTEN (a negative regulator of AKT)-positive genetic status and who have high probability of showing resistance to AKT-directed drugs. In addition, this hypothesis finding will have significant impact on current paradigms of the regulation of AKT signaling. Namely, because the known kinase inhibitors can co-target kinases and non-kinase target NQO2 as mentioned above; chemical inhibitors specifically targeting AKT might not have the effectiveness in preventing the emergence of therapy-resistant CRPC especially in the patient population displaying loss of PTEN.

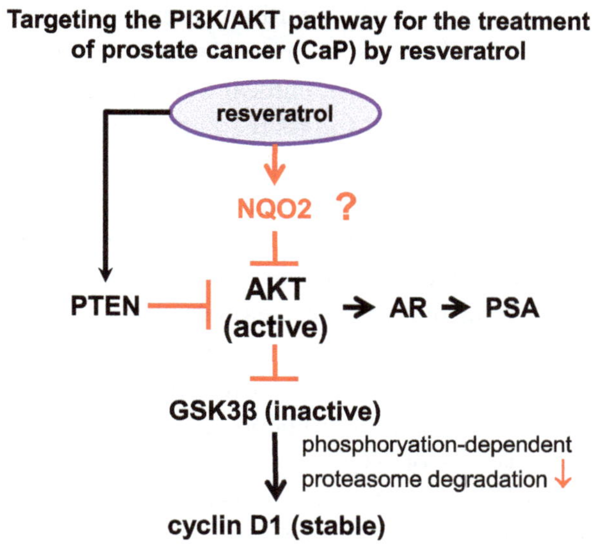

Fig. 5.2. Hypothesis: NQO2-mediated AKT signaling is postulated to be independent of or is coordinated with PTEN status and its control by resveratrol.

Identification the novel interaction between NQO2 and AKT may provide insights on the development of drug resistance and design strategies on how best to co-target dysfunctional AKT with using drugs directed at AKT and/or its off-target inhibitors in order to improve efficacy as well as eliminate resistance. Whether the physiologic role NQO2 plays involves the desensitization and inactivation of hyperactive AKT in PTEN-null CRPC cells and its relevance to the prevention of CaP by resveratrol warrants research in the future.

5.5. Implication of the findings

5.5.1. *Patients*

CaP is a cancer that specifically impacts men. The physical and emotional challenge CaP brings can be particularly devastating when the disease progresses to CRPC, due to the fact that at this clinical stage there is generally no cure. The drug docetaxel does provide extension of life expectancy, but also has significant side effects. The fundamental and perhaps most bewildering challenge all CRPC patients face is the continuing rise in PSA occurring in parallel with castrate levels of testosterone, and independent of the presence or absence of other clinical symptoms, or evidence of metastasis. The focus on AKT and its regulation by binding to NQO2 is particularly relevant for CaP patients diagnosed with CRPC, specifically in men who expresses high AKT despite their PTEN status and who have high probability of showing resistance to AKT-directed drugs. We believe these patients fall into the sub-groups that will benefit from our findings, namely, they fit our postulate that endogenous proteins like NQO2 may regulate and modulate the activation and function of AKT. In this context, resveratrol may prolong life expectancy as a standalone or adjunctive therapy with AKT inhibitors; studies of NQO2 as a modulator of the activation and function of AKT also has the potential to generate biomarkers that can predict and chart the course and establishment of CRPC, and may be developed into an active screen and search for agents with CRPC-sparing effects. Thus it is conceivable that discovery of the control of AKT by NQO2 has high public health impact for CRPC.

5.5.2. *Clinical benefits*

From a prevention point of view, individuals who can benefit from our discovery include healthy men at high risk of CRPC. From the perspective of treatment of CaP, it may be beneficial for men with localized CaP before treatment, or men with biochemical failure after a prior therapy. CRPC patients that develop docetaxel (the standard-of-care for men diagnosed with mCRPC)-resistant disease might also benefit. We envisage that control of AKT by NQO2 may provide a novel foundation for the development of NQO2 inhibitors as added armamentarium for the prevention of CRPC, particularly in a setting of genetically dysregulated AKT and/or PTEN. The finding will also advance cancer prevention by dietary agents, in support that resveratrol and NQO2 both participate in CRPC management by binding and modulating AKT.

In CaP and CRPC carrying a PTEN-deficient genetic status, AKT is maintained in a constitutively active state which consequentially can suppress transcription of androgen-responsive genes — both of these molecular changes are known to facilitate the progression of CaP to CRPC. In support, >70% of late stage CaP exhibits the loss of PTEN, a genetic status which is correlated with the development of CRPC. However, inconsistent results have also been observed. For instance, ~20–40% of androgen responsive localized primary CaP cases show a loss of PTEN while up to 30% of CRPC cases are PTEN-positive. What might be the underlying explanation for these findings, particularly with respect to the control of AKT activation and function?

A hyperactive AKT in the context of CRPC is particularly noteworthy; the robust functioning of AKT impinges on the control of numerous cellular activities driving various cancers including CRPC. As such, inhibition of the AKT kinase activation/function is a clinically relevant approach for the treatment of CRPC and for years considerable efforts have been spent in developing anti-AKT therapeutics with dubious levels of success since the strategy has mostly centered on designing chemicals that competitively inhibit the binding of ATP whose hydrolysis is essential for the activity of AKT. This strategy tends to produce leads with promiscuous effects on other homologous kinases. In comparison, we have introduced a novel perspective by placing at its mechanistic core NQO2 as an effector/modulator of the activation/function of AKT, and furthermore, have added the novel dimension of resveratrol acting as an allosteric modifier targeting the region of AKT involved in protein:protein interaction. These additions are novel and unique. The protein-protein interaction site of AKT with NQO2 should have a conformation or topology unlike the ATP-binding site, fine-tuning of this site by phytochemicals is equally novel. The information to be obtained using this conceptual and mechanistic strategy is significant in considering the targeting of AKT for CRPC therapy since it provides an "inside-out" responsive *vs* an "outside-in" provocative strategy for CaP prevention/therapy.

5.5.3. *Public health impact*

For years population-based studies have explored the association between diet and CaP, with mixed results. A more consistent observation of prominent public health significance is the systematically positive correlation found to exist between the intake/consumption of fruits and vegetables and the reduced risk of cancers at various anatomical sites including CaP, leading to the widely accepted concept that specific ingredients, most notably phytochemicals derived from fruits and vegetables, have efficacy both in prevention and therapy of malignant tumors. Accordingly, focus in recent years has shifted to the identification of bioactive phytochemicals as well as the elucidation of mechanism and cellular targets of the identified phytochemicals. This strategy has been successfully applied in our laboratory to the study of the anti-CaP properties by resveratrol we described in this chapter.

Our studies have immediate public health impact. Although localized CaP responds readily to androgen-ablation therapy, the remission is only temporary for a large percentage of responders for whom the apparent quiescent interlude eventually reappears in the form of the CRPC, characterized by high aggressiveness, metastasis and eventually death. Thus, discovery of novel preventive and therapeutic strategies for CRPC, a relevant objective of this application, is imperative. The knowledge gained and leads generated may significantly improve the overall survival rate of CRPC. They may uncover insights that could be explored and developed into tools to improve diagnosis and therapeutic intervention modalities that can save lives.

Discovery of the control of AKT by NQO2 has several objective implications. A hyperactive AKT phosphorylates and inactivates a multitude of intracellular protein substrates playing an essential role in the control of apoptosis in various cancers including CRPC. Thus, inhibition of the kinase activity of AKT is a clinically relevant approach for the treatment of CRPC. In comparison, we focus on cellular protein NQO2 that modulates the activation/function of AKT, and directs emphasis on phytochemicals possibly acting allosterically on the AKT protein-protein interaction domain adjacent to the ATP site. Both approaches are unique. The protein-protein interaction site in AKT has a conformation/topology unlike the ATP-binding site; thus, elucidation on how AKT:NQO2 interacts will produce new therapeutic leads and targets. Secondly, phytochemicals such as resveratrol are low-specificity non-peptide substrate inhibitory mimetics that are less prone to the development of AKT mutant and instead could generate a profile applicable to the identification of patients likely to be resistant to AKT-directed therapeutics. Since CaP has long latency before CRPC becomes established, the development of non-toxic, easily administered, and cancer selective preventive/therapeutic measures is considered imperative.

5.6. Summary and perspectives

In its traditional biochemical classification, NQO2 is considered a detoxification enzyme that acts to confer protection against the biotransformation of electrophilic quinone metabolites that have the potential of reacting with DNA to generate apurinic, mutation prone genomic DNA. As explained above, we have advanced an innovative hypothesis in the control of AKT by NQO2.

We advance the thesis that the level and expression of endogenous protein, viz., NQO2 with AKT modulatory activity, acting in synchrony or synergy/antagonism with exogenously administered chemical, e.g., resveratrol, specifically affects the PI3K/AKT activation and signaling events, to restrict CRPC. It is our hypothesis that this is a strategy that might restrict CRPC, or conceivably can reverse CRPC to an androgen depletion responsive state, amenable to androgen ablation therapy. In addition, it could produce leads and targets amenable to the development and design of new anti-CRPC therapeutics applicable to other diseases of public health interest where an over-reactive AKT is also critically involved.

References

1. Jang M, Cai L, Udeani GO, Slowing KV, Thomas CF, Beecher CW, Fong HH, Farnsworth NR, Kinghorn AD, Mehta RG, Moon RC, Pezzuto JM. Cancer chemopreventive activity of resveratrol, a natural product derived from grapes. *Science* 1997; **275**: 218–20.
2. Hsieh TC, Wu JM. Differential effects on growth, cell cycle arrest, and induction of apoptosis by resveratrol in human prostate cancer cell lines. *Exp Cell Res* 1999; **249**: 109–15.

3. Hsieh TC, Wu JM. Grape-derived chemopreventive agent resveratrol decreases prostate-specific antigen (PSA) expression in LNCaP cells by an androgen receptor (AR)-independent mechanism. *Anticancer Res* 2000; **20**: 225–8.
4. Mitchell SH, Zhu W, Young CY. Resveratrol inhibits the expression and function of the androgen receptor in LNCaP prostate cancer cells. *Cancer Res* 1999; **59**: 5892–5.
5. Ratan HL, Steward WP, Gescher AJ, Mellon JK. Resveratrol–a prostate cancer chemopreventive agent? *Urol Oncol* 2002; **7**: 223–7.
6. Sheth S, Jajoo S, Kaur T, Mukherjea D, Sheehan K, Rybak LP, Ramkumar V. Resveratrol reduces prostate cancer growth and metastasis by inhibiting the Akt/MicroRNA-21 pathway. *PLoS One* 2012; **7**: e51655.
7. Wang Z, Hsieh TC, Zhang Z, Ma Y, Wu JM. Identification and purification of resveratrol targeting proteins using immobilized resveratrol affinity chromatography. *Biochem Biophys Res Commun* 2004; **323**: 743–9.
8. Buryanovskyy L, Fu Y, Boyd M, Ma Y, Hsieh TC, Wu JM, Zhang Z. Crystal structure of quinone reductase 2 in complex with resveratrol. *Biochemistry* 2004; **43**: 11417–26.
9. Jaiswal AK, Burnett P, Adesnik M, McBride OW. Nucleotide and deduced amino acid sequence of a human cDNA (NQO2) corresponding to a second member of the NAD(P)H:quinone oxidoreductase gene family. Extensive polymorphism at the NQO2 gene locus on chromosome 6. *Biochemistry* 1990; **29**: 1899–906.
10. Wu K, Knox R, Sun XZ, Joseph P, Jaiswal AK, Zhang D, Deng PS, Chen S. Catalytic properties of NAD(P)H:quinone oxidoreductase-2 (NQO2), a dihydronicotinamide riboside dependent oxidoreductase. *Arch Biochem Biophys* 1997; **347**: 221–8.
11. Long DJ, 2nd, Jaiswal AK. NRH:quinone oxidoreductase2 (NQO2). *Chem Biol Interact* 2000; **129**: 99–112.
12. Vella F, Ferry G, Delagrange P, Boutin JA. NRH:quinone reductase 2: an enzyme of surprises and mysteries. *Biochem Pharmacol* 2005; **71**: 1–12.
13. Hsieh TC, Yang CJ, Lin CY, Lee YS, Wu JM. Control of stability of cyclin D1 by quinone reductase 2 in CWR22Rv1 prostate cancer cells. *Carcinogenesis* 2012; **33**: 670–7.
14. Mailliet F, Ferry G, Vella F, Thiam K, Delagrange P, Boutin JA. Organs from mice deleted for NRH:quinone oxidoreductase 2 are deprived of the melatonin binding site MT3. *FEBS Lett* 2004; **578**: 116–20.
15. Mailliet F, Ferry G, Vella F, Berger S, Coge F, Chomarat P, Mallet C, Guenin SP, Guillaumet G, Viaud-Massuard MC, Yous S, Delagrange P, *et al.* Characterization of the melatoninergic MT3 binding site on the NRH:quinone oxidoreductase 2 enzyme. *Biochem Pharmacol* 2005; **71**: 74–88.
16. Long DJ, 2nd, Iskander K, Gaikwad A, Arin M, Roop DR, Knox R, Barrios R, Jaiswal AK. Disruption of dihydronicotinamide riboside:quinone oxidoreductase 2 (NQO2) leads to myeloid hyperplasia of bone marrow and decreased sensitivity to menadione toxicity. *J Biol Chem* 2002; **277**: 46131–9.

17. Iskander K, Jaiswal AK. Quinone oxidoreductases in protection against myelogenous hyperplasia and benzene toxicity. *Chem Biol Interact* 2005; **153–154**: 147–57.
18. Iskander K, Li J, Han S, Zheng B, Jaiswal AK. NQO1 and NQO2 regulation of humoral immunity and autoimmunity. *J Biol Chem* 2006; **281**: 30917–24.
19. Ahn KS, Gong X, Sethi G, Chaturvedi MM, Jaiswal AK, Aggarwal BB. Deficiency of NRH:quinone oxidoreductase 2 differentially regulates TNF signaling in keratinocytes: up-regulation of apoptosis correlates with down-regulation of cell survival kinases. *Cancer Res* 2007; **67**: 10004–11.
20. Nolan KA, Dunstan MS, Caraher MC, Scott KA, Leys D, Stratford IJ. In silico screening reveals structurally diverse, nanomolar inhibitors of NQO2 that are functionally active in cells and can modulate NF-kappaB signaling. *Mol Cancer Ther* 2012; **11**: 194–203.
21. Gong X, Kole L, Iskander K, Jaiswal AK. NRH:quinone oxidoreductase 2 and NAD(P)H:quinone oxidoreductase 1 protect tumor suppressor p53 against 20s proteasomal degradation leading to stabilization and activation of p53. *Cancer Res* 2007; **67**: 5380–8.
22. Khutornenko AA, Roudko VV, Chernyak BV, Vartapetian AB, Chumakov PM, Evstafieva AG. Pyrimidine biosynthesis links mitochondrial respiration to the p53 pathway. *Proc Natl Acad Sci U S A* 2010; **107**: 12828–33.
23. Brehmer D, Godl K, Zech B, Wissing J, Daub H. Proteome-wide identification of cellular targets affected by bisindolylmaleimide-type protein kinase C inhibitors. *Mol Cell Proteomics* 2004; **3**: 490–500.
24. Duncan JS, Gyenis L, Lenehan J, Bretner M, Graves LM, Haystead TA, Litchfield DW. An unbiased evaluation of CK2 inhibitors by chemoproteomics: characterization of inhibitor effects on CK2 and identification of novel inhibitor targets. *Mol Cell Proteomics* 2008; **7**: 1077–88.
25. Bantscheff M, Eberhard D, Abraham Y, Bastuck S, Boesche M, Hobson S, Mathieson T, Perrin J, Raida M, Rau C, Reader V, Sweetman G, *et al.* Quantitative chemical proteomics reveals mechanisms of action of clinical ABL kinase inhibitors. *Nat Biotechnol* 2007; **25**: 1035–44.
26. Rix U, Hantschel O, Durnberger G, Remsing Rix LL, Planyavsky M, Fernbach NV, Kaupe I, Bennett KL, Valent P, Colinge J, Kocher T, Superti-Furga G. Chemical proteomic profiles of the BCR-ABL inhibitors imatinib, nilotinib, and dasatinib reveal novel kinase and nonkinase targets. *Blood* 2007; **110**: 4055–63.
27. Sexton E, Van Themsche C, LeBlanc K, Parent S, Lemoine P, Asselin E. Resveratrol interferes with AKT activity and triggers apoptosis in human uterine cancer cells. *Mol Cancer* 2006; **5**: 45.
28. Aziz MH, Nihal M, Fu VX, Jarrard DF, Ahmad N. Resveratrol-caused apoptosis of human prostate carcinoma LNCaP cells is mediated via modulation of phosphatidylinositol 3'-kinase/Akt pathway and Bcl-2 family proteins. *Mol Cancer Ther* 2006; **5**: 1335–41.
29. Benitez DA, Pozo-Guisado E, Clementi M, Castellón E, Fernandez-Salguero PM. Non-genomic action of resveratrol on androgen and oestrogen receptors

in prostate cancer: modulation of the phosphoinositide 3-kinase pathway. *Br J Cancer* 2007; **96**: 1595–604.
30. Hsieh TC, Lin CY, Bennett DJ, Wu E, Wu JM. Biochemical and cellular evidence demonstrating AKT-1 as a binding partner for resveratrol targeting protein NQO2. *PLoS One* 2014; **9**: e101070.
31. Yan C, Dufour M, Siegel D, Reigan P, Gomez J, Shieh B, Moody CJ, Ross D. Indolequinone inhibitors of NRH:quinone oxidoreductase 2. Characterization of the mechanism of inhibition in both cell-free and cellular systems. *Biochemistry* 2011; **50**: 6678–88.
32. Liu P, Cheng H, Roberts TM, Zhao JJ. Targeting the phosphoinositide 3-kinase pathway in cancer. *Nat Rev Drug Discov* 2009; **8**: 627–44.
33. Kumar CC, Madison V. AKT crystal structure and AKT-specific inhibitors. *Oncogene* 2005; **24**: 7493–501.
34. Bedolla R, Prihoda TJ, Kreisberg JI, Malik SN, Krishnegowda NK, Troyer DA, Ghosh PM. Determining risk of biochemical recurrence in prostate cancer by immunohistochemical detection of PTEN expression and Akt activation. *Clin Cancer Res* 2007; **13**: 3860–7.
35. Ghosh PM, Malik SN, Bedolla RG, Wang Y, Mikhailova M, Prihoda TJ, Troyer DA, Kreisberg JI. Signal transduction pathways in androgen-dependent and -independent prostate cancer cell proliferation. *Endocr Relat Cancer* 2005; **12**: 119–34.
36. Wang Y, Romigh T, He X, Orloff MS, Silverman RH, Heston WD, Eng C. Resveratrol regulates the PTEN/AKT pathway through androgen receptor-dependent and -independent mechanisms in prostate cancer cell lines. *Hum Mol Genet* 2010; **19**: 4319–29.
37. Blando J, Portis M, Benavides F, Alexander A, Mills G, Dave B, Conti CJ, Kim J, Walker CL. PTEN deficiency is fully penetrant for prostate adenocarcinoma in C57BL/6 mice via mTOR-dependent growth. *Am J Pathol* 2009; **174**: 1869–79.
38. Robinson D, Van Allen EM, Wu YM, Schultz N, Lonigro RJ, Mosquera JM, Montgomery B, Taplin ME, Pritchard CC, Attard G, Beltran H, Abida W, *et al.* Integrative clinical genomics of advanced prostate cancer. *Cell* 2015; **161**: 1215–28.
39. Wendel HG, De Stanchina E, Fridman JS, Malina A, Ray S, Kogan S, Cordon-Cardo C, Pelletier J, Lowe SW. Survival signalling by Akt and eIF4E in oncogenesis and cancer therapy. *Nature* 2004; **428**: 332–7.
40. DeFeo-Jones D, Barnett SF, Fu S, Hancock PJ, Haskell KM, Leander KR, McAvoy E, Robinson RG, Duggan ME, Lindsley CW, Zhao Z, Huber HE, *et al.* Tumor cell sensitization to apoptotic stimuli by selective inhibition of specific Akt/PKB family members. *Mol Cancer Ther* 2005; **4**: 271–9.
41. Ohigashi T, Mizuno R, Nakashima J, Marumo K, Murai M. Inhibition of Wnt signaling downregulates Akt activity and induces chemosensitivity in PTEN-mutated prostate cancer cells. *Prostate* 2005; **62**: 61–8.

Chapter 6

The Effects of Resveratrol and Its Analogues on the Sirt5-GLS Axis-Mediated Glutamine Metabolic Reprogramming in Cancer Cells

Tze-chen Hsieh*, Jaysen Zhang*, Dylan J. Bennett*, John M. Pezzuto†, Mark Cushman‡ and Joseph M. Wu*

**Department of Biochemistry and Molecular Biology, New York Medical College, Valhalla, New York 10595, USA*

†Arnold & Marie Schwartz College of Pharmacy and Health Sciences, Long Island University, Brooklyn, New York 11201, USA

‡Department of Medicinal Chemistry and Molecular Pharmacology, College of Pharmacy, and The Purdue Center for Cancer Research, Purdue University, West Lafayette, IN 47907, USA

Gain-of-survival advantage, a hallmark of many cancers, requires metabolic adaptation to glutamine. The preference for glutamine by tumor cells is attributed to glutaminase (GLS), the enzyme that catalyzes the deamidation of glutamine to glutamate, which serves to fuel the tricarboxylic acid (TCA) cycle reactions in the mitochondria. Recent studies show Sirtuin 5 (Sirt5) interacts with GLS effectively inhibiting its activity by desuccinylation. Glutamine metabolism is also attenuated in Sirt5-overexpressing and increased in Sirt5-silenced cells. Accordingly, it may be proposed that activators of Sirt5 could suppress the addictive preference of tumor cells for glutamine by inhibition of GLS.

This chapter will: (i) review the role that Sirt5 plays in the control of glutamine metabolic reprogramming in cancer cell proliferation via the GLS, (ii) analyze the effect and impact of resveratrol or suramin, a known activator and inhibitor of Sirt5, respectively, on proliferation and modulation of preferred utilization of glucose and glutamine in HeLa cells, (iii) summarize pilot studies showing the interaction/binding of resveratrol, its metabolites, and analogues to Sirt5 using a combination of *in silico* and biochemical approaches, (iv) describe the mechanistic insights into the effects and binding of resveratrol synthetic analogues to Sirt5, and (v) present the hypothesis linking the Sirt5→GLS axis with glutamine metabolic reprogramming in cancer.

6.1. Introduction

A primary focal point of cancer research for the past several decades has been on defining and characterizing DNA mutations. This approach has been supported by evidence that cancer has a genetic basis.[1,2] The identified hallmarks of cancer are well aligned with adaptive metabolic changes for generating ATP to meet the energy demands required for tumor cells to survive and thrive in a nutrient-adverse environment.[3–8] The coupling of gene:metabolism circuits in malignant tumors is supported by the evidence that dysregulated cancer cell growth is coordinated with the required metabolic alterations.

In the 1920s, Warburg first recognized that cancer cells synthesize ATP principally via glucose and aerobic glycolysis, in paradoxical coexistence with an ample oxygen supply.[9] This phenomenon, frequently referred to as the Warburg effect and more recently coined "glucose addiction," requires metabolic reprogramming and is considered a hallmark of cancer.[4] A similar disproportionate dependence on high glutamine is also found in many cancer cell types;[10,11] comparatively, glutamine metabolic utilization occurs at a significantly lower level in normal cells.[12,13] The reliance of malignant tumors on high glutamine is linked to immortalized replication, unrestricted proliferation, invasion, metastasis and deregulated energetics — the collective gain-of-survival advantage that is presumably fueled by glutamine metabolism generating an ample supply of nitrogen for purine and pyrimidine nucleotide and amino acid biosynthesis.

Additionally, glutamine metabolism drives the use of citrate and isocitrate in the TCA cycle for lipid synthesis.[11,14,15] In other words, malignant cells are endowed with the ability to reprogram glutamine metabolism for energy production and biosynthesis via anaplerotic reactions.[16] Separately, glutamine is also a key amino acid for maintaining the integrity of the mitochondria.[12,17] Thus, glutamine metabolism represents an appealing target for cancer chemotherapy.[18,19]

6.2. Sirt5 plays a role in control of glutamine metabolic reprogramming

Sirts are enzymes that catalyze deacetylation of specific lysines of proteins in the nucleus, cytosol, and mitochondria, by an NAD^+ (nicotinamide adenine dinucleotide)-dependent reaction in which NAD^+ acts as a co-factor and is hydrolyzed to yield 2′-*O*-acetyl-ADP-ribose and nicotinamide.[20–23] Initial studies on Sirts have focused on lysine deacetylation of nuclear histones; subsequently, Sirts have also been studied for their post-transcription modification (PTM) activity of enzymes/proteins involved in metabolic control.[24–28]

Seven Sirts (1–7), with different subcellular locations, have been identified in mammalian cells. Human Sirts 1, 6 and 7 are found in the nucleus; Sirt2 occurs primarily in the cytosol; and Sirts 3 to 5 are localized in the mitochondria. Only Sirts 1, 2, and 3 have pronounced deacetylase activity; by contrast, Sirts 4 to 7 have weak to no detectable deacetylase activity.[28,29] Accumulating evidence indicates that Sirts are involved in the regulation of cellular events and biological processes including gene regulation and metabolic control, DNA damage repair, cell survival and apoptosis, and the aging process.[26–33] Moreover, they play a role in the pathogenesis of various diseases including cancer, diabetes, and neurodegenerative disorders.[22,31] To date, all Sirts except Sirt5 have demonstrated activity in tumorigenesis, either as a promoter or a suppressor.[34] Thus, it may be surmised that activators/inhibitors of Sirts may serve a dual purpose, both as tools to explore the roles that Sirts play in cell physiology and pathology and as targets for the development of drug candidates for the treatment of human diseases.

The mitochondrial Sirt5[35] has low deacetylase activity and a narrow protein substrate specificity.[36–38] However, quite distinct from Sirts 4, 6 and 7, Sirt5 displays robust NAD^+-dependent demalonylase, desuccinylase and deglutarylase activity capable of removing malonyl, succinyl and glutaryl groups from lysines of mitochondrial-located metabolic enzymes, such as isocitrate dehydrogenase 2 and glucose 6-phosphate dehydrogenase,[39] serine hydroxymethyltransferase, glyceraldehyde-3-phosphate dehydrogenase, malate dehydrogenase 2, citrate synthase, and carbamoyl phosphate synthetase 1 (CPS1).[27,28,40–44] These novel de-PTM reactions possibly have cellular and metabolic implications. Examples include: CPS1 activation by desuccinylation and deglutarylation may contribute to the control of blood ammonia levels in relation to changing NAD^+ levels during fasting; activation of superoxide dismutase 1 (SOD1) by desuccinylation and deglutarylation may lead to a reduction in formation of reactive oxygen species; and activation via Sirt5-mediated desuccinylation of 3-hydroxy-3-methylglutaryl-CoA synthase 2 (HMGCS2) may result in modulation of ketogenesis.[39,44–47] Counteracting the desuccinylation of proteins is the recent report of protein succinylation by an α-ketoglutarate dehydrogenase complex-dependent mechanism in neurons and neuronal cell line.[48,49]

Sirt5 was recently reported to regulate ammonia levels by controlling glutamine metabolism via interaction with and induced desuccinylation of GLS.[50] As support, glutamine metabolism was reduced in Sirt5-overexpressing cells and increased in wild-type cells treated with the Sirt5 inhibitor MC3482.[50] These results suggest that Sirt5 may function as a switch to control glutamine addiction via the Sirt5-GLS axis. We hypothesize that agents capable of upregulating/activating Sirt5 can block or attenuate GLS-mediated glutamine metabolic reprogramming in tumor cells (Figure 6.1), suggesting that GLS may be considered as an "Achilles heel" and a novel target for cancer chemotherapy.[18,19]

6.3. Pilot studies on the effects of a Sirt5 activator (resveratrol) or inhibitor (suramin) on the proliferation and utilization of glutamine and glucose in HeLa cells

We first determined Sirt5 enzyme activity with/without addition of activator (resveratrol) or inhibitor (suramin or nicotinamide)[24] using

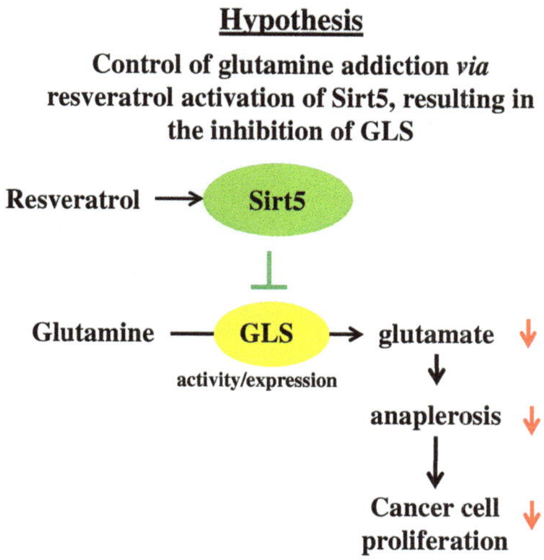

Fig. 6.1. A schematic illustrating the effects of resveratrol on the Sirt5-GLS axis. The activation of Sirt5 by resveratrol inhibits GLS resulting in suppression of cancer cell proliferation mediated by glutamine metabolic reprogramming. It may be hypothesized that chemical activators and inhibitors of Sirt5 function as on/off switch of GLS activity affecting the conversion of glutamine to glutamate.

the FLUOR DE LYS® Sirt5 fluorometric drug discovery assay kit (Enzo Life Sciences, Inc., Farmingdale, NY). Human recombinant Sirt5 was incubated with the FLUOR DE LYS®-succinyl substrate, co-substrate NAD^+, and resveratrol or suramin, each at 200 μM. The lysyl desuccinylase activity of Sirt5 was determined by the changes in emitted fluorescence. A blank (no addition of Sirt5) was included to determine the background emission. The assays were performed in triplicate. Figure 6.2A shows that suramin reduced the emission while resveratrol increased the emission, confirming that they act as an inhibitor and activator of Sirt5, respectively.[24,36]

To obtain information on whether activation/inhibition of Sirt5 by resveratrol/suramin may be linked to cancer cell proliferation and glutamine utilization, HeLa cells, shown to preferentially use glutamine instead of glucose for growth,[51] were examined for cell proliferation using the colony formation assay.[52,53] HeLa cells showed poor growth on co-depleted glucose/glutamine RPMI-1640; both cell proliferation and attachment on culture dishes improved significantly

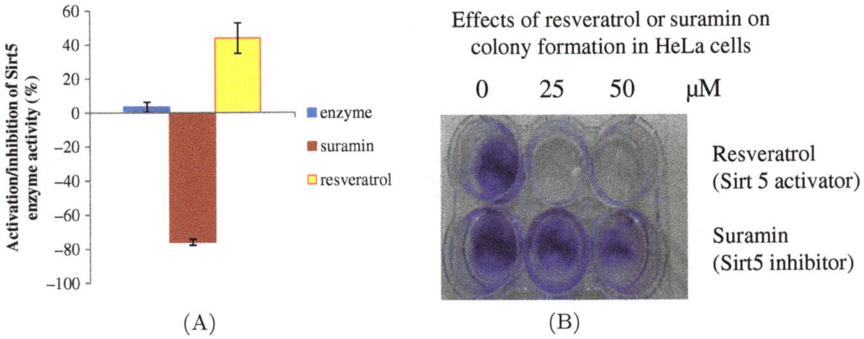

Fig. 6.2. Chemical activation or inhibition of Sirt5 leading to a decrease or increase of GLS, and subsequently impacting on GLS-mediated glutamine metabolism in cancer cells, according to the scheme: chemical modulator of Sirt5 → GLS on/off switch → glutamine metabolism/dependence → cancer cell growth control. **A.** Effects of suramin or resveratrol on SIRT5 enzyme activity compared to enzyme alone. The FLUOR DE LYS® SIRT5 fluorometric drug discovery assay kit (Enzo Life Sciences, Inc., Farmingdale, NY) was used to test the effects of resveratrol or suramin, both at 200 μM on Sirt5 activity. The procedures followed the instructions provided in the kit. The assay was performed by incubating human recombinant Sirt5 with the FLUOR DE LYS®-Succinyl substrate together with the co-substrate NAD^+, with/without addition of resveratrol, a known activator, or suramin, a known inhibitor. The lysyl desuccinylase activity was determined by the change in fluorescent emissions comparing the samples containing Sirt5 alone with samples containing Sirt5 and the added activator of inhibitor. A blank (without Sirt5) was included for determination of the background which was subtracted from all measurements to give the net emission resulting from Sirt5 activity. The assays were performed in triplicate. The bars show the average ± SD (n = 3). **B.** Effects of Sirt5 activator (resveratrol) or inhibitor (suramin) on colony formation in HeLa cells. HeLa cells were maintained in RPMI 1640 at a density of 2000 cells/ml and 2 ml/well in six-well plates. After 24 h, cells were switched to RPMI-1640 medium supplemented with high-glucose (4.5 g/L) and high-glutamine (4 mM) and treated with 25 and 50 μM of resveratrol or suramin or 0.2% DMSO for 1 week. Changes in clonogenicity were visualized by staining colonies with 1.25% crystal violet.

with addition of increasing glucose/glutamine to the culture media (data not shown). Effects of resveratrol or suramin on HeLa cells exposed to high glucose (4.5 g/L)/glutamine (4 mM) were investigated. Figure 6.2B shows that colony formation was >95% inhibited by 25 μM resveratrol; comparatively, the same dose of suramin had minimal effects on colony growth. These results are consistent with the hypothesis that chemical activators/inhibitors of Sirt5 may act as an on/off switch of GLS activity and impact GLS-mediated glutamine metabolism in cancer cells. A scheme of this hypothesis is presented in Figure 6.1.

To further study the modulatory effects of resveratrol on the growth dependence of cancer cells on glucose and glutamine; HeLa cells were maintained in culture media containing 0, 2 and 4 mM glutamine and varying glucose concentrations of 0, 1, 2 and 4.5 g/L. Colony formation was monitored. Poor colony formation occurred in glucose-depleted RPMI-1640, regardless of whether 0, 2, or 4 mM glutamine was added to the culture media. However, supplementation of 1, 2, or 4 g/L glucose resulted in a glutamine dose-dependent increase in colonies formed (Figure 6.3). Interestingly, the observed glucose- and glutamine-dependent increase in colony formation was significantly inhibited by the addition of 25 μM resveratrol, except when cells were maintained in glucose/glutamine co-depleted state (Figure 6.3). These results suggest that resveratrol in the culture media attenuates the utilization of glucose or glutamine by HeLa cells, supporting the hypothesis that resveratrol plays a role on control of gene:metabolism circuits by cancer cells.

6.3.1. *Preliminary studies of effects and binding of resveratrol or its metabolites/analogues to Sirt5 using in silico and biochemical approaches*

Several challenges offsetting the potential health benefits of resveratrol have been identified. These include its poor bioavailability, low potency, and rapid metabolism.[54,55] Despite these lingering concerns, ample evidence, ranging from *in vitro* studies, animal experiments, and limited clinical trials, support its disease-prevention and anti-tumorigenic potential. For example, in colon cancer patients orally administered resveratrol, there was a significant correlation between the reduction in tumor proliferation and increases in resveratrol and its metabolites in resected colorectal tissues as documented by HPLC/UV/mass spectrometry.[56] These results suggest that resveratrol and its metabolites are capable of reaching the responsive

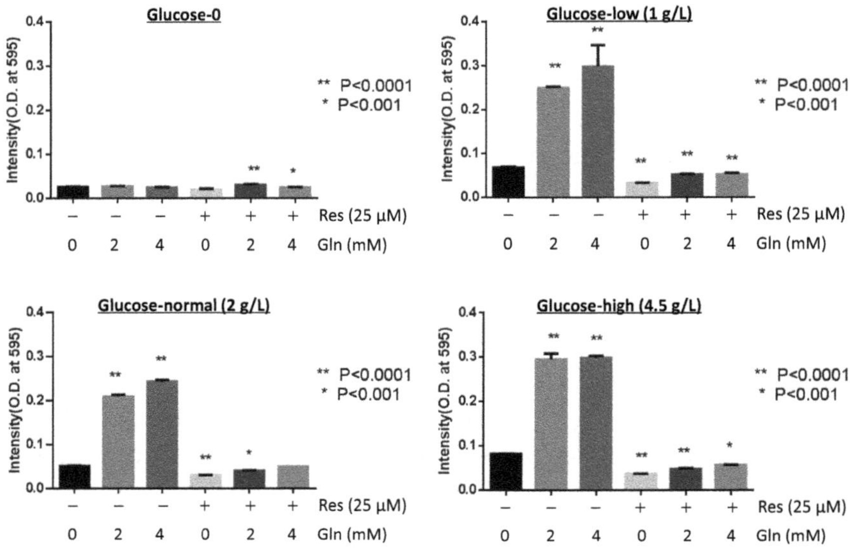

Fig. 6.3. The control of HeLa cell growth by resveratrol (Res) at varying concentrations of glucose or glutamine (Gln), using the colony formation assay. Briefly, HeLa cells maintained in RPMI 1640 were seeded at a density of 2000 cells/ml and 2 ml/well in six-well plates. After 24 h, cells were switched to RPMI-1640 medium containing varying concentrations of glucose and glutamine and treated with 25 μM of resveratrol or 0.2% DMSO for 1 week. Changes in clonogenicity were visualized by staining colonies with 1.25% crystal violet. The bars show average ± SD (n = 3).

target tissues in sufficiently high, anti-carcinogenic levels.[56,57] It is also worth noting that most studies to date have used free resveratrol and not its metabolites despite data showing that the administered resveratrol is rapidly converted to its glucuronated and sulfated derivatives by glucuronosyl transferase and sulfotransferase in the intestine and liver.[58–62] The synthesis and biological evaluation of five resveratrol sulfates have been reported.[63]

Control of Sirt5 activity by resveratrol metabolites is an open question. We used 'docking' to test the theoretical binding of small molecules to their cognate protein targets. In the docking analyses, UCSF DOCK6.3 programs were used to determine interaction between resveratrol and its derivatives with Sirt5. The resveratrol derivatives in the mol2 format were generated using the UCSF

ZINC database or created using MarvinSketch from ChemAxon. The binding affinities of resveratrol derivatives were computed using DOCK6.3 and the results were analyzed using the UCSF program Chimera and viewed by Chimera ViewDock tool. The strength of binding is represented by the calculated grid score (GS), as the summation of E_{VDW} — attributed to VDW (van der Waals interaction) and ES (electrostatic interactions), respectively. The hydrogen bonds formed between the tested chemicals and specific amino acid residues in the docked protein are visualized and determined using Chimera ViewDock.[64] VDW and ES between the ligands (resveratrol and its derivatives) and receptor (Sirt5), generated on the basis of 25,000 possible binding conformations, were computed and presented as binding energy scores in kilocalories per mole. Ligands with low energy scores are presumed to possess a higher affinity for the receptor. All dockings were performed in a Linux-based shell program.

The published crystal structure of Sirt5 in complex with resveratrol and Fluor-de-Lys peptide was obtained from the RSCB Protein Data Bank (PDB) and used as a reference receptor (PDB code: 4HDA).[24] The possible interactions of Sirt5 with piceatannol, resveratrol glucuronide metabolites, and piceid were shown in Figure 6.4A. The docking output and the number of hydrogen bonds are shown in Figure 6.4B. The calculated binding energies show a ranking order: resveratrol-3-*O*-glucuronide> piceid > resveratrol-4′-*O*-glucuronide > piceatannol > resveratrol (Figure 6.4B). These results suggest that resveratrol glucuronides can interact with the same target protein as resveratrol, implying that resveratrol glucuronides may be comparable to resveratrol in activating Sirt5. We therefore tested effects of resveratrol glucuronides on Sirt5 activity, for comparison with resveratrol (Sirt5 activator) and suramin (Sirt5 inhibitor). The results showed that 200 μM resveratrol-4′-*O*-glucuronide displayed equal potency as resveratrol in activating Sirt5. However, resveratrol-3-*O*-glucuronide was less effective in activating Sirt5 (Figure 6.4C) despite showing the best grid score for Sirt5 in docking (Figure 6.4B). These results are consistent with the known limitations of molecular mechanics-based molecular modeling.

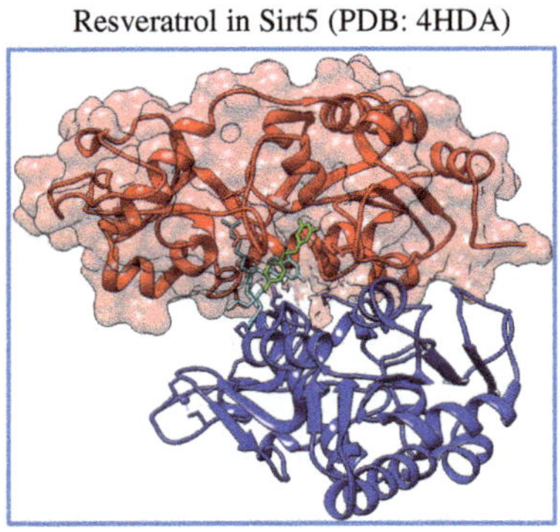

(A)

Name	Grid Score (kcal/mol)	VDW	ES	H-Bonds
resveratrol (Res)	−42.06	−36.09	−5.97	2
piceatannol	−42.78	−37.06	−5.72	0
resveratrol-4′-*O*-glucuronide	−53.99	−53.99	0	0
resveratrol-3-*O*-glucuronide	−56.35	−55.46	−0.88	1
piceid	−54.94	−53.51	−1.43	2

(B)

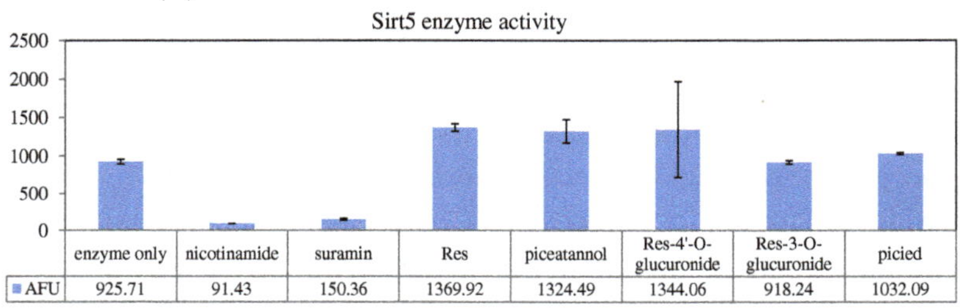

(C)

Fig. 6.4. **A.** Sirt5-resveratrol complex (PDB ID: 4HDA). The model of resveratrol (shown in green stick) interacting with a lysine residue (shown in cyan stick) of Sirt5. **B.** Output of the grid score generated by DOCK 6.3 program for binding of resveratrol or its metabolites to Sirt5. The number of H-bonds is calculated using the "view dock" accessory program in Chimera. **C.** Effects of resveratrol (Res), its metabolites and suramin on Sirt5 enzyme activity. The bars show the average ± SD (n = 3).

In addition to resveratrol metabolites, we also tested chemically modified resveratrol derivatives and analogues for their effects on Sirt5 activity compared to resveratrol, suramin, and nicotinamide (obtained from Enzo Life Sciences, Inc., Farmingdale, NY, USA). The resveratrol analogues and derivatives (structures shown in Figure 6.5) used were: (*E*)-4-(3,5-dimethoxystyryl)aniline (**1**), (*E*)-4-(3,4,5-trimethoxystyryl)aniline (**2**), (*E*)-4-(3,4-dimethoxystyryl)aniline (**3**), (*E*)-4-(2-(benzo[*d*][1,3]dioxole-5-yl)vinyl)aniline (**4**), (*E*)-5-(4-aminostyryl) benzene-1,3-diol (**5**), (*E*)-4-(2,3,4-trimethoxystyryl)aniline (**6**), (*E*)-4-(3,5-dimethoxystyryl)phenol (**7**), (*E*)-4-(3,5-dinitrostyryl) phenyl acetate (**8**), (*E*)-5-(4-nitrostyryl)benzo[*d*][1,3]dioxole (**9**),

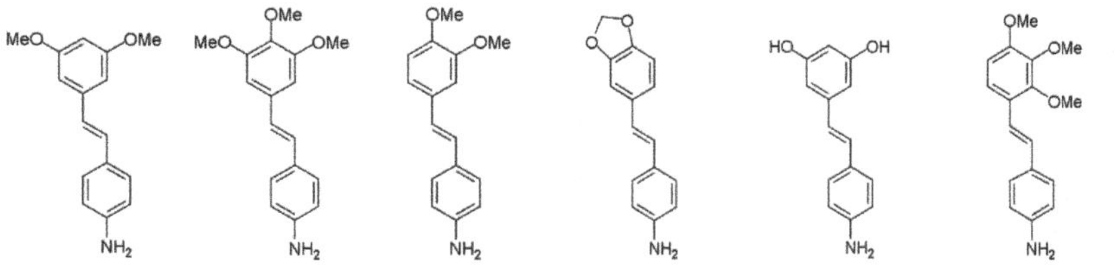

(*E*)-4-(3,5-dimethoxystyryl) aniline (**1**)

(*E*)-4-(3,4,5-trimethoxystyryl) aniline (**2**)

(*E*)-4-(3,4-dimethoxystyryl) aniline (**3**)

(*E*)-4-(2-(benzo [d][1,3]dioxole-5-yl)vinyl)aniline (**4**)

(*E*)-5-(4-aminostyryl) benzene-1,3-diol (**5**)

(*E*)-4-(2,3,4-trimethoxystyryl) aniline (**6**)

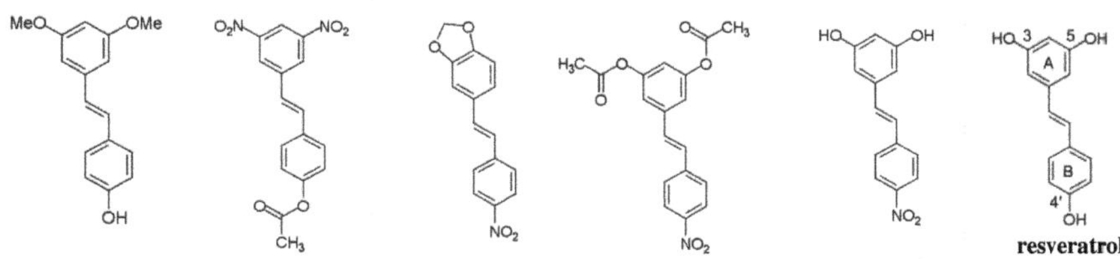

resveratrol

(*E*)-4-(3,5-dimethoxystyryl) phenol (**7**)

(*E*)-4-(3,5-dinitrostyryl) phenyl acetate (**8**)

(*E*)-5-(4-nitrostyryl)benzo [d][1,3]dioxole (**9**)

(*E*)-5-(4-nitrostyryl)-1,3-phenylene diacetate (**10**)

(*E*)-5-(4-nitrostyryl) benzene-1,3-diol (**11**)

Fig. 6.5. Chemical structure of resveratrol analogues and resveratrol tested for modulation of Sirt5 activity.

(*E*)-5-(4-nitrostyryl)-1,3-phenylene diacetate (**10**), and (*E*)-5-(4-nitrostyryl)benzene-1,3-diol (**11**).[65] Of the eleven resveratrol derivatives tested, ten had the 4′-hydroxy group changed to 4′-amino, 4′-acetoxy, or 4′-nitro groups, and one had the 4′-hydroxy group left intact. The 3,5-dihydroxybenzene portion was either left unchanged or changed to other substituted or fused benzene rings.

Based on results of the Sirt5 enzymatic assay, all six 4′-amino resveratrol derivatives tested (**1**–**6**, see Figure 6.5) were found to be more potent in activating Sirt5 than resveratrol. Figure 6.6 shows that compounds **1**–**6** when tested at 200 μM displayed higher potency in activating Sirt5 than resveratrol. The most potent of them is **4**, which quadrupled the activity increase compared with resveratrol at the same dose (209% increase *vs* 48% increase). This represents the most effective Sirt5 activator reported thus far. (*E*)-4-(3,5-Dimethoxystyryl)phenol (**7**), the resveratrol derivative in which the 3,5-dihydoxy groups were converted to 3,5-dimethoxy groups but the 4′-hydroxy group was left intact, activated Sirt5 about the same

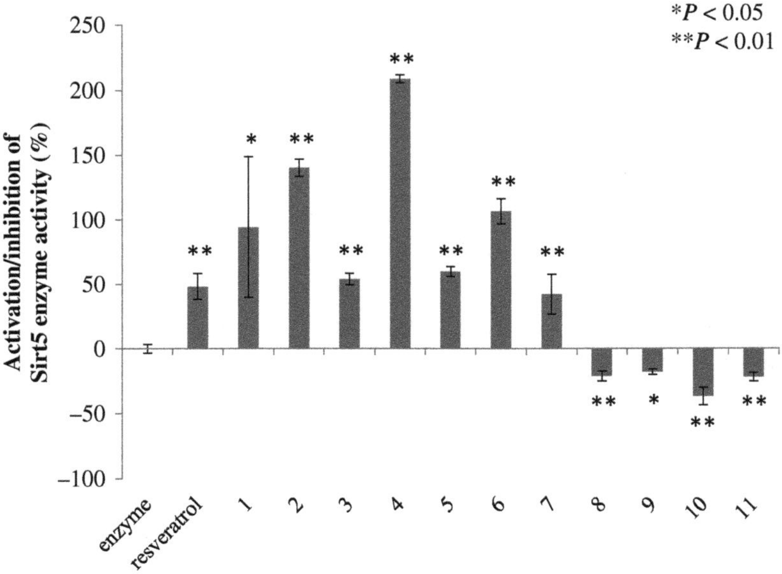

Fig. 6.6. Effects of resveratrol and resveratrol analogues (compounds **1**–**11**) on Sirt5 enzyme activity. The FLUOR DE LYS® SIRT5 fluorometric drug discovery assay kit (Enzo Life Sciences, Inc., Farmingdale, NY) was used to test the effects of resveratrol or its analogues on Sirt5 activity. The assay was performed by incubating human recombinant Sirt5 with the FLUOR DE LYS®-Succinyl substrate together with the co-substrate NAD^+, with/without addition of resveratrol or resveratrol analogues, all used at 200 μM. The lysyl desuccinylase activity of the recombinant human Sirt5 was determined by the changes in fluorescent emissions as detailed in Figure 6.2. The assays were performed in triplicate. The bars show the average ± SD (n = 3); $^*P < 0.05$, and $^{**}P < 0.01$ (calculated using unpaired t-test) when compared to enzyme only.

amount as resveratrol did. Derivatives that replace the corresponding 4′-hydroxy group in resveratrol with nitro and acetoxy groups (**8**–**11**, see Figure 6.5) were Sirt5 inhibitors. As shown in Figure 6.6, these compounds inhibited Sirt5 when tested at 200 μM.

We further used the docking program UCSF DOCK6.3 to investigate the possible binding between Sirt5 and all resveratrol analogues. The predicted binding affinities to Sirt5 are shown in Figure 6.7. The analogues and resveratrol were calculated to bind to the same area of Sirt5 in roughly the same orientation but with a different rank order of binding affinities than predicted: resveratrol's score is

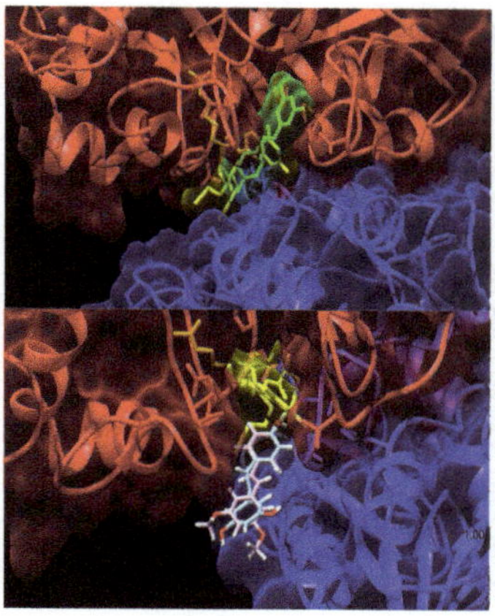

(A)

Name	Grid Score (kcal/mol)	VDW	Electrostatic
Compound 1	−41.999538	−43.42186	1.42232
Compound 2	−45.375465	−39.154789	−6.220675
Compound 3	−44.338474	−43.196526	−1.141949
Compound 4	−39.987591	−37.529282	−2.458308
Compound 5	−37.619747	−36.246651	−1.373095
Compound 6	−44.381565	−42.223106	−2.158457
Compound 7	−41.21143	−39.615467	−1.595963
Compound 8	−53.574028	−50.706337	−2.867692
Compound 9	−44.483223	−41.848473	−2.634749
Compound 10	N.D.	N.D.	N.D.
Compound 11	−44.353962	−39.51754	−4.836421

(B)

Fig. 6.7. **A.** Binding of resveratrol and its analogues inside the monomers of Sirt5 (red and blue) using PDB ID: 4HDA. The top image shows an X-ray crystal structure in which resveratrol (green) is keeping the lysine substrate (yellow) inside the Sirt5 active site located in one of the Sirt5 monomers (red). The bottom image shows our computer model calculated using UCSF DOCK6.3. Resveratrol (light blue) and compound **7** (light pink) are predicted to bind with a similar binding conformation. **B.** Output data of all the analogues of resveratrol that were tested. Their grid scores represent kcal/mol for the reaction.

−42.1 kcal/mol (Figure 6.4) while its derivatives' scores range from −37.6 to −53.6 kcal/mol (Figure 6.7B). Thus, computer modeling suggests that these analogues can bind to Sirt5 at the same site as resveratrol, with similar or better affinities. Biochemical studies show that modifications at its 4′-position could augment or diminish Sirt5 activity. The replacement of the 4′-hydroxy on resveratrol with 4′-amino resulted in an increase in the ability to activate Sirt5 activity up to fourfold, while Sirt5 activity was inhibited when 4′-hydroxy was replaced by 4′-nitro or 4′-acetoxy groups. Thus, the evidence indicates that differences in interaction between the 4′-substituents of the derivatives with the same binding area on Sirt5 are responsible for the differences observed in the activation or inhibition of Sirt5 activity. These novel findings require further analysis of the mechanism contributing to the dichotomous regulation of Sirt5 activity in order to develop drugs targeting Sirt5-mediated human disorders.

6.4. Mechanistic insights into the effects and binding of resveratrol synthetic analogues to Sirt5

Due to the multifaceted role played by Sirt5 in normal and diseased states, both inhibitors and activators of Sirt5 have been developed. As an example, suramin is an inhibitor of some Sirts including Sirt5. Suramin acts by a mechanism targeting the blockade of the active site of Sirt5,[36] which may contribute to the augmentation of its ability to enhance paclitaxel in the treatment of breast cancer.[66] Resveratrol (3,4′,5-trihydroxystilbene), a natural grape polyphenol shown to reduce tumorigenesis,[67] increase stress-resistance,[38] and extend lifespan,[24] is a known activator of Sirt5.[38] Derivatives and analogues of resveratrol have been developed to address its low solubility and bioavailability in humans.[68,69] The structural modifications have involved changing 4′-hydroxy group to 4′-amino, 4′-acetoxy, or 4′-nitro groups, while other analogues have preserved the 4′-hydroxy group. Alternatively, other synthetic analogues leave the 3,5-dihydroxybenzene portion unchanged or modify it to other substituted or fused benzene rings.[69–72]

In this study the structural requirements for several resveratrol analogues have been evaluated in pilot experiments, in attempts to obtain information for understanding the role that Sirt5 plays in normal and diseased states as well as to glean insights on how to improve activation or inhibition of Sirt5 activity for development as target-specific pharmaceutical agents. Several results are noteworthy. Our computer modeling analyses showed that all 11 resveratrol derivatives tested are calculated to bind at the same site as resveratrol with similar or better binding affinity (Figure 6.7A and B). However, determination of their effects on Sirt5 enzyme activity changes revealed that there are distinct differences among these derivatives. The molecular explanations for the observed Sirt5 activity changes will require detailed investigations in the future. X-ray crystallography and molecular modeling studies have provided some insights on the binding mode of Sirt5 to its inhibitor, suramin, and activator, resveratrol, respectively.[36,38] Two Sirt5 monomers form an asymmetric dimer with an active site forming a cleft in between. Suramin inhibits Sirt5 activity by blocking the active site by virtue of its large size.[36,38] Resveratrol binds at the 'docking patch' in close proximity to the opening/closing gate of the active site. This effectively traps the substrate in the active site, facilitating timely, rapid deacetylation of substrate's lysine side chain.[24] According to this model, the 3- and 5- positions in ring A of resveratrol are exposed to the solvent, allowing modifications to be relatively well tolerated. The 4′-position on ring B is crucial for the interaction with the substrate and therefore should exhibit greater sensitivity to structural modifications. Results in the Sirt5 enzyme activity assays support these anticipated outcomes.

By modifying functional groups of resveratrol, one can modulate resveratrol derivatives' activity on Sirt5. Thus, the 3,5-dihydroxybenzene A ring of the resveratrol may be changed in various ways without significantly changing activity, as shown by direct comparison with analogues modified on the 4′-functional group. The 4′-hydroxylbenzene B ring, however, seems to be more critical for determining the activity of this series of derivatives. These observations are in line with the predictions revealed by the X-ray crystallography structural model analysis.[24] Replacing the 4′-hydroxy group with a 4′-amino group preserves and probably enhances the Sirt5-activating capability of resveratrol, as shown by a direct comparison of compound **5** with resveratrol. We surmise that both the 4′-hydroxy and 4′-amino derivatives act as effective Sirt5 activators since both can serve as hydrogen-bond donors at the 4′-position. However, additional modifications at the 4′-position and X-ray crystallography studies are needed to confirm this hypothesis. A 4′-nitro or 4′-acetoxy group reverses resveratrol's function from a Sirt5 activator to an inhibitor, as shown by a direct comparison of compound **11** vs. resveratrol. These inhibitors most likely act differently than suramin. Whereas suramin, by its large size, inhibits Sirt5 activity by restricting access of the substrate to the Sirt5 binding site, our computer modeling studies suggest that the inhibitory resveratrol derivatives bind to Sirt5 with a substrate. Thus, their inhibition of Sirt5 may be attributed to a lack of a hydrogen-bond donor in the 4′-position. Of note, our preliminary results also support previous studies showing that five derivatives of resveratrol (the parent molecule resveratrol is a Sirt1 activator) where the 4′-hydroxy was replaced with five different hydrophobic groups including a 4′-acetoxy group act as Sirt1 activators with varying degrees of activity.[73]

6.4.1. *Perspectives on binding of resveratrol and its synthetic analogues to Sirt5*

Here, we have provided results of pilot experiments showing that by modifying the 4′-hydroxyl group in resveratrol, it is possible to enhance resveratrol's role as an activator of Sirt5, or alternatively, convert its role from a Sirt5 activator to an inhibitor. What accounts for the higher activity observed with the 4′-amino derivatives and the inversion of the activity shown by the 4′-acetoxy and 4′-nitro derivatives requires further investigation.

Our computer modeling analysis suggests that structural modifications of the resveratrol molecule are allowed for modulating Sirt5 activity; thus, further modifications of both A and B rings should be explored. X-ray crystal structures of Sirt5 bound to an activating or inhibitory resveratrol derivative should be performed to better understand the mechanism behind Sirt5 activity control, particularly with respect to the interactions between the B ring and Sirt5 responsible for activation and inhibition. Elucidation of the molecular details by which Sirt5 activation or inhibition is achieved would lay the mechanistic foundation for the development of new pharmaceutical agents to control glutamine metabolic reprogramming in cancer.

6.5. A hypothesis linking the Sirt5 → GLS axis with glutamine metabolic rewiring in cancer

Cancer cells show increased dependence on glutamine metabolism for their growth and proliferation. Therefore, control of glutamine metabolism has attracted attention as a target for cancer chemotherapy.[10,11,14,15] The mitochondrial enzyme GLS plays a pivotal role in the control of glutamine metabolism; thus, controlling the expression and/or activity of GLS represents an attractive approach in cancer chemotherapy.[18,38,74] GLS catalyzes the conversion of glutamine to glutamate and the release of ammonia. Mitochondrial Sirt5 controls ammonia levels by interacting with and desuccinylating GLS, showing the link of regulation of glutamine metabolism and GLS control by Sirt5.[50]

In this chapter, we show for the first time that resveratrol, a Sirt5 activator; inhibits glutamine-dependent HeLa cell growth (Figure 6.2), providing support for the tenet of chemical control of Sirt5 in GLS/glutamine addiction (Figure 6.1). As a final note, another important aspect of our hypothesis not addressed in this book chapter concerns the control of GLS expression by the proto-oncogene protein c-Myc via the c-Myc–miR-23a/b–GLS axis.[75] In this axis, c-Myc suppresses miR-23a/b, which suppresses GLS expression. The activity of GLS is known to be controlled by the mitochondrial enzyme Sirt5[50] whereas mitochondrial glutaminolysis is associated with c-Myc.[75,76] It is currently unknown and yet to be determined whether Sirt5 is involved in c-Myc-mediated glutamine addiction in cancer cells. Moreover, hitherto the link between c-Myc and Sirt5 has not been established.

Resveratrol derivatives and analogues may be considered a novel approach to control glutamine metabolic reprogramming in cancer. Our goal is to develop more potent chemical agents to reprogram glutamine utilization in cancer cells by control of glutamine metabolism via the Sirt5–GLS axis. In pilot experiments we showed that some resveratrol derivatives are more potent in desuccinylation than resveratrol in the control Sirt5 enzyme activity. These can be used to test Sirt5-mediated control of GLS/glutamine addiction in HeLa and other cancer cell types to elucidate details on the crucial role that Sirt5 plays in the control of GLS-mediated anaplerosis and cell proliferation (Figure 6.8).

It should be noted that ample evidence already exists on the inhibition of glucose transport and metabolism by resveratrol.[77,78] Future work on resveratrol analogues should focus on the identification of compounds that control both glucose utilization via aerobic glycolysis and glutamine metabolic reprogramming via anaplerosis. This may provide insight into the metabolic dysregulation that is a hallmark of cancer. These studies may also offer some insights on

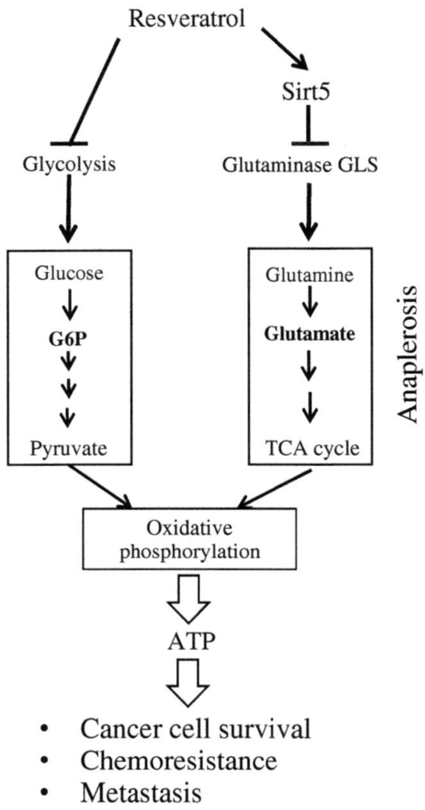

Fig. 6.8. A hypothesis showing resveratrol exerting dual control of glucose utilization via aerobic glycolysis and glutamine metabolic reprogramming targeting control of the Sirt5-GLS axis and anaplerosis as combined dysregulated hallmark metabolic features of cancer.

the detailed mechanism of attenuation of GLS activity through the Sirt5–GLS axis by small molecules.

Acknowledgments

Jaysen Zhang was a participant in STAR 2014–16 (Summer Trainees in Academic Research), a summer program at New York Medical College aimed at enhancing the research experience of students. He was a recipient of the semifinalist Siemens competition in 2015 and 2016 as well as a recipient of the semifinalist Regeneron STS (previous Intel) competition award, 2016 for his research, some of which is included in this book chapter.

References

1. Stoler AB. Genes and cancer. *Br Med Bull* 1991; **47**: 64–75.
2. Pearson PL, Van der Luijt RB. The genetic analysis of cancer. *J Intern Med* 1998; **243**: 413–7.
3. Hanahan D, Weinberg RA. The hallmarks of cancer. *Cell* 2000; **100**: 57–70.
4. Hanahan D, Weinberg RA. Hallmarks of cancer: the next generation. *Cell* 2011; **144**: 646–74.
5. Chaffer CL, Weinberg RA. How does multistep tumorigenesis really proceed? *Cancer Discov* 2015; **5**: 22–4.
6. Thompson CB, Palm W. Reexamining how cancer cells exploit the body's metabolic resources. *Cold Spring Harb Symp Quant Biol* 2016; **81**: 67–72.
7. Zhang J, Pavlova NN, Thompson CB. Cancer cell metabolism: the essential role of the nonessential amino acid, glutamine. *EMBO J* 2017; **36**: 1302–15.
8. Palm W, Thompson CB. Nutrient acquisition strategies of mammalian cells. *Nature* 2017; **546**: 234–42.
9. Warburg O, Wind F, Negelein E. The metabolism of tumors in the body. *J Gen Physiol* 1927; **8**: 519–30.
10. Ward PS, Thompson CB. Metabolic reprogramming: a cancer hallmark even Warburg did not anticipate. *Cancer Cell* 2012; **21**: 297–308.
11. Stine ZE, Walton ZE, Altman BJ, Hsieh AL, Dang CV. MYC, metabolism, and cancer. *Cancer Discov* 2015; **5**: 1024–39.
12. Wise DR, Thompson CB. Glutamine addiction: a new therapeutic target in cancer. *Trends Biochem Sci* 2010; **35**: 427–33.
13. Lu W, Pelicano H, Huang P. Cancer metabolism: is glutamine sweeter than glucose? *Cancer Cell* 2010; **18**: 199–200.
14. Altman BJ, Stine ZE, Dang CV. From Krebs to clinic: glutamine metabolism to cancer therapy. *Nat Rev Cancer* 2016; **16**: 619–34.
15. Ochoa-Ruiz E, Diaz-Ruiz R. Anaplerosis in cancer: another step beyond the Warburg effect, *Am J Mol Biol*, 2012; **2**: 291–303.
16. DeBerardinis RJ, Mancuso A, Daikhin E, Nissim I, Yudkoff M, Wehrli S, Thompson CB. Beyond aerobic glycolysis: transformed cells can engage in glutamine metabolism that exceeds the requirement for protein and nucleotide synthesis. *Proc Natl Acad Sci U S A* 2007; **104**: 19345–50.
17. Jeong SM, Hwang S, Park K, Yang S, Seong RH. Enhanced mitochondrial glutamine anaplerosis suppresses pancreatic cancer growth through autophagy inhibition. *Sci Rep* 2016; **6**: 30767.
18. Wang JB, Erickson JW, Fuji R, Ramachandran S, Gao P, Dinavahi R, Wilson KF, Ambrosio AL, Dias SM, Dang CV, Cerione RA. Targeting mitochondrial glutaminase activity inhibits oncogenic transformation. *Cancer Cell* 2010; **18**: 207–19.

19. Hensley CT, Wasti AT, DeBerardinis RJ. Glutamine and cancer: cell biology, physiology, and clinical opportunities. *J Clin Invest* 2013; **123**: 3678–84.
20. Imai S, Armstrong CM, Kaeberlein M, Guarente L. Transcriptional silencing and longevity protein Sir2 is an NAD-dependent histone deacetylase. *Nature* 2000; **403**: 795–800.
21. Lin SJ, Defossez PA, Guarente L. Requirement of NAD and SIR2 for life-span extension by calorie restriction in Saccharomyces cerevisiae. *Science* 2000; **289**: 2126–8.
22. Finkel T, Deng CX, Mostoslavsky R. Recent progress in the biology and physiology of sirtuins. *Nature* 2009; **460**: 587–91.
23. Sauve AA, Youn DY. Sirtuins: NAD(+)-dependent deacetylase mechanism and regulation. *Curr Opin Chem Biol* 2012; **16**: 535–43.
24. Gertz M, Nguyen GT, Fischer F, Suenkel B, Schlicker C, Fränzel B, Tomaschewski J, Aladini F, Becker C, Wolters D, Steegborn C. A molecular mechanism for direct sirtuin activation by resveratrol. *PLoS One* 2012; **7**: e49761.
25. Nakagawa T, Guarente L. Sirtuins at a glance. *J Cell Sci* 2011; **124**: 833–8.
26. Newman JC, He W, Verdin E. Mitochondrial protein acylation and intermediary metabolism: regulation by sirtuins and implications for metabolic disease. *J Biol Chem* 2012; **287**: 42436–43.
27. Kumar S, Lombard DB. Mitochondrial sirtuins and their relationships with metabolic disease and cancer. *Antioxid Redox Signal* 2015; **22**: 1060–77.
28. Parihar P, Solanki I, Mansuri ML, Parihar MS. Mitochondrial sirtuins: emerging roles in metabolic regulations, energy homeostasis and diseases. *Exp Gerontol* 2015; **61**: 130–41.
29. Sebastián C, Mostoslavsky R. The role of mammalian sirtuins in cancer metabolism. *Semin Cell Dev Biol* 2015; **43**: 33–42.
30. Jeong SM, Haigis MC. Sirtuins in cancer: A balancing act between genome stability and metabolism. *Mol Cells* 2015; **38**: 750–8.
31. Baur JA, Ungvari Z, Minor RK, Le Couteur DG, de Cabo R. Are sirtuins viable targets for improving healthspan and lifespan? *Nat Rev Drug Discov* 2012; **11**: 443–61.
32. Chalkiadaki A, Guarente L. The multifaceted functions of sirtuins in cancer. *Nat Rev Cancer* 2015; **15**: 608–24.
33. George J, Ahmad N. Mitochondrial sirtuins in cancer: Emerging roles and therapeutic potential. *Cancer Res* 2016; **76**: 2500–6.
34. Yuan H, Su L, Chen WY. The emerging and diverse roles of sirtuins in cancer: a clinical perspective. *Onco Targets Ther* 2013; **6**: 1399–416.
35. Michishita E, Park JY, Burneskis JM, Barrett JC, Horikawa I. Evolutionarily conserved and nonconserved cellular localizations and functions of human SIRT proteins. *Mol Biol Cell* 2005; **16**: 4623–35.
36. Schuetz A, Min J, Antoshenko T, Wang CL, Allali-Hassani A, Dong A, Loppnau P, Vedadi M, Bochkarev A, Sternglanz R, Plotnikov AN. Structural basis of inhibition of the human NAD+-dependent deacetylase SIRT5 by suramin. *Structure* 2007; **15**: 377–89.

37. Schlicker C, Gertz M, Papatheodorou P, Kachholz B, Becker CF, Steegborn C. Substrates and regulation mechanisms for the human mitochondrial sirtuins Sirt3 and Sirt5. *J Mol Biol* 2008; **382**: 790–801.
38. Gertz M, Steegborn C. Function and regulation of the mitochondrial sirtuin isoform Sirt5 in Mammalia. *Biochim Biophys Acta* 2010; **1804**: 1658–65.
39. Zhou L, Wang F, Sun R, Chen X, Zhang M, Xu Q, Wang Y, Wang S, Xiong Y, Guan KL, Yang P, Yu H, *et al.* SIRT5 promotes IDH2 desuccinylation and G6PD deglutarylation to enhance cellular antioxidant defense. *EMBO Rep* 2016; **17**: 811–22.
40. Rardin MJ, He W, Nishida Y, Newman JC, Carrico C, Danielson SR, Guo A, Gut P, Sahu AK, Li B, Uppala R, Fitch M, *et al.* SIRT5 regulates the mitochondrial lysine succinylome and metabolic networks. *Cell Metab* 2013; **18**: 920–33.
41. Buler M, Aatsinki SM, Izzi V, Uusimaa J, Hakkola J. SIRT5 is under the control of PGC-1α and AMPK and is involved in regulation of mitochondrial energy metabolism. *FASEB J* 2014; **28**: 3225–37.
42. Tan M, Peng C, Anderson KA, Chhoy P, Xie Z, Dai L, Park J, Chen Y, Huang H, Zhang Y, Ro J, Wagner GR, *et al.* Lysine glutarylation is a protein posttranslational modification regulated by SIRT5. *Cell Metab* 2014; **19**: 605–17.
43. Rauh D, Fischer F, Gertz M, Lakshminarasimhan M, Bergbrede T, Aladini F, Kambach C, Becker CF, Zerweck J, Schutkowski M, Steegborn C. An acetylome peptide microarray reveals specificities and deacetylation substrates for all human sirtuin isoforms. *Nat Commun* 2013; **4**: 2327.
44. Nishida Y, Rardin MJ, Carrico C, He W, Sahu AK, Gut P, Najjar R, Fitch M, Hellerstein M, Gibson BW, Verdin E. Sirt5 regulates both cytosolic and mitochondrial protein malonylation with glycolysis as a major target. *Mol Cell* 2015; **59**: 321–32.
45. Du J, Zhou Y, Su X, Yu JJ, Khan S, Jiang H, Kim J, Woo J, Kim JH, Choi BH, He B, Chen W, et al. Sirt5 is a NAD-dependent protein lysine demalonylase and desuccinylase. *Science* 2011; **334**: 806–9.
46. Peng C, Lu Z, Xie Z, Cheng Z, Chen Y, Tan M, Luo H, Zhang Y, He W, Yang K, Zwaans BM, Tishkoff D, *et al.* The first identification of lysine malonylation substrates and its regulatory enzyme. *Mol Cell Proteomics* 2011; **10**: M111.012658.
47. Park J, Chen Y, Tishkoff DX, Peng C, Tan M, Dai L, Xie Z, Zhang Y, Zwaans BM, Skinner ME, Lombard DB, Zhao Y. SIRT5-mediated lysine desuccinylation impacts diverse metabolic pathways. *Mol Cell* 2013; **50**: 919–30.
48. Gibson GE, Xu H, Chen HL, Chen W, Denton TT, Zhang S. Alpha-ketoglutarate dehydrogenase complex-dependent succinylation of proteins in neurons and neuronal cell lines. *J Neurochem* 2015; **134**: 86–96.
49. McKenna MC, Rae CD. A new role for α-ketoglutarate dehydrogenase complex: regulating metabolism through post-translational modification of other enzymes. *J Neurochem* 2015; **134**: 3–6.

50. Polletta L, Vernucci E, Carnevale I, Arcangeli T, Rotili D, Palmerio S, Steegborn C, Nowak T, Schutkowski M, Pellegrini L, Sansone L, Villanova L, *et al.* SIRT5 regulation of ammonia-induced autophagy and mitophagy. *Autophagy* 2015; **11**: 253–70.
51. Reitzer LJ, Wice BM, Kennell D. Evidence that glutamine, not sugar, is the major energy source for cultured HeLa cells. *J Biol Chem* 1979; **254**: 2669–76.
52. Elangovan S, Hsieh TC, Wu JM. Growth inhibition of human MDA-MB-231 breast cancer cells by delta-tocotrienol is associated with loss of cyclin D1/CDK4 expression and accompanying changes in the state of phosphorylation of the retinoblastoma tumor suppressor gene product. *Anticancer Res* 2008; **28**: 2641–7.
53. Hsieh TC, Bennett DJ, Lee YS, Wu E, Wu JM. In silico and biochemical analyses identify quinone reductase 2 as a target of piceatannol. *Curr Med Chem* 2013; **20**: 4195–202.
54. Calamini B, Ratia K, Malkowski MG, Cuendet M, Pezzuto JM, Santarsiero BD, Mesecar AD. Pleiotropic mechanisms facilitated by resveratrol and its metabolites. *Biochem J* 2010; **429**: 273–82.
55. Hsieh TC, Wu JM. Resveratrol: Biological and pharmaceutical properties as anticancer molecule. *Biofactors* 2010; **36**: 360–9.
56. Patel KR, Brown VA, Jones DJ, Britton RG, Hemingway D, Miller AS, West KP, Booth TD, Perloff M, Crowell JA, Brenner DE, Steward WP, et al. Clinical pharmacology of resveratrol and its metabolites in colorectal cancer patients. *Cancer Res* 2010; **70**: 7392–9.
57. Patel KR, Andreadi C, Britton RG, Horner-Glister E, Karmokar A, Sale S, Brown VA, Brenner DE, Singh R, Steward WP, Gescher AJ, Brown K. Sulfate metabolites provide an intracellular pool for resveratrol generation and induce autophagy with senescence. *Sci Transl Med* 2013; **5**: 205ra133.
58. Meng X, Maliakal P, Lu H, Lee MJ, Yang CS. Urinary and plasma levels of resveratrol and quercetin in humans, mice, and rats after ingestion of pure compounds and grape juice. *J Agric Food Chem* 2004; **52**: 935–42.
59. Wenzel E, Soldo T, Erbersdobler H, Somoza V. Bioactivity and metabolism of trans-resveratrol orally administered to Wistar rats. *Mol Nutr Food Res* 2005; **49**: 482–94.
60. Bode LM, Bunzel D, Huch M, Cho GS, Ruhland D, Bunzel M, Bub A, Franz CM, Kulling SE. In vivo and in vitro metabolism of trans-resveratrol by human gut microbiota. *Am J Clin Nutr* 2013; **97**: 295–309.
61. Rabassa M, Zamora-Ros R, Urpi-Sarda M, Andres-Lacueva C. Resveratrol metabolite profiling in clinical nutrition research–from diet to uncovering disease risk biomarkers: epidemiological evidence. *Ann N Y Acad Sci* 2015; **1348**: 107–15.
62. Zhou J, Zhou M, Yang FF, Liu CY, Pan RL, Chang Q, Liu XM, Liao YH. Involvement of the inhibition of intestinal glucuronidation in enhancing the oral bioavailability of resveratrol by labrasol containing nanoemulsions. *Mol Pharm* 2015; **12**: 1084–95.

63. Hoshino J, Park EJ, Kondratyuk TP, Marler L, Pezzuto JM, van Breemen RB, Mo S, Li Y, Cushman M. Selective synthesis and biological evaluation of sulfate-conjugated resveratrol metabolites. *J Med Chem* 2010; **53**: 5033–43.
64. Pettersen EF, Goddard TD, Huang CC, Couch GS, Greenblatt DM, Meng EC, Ferrin TE. UCSF Chimera–a visualization system for exploratory research and analysis. *Journal of computational chemistry* 2004; **25**: 1605–12.
65. Sun B, Hoshino J, Jermihov K, Marler L, Pezzuto JM, Mesecar AD, Cushman M. Design, synthesis, and biological evaluation of resveratrol analogues as aromatase and quinone reductase 2 inhibitors for chemoprevention of cancer. *Bioorg Med Chem* 2010; **18**: 5352–66.
66. Lustberg MB, Pant S, Ruppert AS, Shen T, Wei Y, Chen L, Brenner L, Shiels D, Jensen RR, Berger M, Mrozek E, Ramaswamy B, et al. Phase I/II trial of non-cytotoxic suramin in combination with weekly paclitaxel in metastatic breast cancer treated with prior taxanes. *Cancer Chemother Pharmacol* 2012; **70**: 49–56.
67. Villalba JM, Alcaín FJ. Sirtuin activators and inhibitors. *Biofactors* 2012; **38**: 349–59.
68. Balcerczyk A, Pirola L. Therapeutic potential of activators and inhibitors of sirtuins. *Biofactors* 2010; **36**: 383–93.
69. Mazué F, Colin D, Gobbo J, Wegner M, Rescifina A, Spatafora C, Fasseur D, Delmas D, Meunier P, Tringali C, Latruffe N. Structural determinants of resveratrol for cell proliferation inhibition potency: experimental and docking studies of new analogs. *Eur J Med Chem* 2010; **45**: 2972–80.
70. Kondratyuk TP, Park EJ, Marler LE, Ahn S, Yuan Y, Choi Y, Yu R, van Breemen RB, Sun B, Hoshino J, Cushman M, Jermihov KC, et al. Resveratrol derivatives as promising chemopreventive agents with improved potency and selectivity. *Mol Nutr Food Res* 2011; **55**: 1249–65.
71. Oskarsson A, Spatafora C, Tringali C, Andersson Å. Inhibition of CYP17A1 activity by resveratrol, piceatannol, and synthetic resveratrol analogs. *Prostate* 2014; **74**: 839–51.
72. Lundqvist J, Tringali C, Oskarsson A. Resveratrol, piceatannol and analogs inhibit activation of both wild-type and T877A mutant androgen receptor. *J Steroid Biochem Mol Biol* 2017; **174**: 161–8.
73. Yang H, Baur JA, Chen A, Miller C, Adams JK, Kisielewski A, Howitz KT, Zipkin RE, Sinclair DA. Design and synthesis of compounds that extend yeast replicative lifespan. *Aging Cell* 2007; **6**: 35–43.
74. Xiang Y, Stine ZE, Xia J, Lu Y, O'Connor RS, Altman BJ, Hsieh AL, Gouw AM, Thomas AG, Gao P, Sun L, Song L, *et al.* Targeted inhibition of tumor-specific glutaminase diminishes cell-autonomous tumorigenesis. *J Clin Invest* 2015; **125**: 2293–306.
75. Gao P, Tchernyshyov I, Chang TC, Lee YS, Kita K, Ochi T, Zeller KI, De Marzo AM, Van Eyk JE, Mendell JT, Dang CV. c-Myc suppression of miR-23a/b enhances mitochondrial glutaminase expression and glutamine metabolism. *Nature* 2009; **458**: 762–5.

76. Wise DR, DeBerardinis RJ, Mancuso A, Sayed N, Zhang XY, Pfeiffer HK, Nissim I, Daikhin E, Yudkoff M, McMahon SB, Thompson CB. Myc regulates a transcriptional program that stimulates mitochondrial glutaminolysis and leads to glutamine addiction. *Proc Natl Acad Sci U S A* 2008; **105**: 18782–7.
77. Azevedo C, Correia-Branco A, Araújo JR, Guimarães JT, Keating E, Martel F. The chemopreventive effect of the dietary compound kaempferol on the MCF-7 human breast cancer cell line is dependent on inhibition of glucose cellular uptake. *Nutr Cancer* 2015; **67**: 504–13.
78. León D, Uribe E, Zambrano A, Salas M. Implications of resveratrol on glucose uptake and metabolism. *Molecules* 2017; **22**.

Chapter 7

Resveratrol and Cancer Cell Biology

Chandra K. Singh, Gagan Chhabra and Nihal Ahmad

Department of Dermatology, University of Wisconsin, Madison, WI, USA
William S. Middleton VA Medical Center, Madison, WI, USA

Recent advances in cancer biology have resulted in a near-complete information regarding the major pathways involved in cancer development and progression. This has resulted in the identification of newer drug-targets and designing of novel drugs for cancer management. However, cancer still remains a global challenge affecting millions of people worldwide. Therefore a quest for an effective cure for cancer continues. The grape antioxidant resveratrol has shown considerable promise towards cancer management, due to its pleiotropic mechanistic ability in affecting multiple cancer-related pathways. In this chapter, we have discussed the mechanistic action(s) of resveratrol against the hallmark characteristics of cancer. Studies have shown how resveratrol is capable of challenging cancer cells against independence in growth signals, promoting sensitivity to anti-growth signals, promoting apoptosis, limiting replicative potential, discontinuing angiogenesis, inhibiting invasion and metastasis, inhibiting metabolic reprogramming and promoting immune destruction. Resveratrol's actions against all these processes make it very promising in cancer management.

7.1. Introduction

Cancer is not a single disease, but a conglomeration of a whole family of diseases sharing a common characteristic of having aberrant uncontrolled cells with seriously dysregulated normal cell growth and division cycle. This makes cancer cells independent, motile and ultimately capable of traveling a long distance to other organs. Continuous and expanding research efforts are ongoing to unravel why a normal cell abruptly rebels against the programmed regulation, dividing frantically, corrupting neighbor cells and eventually compromising the functionality of the organ and collapsing the whole body. As per the recent estimate of American Cancer Society, 1,735,350 new cases of cancer are expected to be diagnosed in 2018, and 609,640 cancer-related deaths are expected this year in the United States. The economic burden of cancer is too high for the suffering person as well as for the whole society.[1] Since cancer is a group of diseases, a single drug or agent is not expected to be successful in curing cancer. Intense ongoing research has resulted into the development of a number of new therapies, including targeted drugs, biologics, radiation modalities, and surgical options. Similarly, naturally occurring and dietary agents have also made to the list of modalities that have shown considerable promise in cancer management. Interestingly, resveratrol (3,5,4′-trihydroxystilbene), an antioxidant found in red grapes, nuts and berries, and other botanical sources, has shown promise in the management of almost all types of cancer. This is clearly evident from a PubMed search with keyword 'resveratrol and cancer' yielding ∼2900 publications at the beginning of the year 2018.

Resveratrol was first isolated from the roots of white hellebore in 1940, and subsequently in 1963 from Japanese Knotweed (*Polygonum cuspidatum*), which is currently the primary commercial source of resveratrol.[2] Resveratrol started receiving enormous attention in 1997 when a group of researchers led by Dr. John Pezzuto at the University of Illinois, Chicago, reported the anti-initiation, anti-promotion, and anti-progression activities of resveratrol in multiple cancer models.[3] Following this study, a plethora of evidence suggested that resveratrol could be useful in cancer management. Moreover, resveratrol has been shown to be beneficial for a number of other health conditions including aging, cardiovascular disease and metabolic conditions.[4–7] This chapter is devoted to discussing how resveratrol modulates the cellular processes and biology to be effective against cancer.

7.2. Resveratrol and cancer cell biology

Cancer development and progression happens in a series of steps starting from hyperplasia, which appears normal but too many cells resulting due to some loss of control of cell division. Dysplasia, as the next step, is characterized by abnormal changes in the growth and differentiation of cells. Further abnormality allows cells to spread over the affected tissue and lose their original function. These cells now termed as anaplastic are not invasive, as they remain at their original location (in situ), but possess the characteristics to be potentially malignant, if they continue to grow. In the final step, the cancer cells invade neighboring tissue and spread into the bloodstream and other distant sites/organs. The cancer types are categorized into five groups based on the cell type that was primarily malformed. These are defined as carcinomas (result from altered epithelial cells, which cover the exterior of the skin and internal organs), sarcomas (result from changes in mesenchymal cells e.g. muscle, bone, fat, or connective tissue), leukemia (results from malignant white blood cells), lymphoma (cancer of the lymphatic system) and myelomas (cancers of specialized white blood cells that make antibodies). As described by Hanahan and Weinberg in two milestone articles, cancer cells possess eight hallmark characteristics, which keep the sequence of oncogenic events in action for the development and progression of cancer. These are defined as insensitivity to anti-growth signals, independence in growth signals, escaping apoptosis, unlimited replicative potential, continued angiogenesis, activating invasion and metastasis, metabolic reprogramming, and evading immune destruction.[8,9] Here, we have discussed the studies describing the molecular action(s) of resveratrol against the hallmark characteristics of cancers.

7.2.1. *Molecular actions of resveratrol modulating cancer cell independence in growth signals*

The cancer cells are known to modulate the normal cell signaling machinery and tissue microenvironment to gain independence. This happens due to alterations in the components of signaling circuitry that senses and processes the signals released by ligand-activated growth signal receptors and transmembrane receptors called as integrins. Integrins facilitate cell to extracellular matrix (ECM) adhesion and play a key role in cell signaling pathways of transmembrane protein kinases such as receptor tyrosine kinases (RTKs). As a consequence, cells may experience a dysregulated cell growth, division, survival, cellular differentiation, and apoptosis. The SOS-RAS-RAF-MAPK cascade plays a central role in this process.[8,9] Son of Sevenless (SOS) encodes guanine nucleotide exchange factors, which act on the RAS subfamily of small GTPases. RAS family members (HRAS, KRAS and NRAS) are known to regulate several downstream pathways involved in cancer, including the mitogen-activated protein kinases (MAPKs) and phosphatidylinositol 3-kinase (PI3K) signaling pathways.[10] Mutant RAS has been identified as the most common gene in human cancers; with ~25% in almost all and up to 90% in certain types of cancer such as pancreatic cancer.[11] Similarly, RAF family members (A-RAF, B-RAF and C-RAF) are serine/threonine-protein kinases known as bona fide effectors of RAS. Approximately, 40% cases of melanomas are known to possess B-RAF mutation, resulting in constitutive signaling of MAPK pathway.[12]

As cancer cells take over normal growth factor response pathways to become self-sufficient, it's logical to target these delinquent pathways precisely. Interestingly, resveratrol has been shown to modulate several of these targets. We have elected to discuss only a few examples showing the effects of resveratrol on molecules/pathways known to affect the independence of cancer cells from normal growth signaling. In a promising study, Saud and colleagues have demonstrated the effects of resveratrol against colorectal cancer (CRC) in a genetically engineered mouse model knocked out for adenomatous polyposis coli (APC) for sporadic CRC where KRAS is activated specifically in the distal colon.[13] The resveratrol supplementation, in a chemoprevention protocol, was found with ~60% inhibition in tumor development. Fascinatingly, in the remaining mice that developed tumors, KRAS expression was found to be lost. Further, in an intervention protocol, where resveratrol was supplemented after tumor development, a complete remission in 33% of the mice and a 97% decrease in tumor size in the remaining mice were observed. Furthermore, a micro-RNA (miRNA) analysis showed an increased expression of miR-96, which is known to regulate KRAS translation, suggesting that anti-cancer effect of resveratrol is mediated by downregulating KRAS expression.[13] Dietary resveratrol has also been shown to inhibit the self-renewal capacity of pancreatic cancer stem cells (CSCs) derived from human primary tumors and KRAS(G12D) transgenic mice.[14] In this study, resveratrol was found to inhibit the tumor growth and development of PanIN lesions in KRAS(G12D) mice, suggesting a promising potential of resveratrol in the management of pancreatic cancer and possibly other cancers.[14] Another study showed that resveratrol suppressed AKT phosphorylation in Vemurafenib-resistant human metastatic melanoma cells and overcame the resistance of the BRAF inhibitor Vemurafenib, used in the treatment of late-stage melanomas with BRAF mutation.[15] Resveratrol-induced anti-cancer effects have also been shown to be modulated via alterations in MAPK pathway. Interestingly, resveratrol has been shown to activate MAPK in some cells[16,17] and inhibit in others.[18,19] Resveratrol has been shown to inhibit cell proliferation and induce apoptosis in Waldenstrom's macroglobulinemia cells and overcome the growth-promoting activities of the bone marrow microenvironment, which plays a key role in plasma cell malignancies. These anti-proliferative effects of resveratrol were suggested to be mediated via the down-regulation of MAPK, AKT and WNT signaling pathways.[19] The molecular actions of resveratrol modulating cancer cell independence in growth signals have been illustrated in Figure 7.1.

7.2.2. *Molecular actions of resveratrol modulating cancer cell insensitivity to anti-growth signals*

In the event of any damage to normal cell proliferation machinery, the anti-growth signals are evolved to block cell proliferation by pushing the cells in (i) a quiescent (G0) state from where they may return back when the damage is repaired and the extracellular signals permit, or (ii) permanently abandon their proliferative potential via entering into post-mitotic states. At the molecular level, the anti-growth signals are funneled through a tumor suppressor

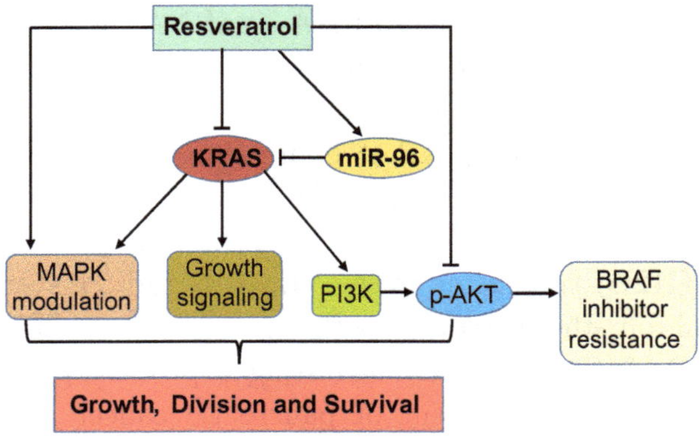

Fig. 7.1. Molecular actions of resveratrol modulating cancer cell independence in growth signals.

protein known as the retinoblastoma protein (pRb) that is encoded by the retinoblastoma gene (Rb) that was identified in 1987.[20] Retinoblastoma binds to and inactivates E2F transcription factors, which regulate cell cycle and DNA replication, including the Cyclins and CDKs specific to G1 and S phases of the cell cycle. Disruption of the retinoblastoma pathway releases E2F transcription factors, resulting increased cell division and cells become insensitive to anti-growth signals.[8,9] Tumor growth factor-beta (TGF-β) is one of the well acknowledged anti-growth molecule preventing the phosphorylation of retinoblastoma. However, in late-stage tumors, instead of working as an anti-growth signal, TGF-β activates epithelial-to-mesenchymal-transition (EMT) to empower cancer cells with stem-cell-like abilities.[8,9,21]

The first evidence that pRb-E2F/DP pathway is a contributor to resveratrol-mediated cell cycle arrest and apoptosis, came from a study reported by Adhami and colleagues.[22] In this study, it was demonstrated that resveratrol decreases hyperphosphorylated pRb with a relative increase in hypophosphorylated pRb, which was accompanied by downregulation of all five E2F (1–5) family members of transcription factors and their heterodimeric partners DP1 and DP2. Another separate study showed that resveratrol imparts cancer chemopreventive effects via modulating cyclin-dependent kinase

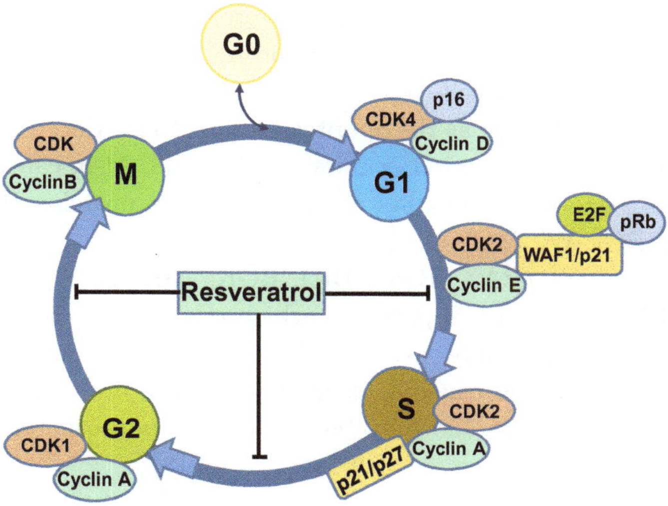

Fig. 7.2. Molecular actions of resveratrol modulating cancer cell insensitivity to anti-growth signals.

(CDK) inhibitor-cyclin-CDK machinery, resulting in a G1-phase cell arrest followed by apoptosis in human epidermoid carcinoma (A431) cells.[23] In this study, resveratrol was found to induce WAF1/p21 and decrease in the levels of cell cycle proteins (cyclin D1, cyclin D2, cyclin E, CDK2, CDK4 and CDK6), resulting in an inhibition of cyclin D1/D2-CDK6, cyclin D1/D2-CDK4, and cyclin E-CDK2 complexes, leading to a G0/G1 phase cell cycle and subsequent apoptosis.[23] Resveratrol has also been shown to inhibit TGF-β2 in squamous cell carcinoma (SCC)/skin in P53+/−/SKH-1 mice.[24] In this study, resveratrol was found to inhibit invasiveness in human A431 SCC cells possibly via AKT-mediated downregulation of TGF-β2.[24] The molecular actions of resveratrol modulating cancer cell insensitivity to anti-growth signals have been illustrated in Figure 7.2.

7.2.3. *Molecular actions of resveratrol modulating cancer cell escaping apoptosis*

All the cells possess a machinery for a programmed self-destruction, in the event of damages to normal cell signaling. This programmed self-destruction of cells is known as apoptosis, which is an orchestrated process controlled by a set of proteins. The most common method used by cancer cells to escape death is the loss of tumor suppressor P53, also known as the apoptosis gatekeeper. More than 50% of all cancers have a mutated or missing P53 gene. Also, cancer cells can produce excessive amounts of anti-apoptotic proteins (BCL-2, BCL-XL, BCL-W, MCL-1, A1) and lower amounts of the pro-apoptotic proteins (BAX, BAD, BID, BOK, BIK, BAK).[8,9] Additionally, PI3K–AKT/PKB pathway, which transmits anti-apoptotic survival signals, is likely involved in mitigating apoptosis in a substantial fraction in cancers. In addition, there are a number of caspases that have been shown to be involved in apoptotic regulation of cells.

Based on a number of studies, resveratrol has been shown to induce apoptosis of cancer cells through P53-dependent as well as P53-independent pathways. For example, Ferraz da Costa *et al.* have shown that resveratrol affects the viability of cancer cell lines expressing constitutively wild-type P53 with greater impact than the one carrying only a partial fragment of the P53 gene.[25] Moreover, resveratrol-induced cytotoxicity in constitutively active wild-type P53 was partly mediated by P53 and involved caspases 9 and 7 activations and PARP cleavage; while limited cell death observed in H1299 non-small lung cancer cells, which possess partially deleted P53, was not accompanied by caspase activation. Interestingly, a transient transfection of a wild-type P53 in H1299 cells was found to make it more responsive to the pro-apoptotic properties of resveratrol, comparable to P53-positive MCF-7 cells.[25] Resveratrol has also been shown to downregulate ΔNp63, which is N-terminal truncated isoform of P63, and known to overexpress in epithelial cancers.[26] Interestingly, in this study resveratrol was found to upregulate P53 as well, however, siRNA mediated P53 suppression did not inhibit resveratrol-induced apoptosis, suggesting P53-independent pathway at play in human nasopharyngeal carcinoma.[26] Resveratrol has been shown to cause activation and nuclear translocation of ERK1/2, resulting phosphorylation of Ser-15 of mutant P53 in DU145 prostate cancer cells. This led to the restoration of wild-type P53

DNA binding activity and apoptosis induction.[27] Cyclooxygenase-2 (COX-2), which is known for its association with tumor cell growth and metastasis, enhanced cellular adhesion and inhibition of apoptosis, plays an important role in P53-dependent apoptosis in cancer cells. Tang *et al.* have found that resveratrol induces COX-2, which facilitates P53-dependent apoptosis in human breast cancer cells.[28,29] Resveratrol has also been shown to downregulate PI3K/AKT/mTOR signaling pathways in mediating apoptosis in glioma cells,[30] suggesting the pleiotropic mechanism of resveratrol against cancer cells. The molecular actions of resveratrol modulating cancer cell escaping apoptosis have been illustrated in Figure 7.3.

7.2.4. *Molecular actions of resveratrol controlling unlimited replicative potential of cancer cells*

Another defining feature of cancer cells is its ability to divide endlessly, without exhaustion, which they achieve by maintaining

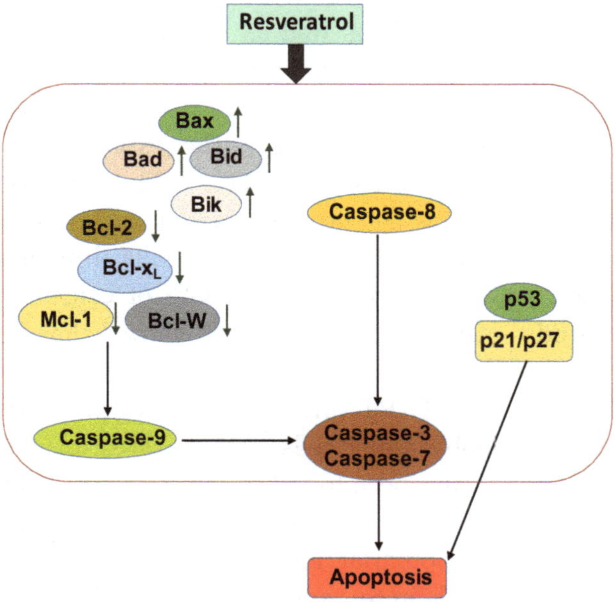

Fig. 7.3. Molecular actions of resveratrol modulating cancer cell escaping apoptosis.

the cellular timekeeper, the telomere. In 90% of cases, they do so by increasing the production of the enzyme telomerase, which functions by adding telomeric DNA to the ends of chromosomes. Many cancer-causing proteins (oncoproteins) are able to activate the production of telomerase, while many tumor suppressor proteins such as P53 can trigger the pathways to inhibit the production of telomerase.[8,9] Additional clues to the importance of telomere maintenance for cancer come from the analysis of mice lacking telomerase function. For example, mice carrying a homozygous knockout of the cell cycle inhibitor $p16^{INK4A}$ are tumor prone, particularly when exposed to carcinogens; the tumors that arise, in these mice, show comparatively elevated telomerase activity.[31] In addition to inducing apoptosis and inhibiting proliferation of cancer cells, the anti-proliferative action of resveratrol is also shown to be associated with the inhibition of telomerase. For example, one of the major mechanisms resveratrol uses to elicit its response is via activation of tumor suppressor P53 protein, which once activated can trigger the transcription of precise target genes known to be involved in growth arrest, senescence, cell differentiation, apoptosis as well as telomere erosion (reviewed in[32]). Zhai *et al.* have shown that resveratrol's anti-proliferative effect against A431 cells was associated with resveratrol's ability to inhibit telomerase and to decrease the expression of human telomerase reverse transcriptase (hTERT).[33] The catalytic subunit of telomerase, hTERT, is intently associated with the activation of telomerase. hTERT mRNA is a potentially immortalization-specific marker, which is expressed in ~85% of cancer cells. In a very recent study, Mirzazadeh *et al.* have shown that resveratrol effectively inhibits cell growth and causes cell death as well as dramatically decreases hTERT mRNA leading to prevention of cell division in human glioblastoma cells.[34] Resveratrol in combination with pterostilbene has also been found to suppress both hTERT gene and its enzymatic activity in triple-negative breast cancer cell lines, suggesting one of the potential mechanisms of resveratrol's over-powering cancer cells indefinite replicative capacity.[35] The molecular network of resveratrol controlling unlimited replicative potential of cancer cells is depicted in Figure 7.4.

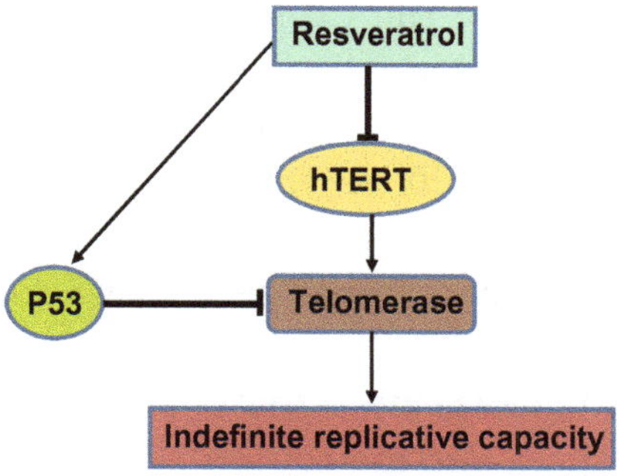

Fig. 7.4. Molecular actions of resveratrol controlling unlimited replicative potential of cancer cells.

7.2.5. *Molecular actions of resveratrol modulating angiogenesis*

A hallmark feature of cancer and its distant organ metastasis is new growth in vascular network, since the cancer cells cannot thrive without an adequate supply of oxygen and nutrients. The process of formation of the new blood vessels is called 'angiogenesis'. Without the additional blood supplied by angiogenesis, tumors cannot grow more than 0.5 mm and also cannot spread, or metastasize, to new tissues. Angiogenesis is regulated by a number of different proteins, which serve as activator or inhibitor molecules. Certain oncogenes in tumor cells are known to cause an increased expression of genes that make angiogenic factors. There are approximately 15 angiogenic factors whose production are regulated by a variety of oncogenes.[8,9,36] The primary angiogenic growth factor, vascular endothelial growth factor (VEGF) binds to VEGF receptors on the surface of endothelial cells, and breakthrough blood vessels, to further grow and form new blood vessels. The angiogenesis-initiating signals are characterized by VEGF and acidic and basic fibroblast growth factors (FGF1/2).[36] The speedy growth of a tumor results in deprivation of an adequate oxygen supply, which is generally experienced by cells found towards the inside of the tumor. These hypoxic cells activate the hypoxia stress response, which is mediated at the cellular level by a transcription factor called hypoxia-inducible factor (HIF). During hypoxia, HIF accumulates and transports into the nucleus to induce the expression of several target genes, including VEGF.[8,9] The activation of oncoproteins such as RAS results in an increase in VEGF production. Cancer cells also stop producing angiogenesis inhibitors like thrombospondin-1 (TSP1), which is under the positive control of P53. The loss of P53 found in more than half of all cancers, has been shown to cause a decrease in TSP1 levels.

Resveratrol's anti-angiogenic activity has been well documented in several model systems. Resveratrol-mediated suppression of glioma growth has been shown to correlate with the inhibition of macroscopic and microscopic angiogenesis.[37] Resveratrol suppresses multiple myeloma angiogenesis by modulating the levels of VEGF, basic FGF, MMP-2 and MMP-9.[38] Resveratrol has been found to inhibit angiogenesis in human breast cancer xenografts *in vivo*, in addition to lowering tumor growth, and increasing apoptotic index.[39] The anti-angiogenic effect of resveratrol has been shown to be mostly mediated via inhibition of hypoxia-induced HIF-1α accumulation and down-regulation of VEGF.[40,41] Wu *et al.* have shown that resveratrol inhibits the migration, adhesion, invasion as well as secretion of MMP-2 and MMP-9 in a highly metastatic Lovo colon carcinoma cell line, cultured under normoxia and hypoxia.[42] Trapp *et al.* have shown that resveratrol stimulates vascular endothelial cell (VEC) growth in monolayer culture but causes a decrease in VEC survival in the three-dimensional tumor co-culture of melanoma-endothelial cells. This happens as a result of shifting the ratio between VEGF expression and matrix TSP1, which is a natural inhibitor of angiogenesis and downstream target of P53 as well.[43,44] The molecular network of resveratrol modulating angiogenesis is illustrated in Figure 7.5.

7.2.6. *Molecular actions of resveratrol modulating invasion and metastasis*

The cancer cell ability to invade and metastasize is the most complicated and a key hallmark of cancer progression. The beginning of metastasis includes invasion, which is facilitated by EMT. Normally, cells migrate during embryonic development and in response to inflammation, where they are recruited to specific sites and these cells

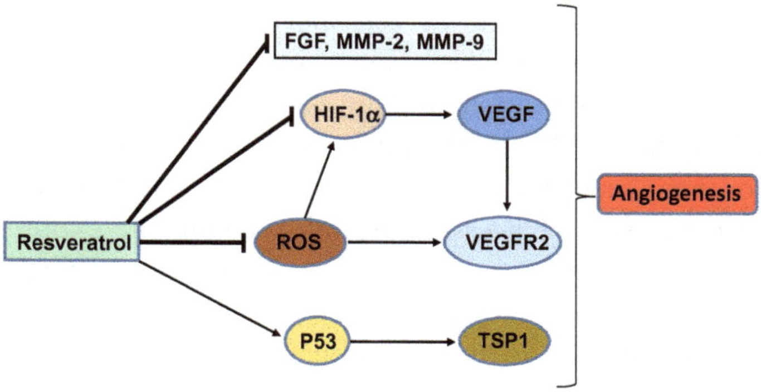

Fig. 7.5. Molecular actions of resveratrol modulating angiogenesis.

can revert back through mesenchymal to epithelial (MET) transition to form epithelial tissues at distant localities. Cancer cells hijack this process to invade and metastasize at distant sites.[8,9] EMT is a critical phenomenon through which epithelial cells lose their cell-cell adherence and gain mesenchymal characteristics, leading to enhanced migration potential and invasiveness of cancer. Immunoglobulins and cadherins facilitate cell-to-cell adhesions, integrins link cells to the ECM, and E-cadherin cement cells to each other. The linking of cells by E-cadherin results in the transmission of anti-growth signals; causing contact inhibition. Predictably, E-cadherin function is lost in migrating cancer cells. On the other hand, N-cadherin function increases in migrating cancer cells to help it to slip through blood vessels. As discussed before, TGF-β acts as an anti-growth factor in the early stages of cancer, however, it enhances malignancy by promoting invasion and metastasis via EMT activation. Indeed, the addition of TGF-β to epithelial cells in culture is a well-established method for inducing the EMT in *in vitro* studies.

Resveratrol has been shown to affect EMT in various cancer types. Resveratrol modulates the expression of EMT-related markers, such as E-cadherin, N-cadherin, vimentin, and MMPs. These EMT-related markers are important for cancer cell motility, invasiveness and metastasis during tumorigenesis. In pancreatic cancer cells, resveratrol has been shown to reduce EMT by inhibiting the levels of phospho-AKT and phospho-NF-κB. Moreover, resveratrol reverses

TGF-β-induced modulations in cell morphology and invasiveness that are characteristic of EMT. Thus, resveratrol suppresses pancreatic cancer cell migration and invasion through the inhibition of the PI3K/AKT/NF-κB signaling pathway.[45] In lung cancer, resveratrol has been found to inhibit TGF-β-induced EMT by enhancing expression of E-cadherin and reducing the expression of vimentin and fibronectin as well as transcription factors SNAIL and SLUG.[46] In pancreatic cancer, resveratrol has been shown to inhibit migration and invasion of cancer stem cells (CSCs) by inhibiting transcriptional regulators ZEB-1, SLUG and SNAIL, which are known to enhance EMT.[14] In breast cancer, resveratrol represses epidermal growth factor (EGF)-induced ERK activation. EGF is reported to induce EMT via extracellular signal-regulated kinase 1/2 (ERK1/2) signaling pathway.[47] Resveratrol has also been shown to inhibit migration and invasion of colorectal cancer by inhibition of integrin and focal adhesion kinase (FAK).[48] Resveratrol treatment of oral squamous cell carcinoma cells showed decreased cell migration, invasion, and EMT-inducing transcription factor.[49] Resveratrol has been found to decrease the hypoxia-induced production of ROS, invasion and migration of pancreatic cancer cells.[50] Another study showed that resveratrol inhibits the hedgehog signaling pathway and gastric cancer invasion and metastasis by inhibiting EMT.[51] The molecular network of resveratrol in modulating invasion and metastasis is illustrated in Figure 7.6.

7.2.7. *Molecular actions of resveratrol modulating metabolic reprogramming*

Metabolism, within a cell, is a crucial process that is comprised of numerous chemical reactions, which produce or consume energy. The major purposes of cellular metabolism processes include, (i) the conversion of food/fuel to energy to run cellular processes, (ii) to synthesize macromolecule, including proteins, lipids, nucleic acids, and carbohydrates, and (iii) the elimination of nitrogenous wastes. The cancer cells consume more glucose than normal cells to speedily produce ATP and metabolic precursors. As discovered by Otto Warburg,[52,53] cancer cells can reprogram their glucose metabolism, and thus their energy production, by limiting their

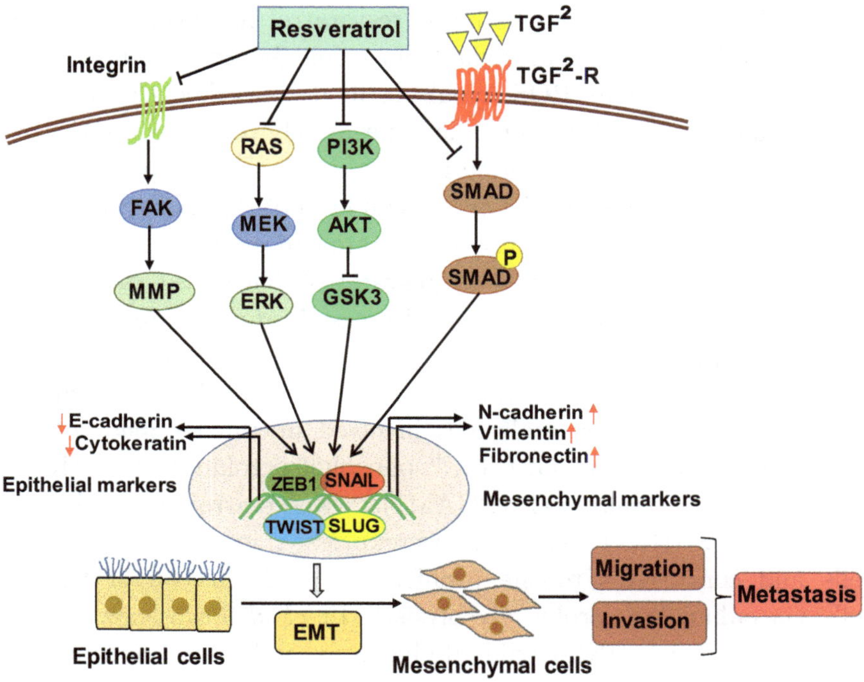

Fig. 7.6. Molecular actions of resveratrol modulating invasion and metastasis.

energy metabolism largely to glycolysis, even in the presence of oxygen, leading to a state called as "aerobic glycolysis". They do so in part by upregulating glucose transporters, notably GLUT1, which substantially increases glucose import into the cytoplasm.[54] Hypoxia-inducible factor (HIF), which is initially activated in response to low oxygen levels in tumors as well as a variety of triggers, such as radiation-induced DNA damage, signaling from growth factors and the presence of pyruvate, plays a central regulatory role in metabolic reprogramming of cancer cells.[9] Similarly, phosphoinositide 3 kinase (PI3K), which is also activated by a variety of survival signals, mediates cell growth and survival, inhibits apoptosis, activates angiogenesis and notably, activates aerobic glycolysis. PI3K also activates AKT, which then activates the protein mTOR.[9] Resveratrol is known to inhibit PI3K/AKT/mTOR pathway in many types of cancers.[30,55] Activated mTOR also activates HIF, further supporting HIF's efforts to switch cancer cell metabolism from normal respiration to aerobic glycolysis.[9] Interestingly, resveratrol has been shown to exert its anti-cancer activities via suppressing glycolytic metabolism, inhibiting intracellular ROS, resulting in decreased levels of HIF-1α accumulation.[56] PI3K signaling is negatively regulated by tumor suppressor PTEN, which is often mutated in cancer. Resveratrol has also been found to restore the expression level of PTEN by targeting oncomiRs of the miR-17 family in prostate cancer.[57]

The anti-cancer activities of resveratrol have also been suggested to occur potentially via decreased glucose uptake and lactate production, as well as via targeting numerous metabolic enzymes and signaling pathways. Resveratrol has been shown to negatively regulate key molecules involved in glucose metabolism e.g. glucose transporter GLUT1,[56] phosphofructokinase-1 (PFK1), phosphoglycerate mutase-1 (PGAM1).[55] Resveratrol also inhibits Hexokinase 2 (HK2), which is overexpressed in most neoplastic cells and contributes to the proliferation and survival of tumor cells via enhancing aerobic glycolysis.[58] Li *et al.* have shown that resveratrol impairs HK2-mediated glycolysis, and inhibit anchorage-dependent and -independent growth of human non-small cell lung cancer (NSCLC) cells. The effect was also related to decreased EGFR and downstream kinases AKT and ERK1/2.[58] In a study by Dai *et al.* resveratrol was shown to induce apoptosis in HCC cells addicted to aerobic glycolysis by reducing HK2, and inhibit tumor growth in mice.[59] In a very recent study, resveratrol was shown to decrease glycolysis as well as pentose phosphate activity reverting metabolic reprogramming in colon cancer cells.[60] Moreover, resveratrol was found to target the pyruvate dehydrogenase (PDH) complex, a key mitochondrial gatekeeper of energy metabolism, as well as to increase oxidative phosphorylation partly via Ca2+ Calmodulin kinase kinase B (CamKKB)/fuel-sensing AMP-activated kinase (AMPK) pathway. Interestingly, in this study, which was done at a concentration as low as 10 μM, no significant decrease in cell viability or an increase of apoptosis was noticed. Still, resveratrol showed the capacity to revert a central feature of cancer cells, metabolic reprogramming.[60] The molecular actions of resveratrol in metabolic reprogramming are depicted in Figure 7.7.

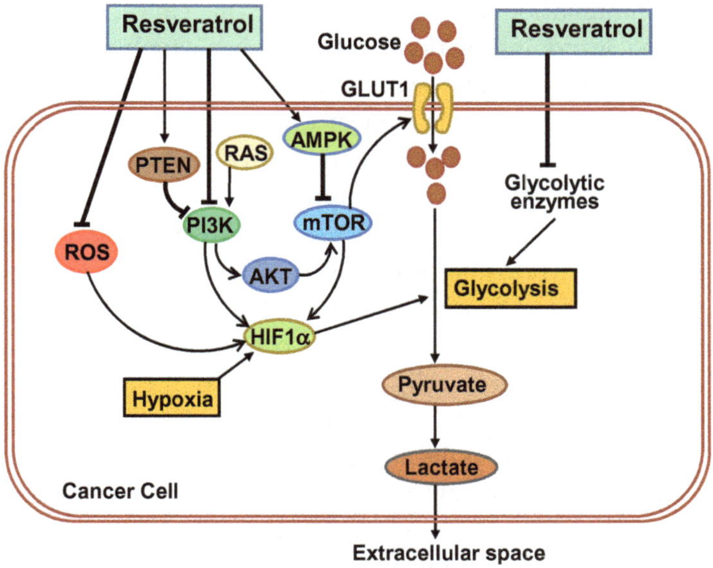

Fig. 7.7. Molecular actions of resveratrol modulating metabolic reprogramming.

7.2.8. *Molecular actions of resveratrol controlling tumor-promoting inflammation and evading immune destruction*

Inflammation is the normal response of body's immune system to damages and injuries. Both inflammation and immune surveillance have been shown to play decisive roles at all the different stages of tumorigenesis. White blood cells, known to protect the body against foreign invaders/infectious diseases, have been seen in tumors for many decades, providing an initial clue that inflammation is associated with cancer. Nuclear factor kappa B (NF-κB), is known as the first responder to inflammation. NF-κB is maintained in an inactive state by IκBα, which is regulated by an IκB kinase. IκBα binds to NF-κB and prevents it from moving into the nucleus of the cell. NF-κB is activated when IκB kinase adds a phosphate group to IκBα to degrade it. NF-κB when free move into the nucleus to trigger wide-ranging genes that allow cell growth promotion, survival, migration and the materialization of new blood vessels. Despite causing increased inflammation and angiogenesis, the cancer cells appear to be able to avoid interaction and/or being capable of disabling components of the immune system that have been dispatched to eliminate them.[9] For example, cancer cells may paralyze infiltrating cytotoxic T lymphocytes (CTL) and natural killer (NK) cells, by secreting TGF-β or other immunosuppressive factors.[61]

In a very recent study, Silva *et al.* have shown that resveratrol decreases surface expression of CD83 and CD86, which are markers for mature dendritic cells (DC) and considered as promising targets for provoking tolerance in inflammatory conditions.[62] Dendritic cells are a functionally diverse cell type known to process antigen to present it on the cell surface to the T cells. In the same study, resveratrol was found to inhibit the effects of TNF-α exposure on dendritic cell phenotype, cytokine secretion, and their capability to stimulate T cells and recruit mesenchymal stem/stromal cells (MSC) through ECM components.[62] Lee-Chang *et al.* have shown that resveratrol efficiently inhibits lung metastasis in mice by inactivating the phosphorylation and acetylation of signal transducer and activator of transcription 3 (STAT3) and disabling tumor-evoked regulatory B cells (tBregs)-induced conversion of Foxp3+ regulatory T cells (Tregs).[63] tBregs are known to play a critical role in breast cancer lung metastasis via TGF-β–dependent conversion of metastasis-promoting Foxp3+ regulatory T cells.[64] The study suggests that resveratrol possesses the ability to control cancer escape via blockage of the tBreg–Treg axis to elicit anti-tumor effector responses.[63] Resveratrol has also been shown to inhibit STAT3 and NF-κB, which are major inflammatory mediators known to play a critical role in pancreatic carcinogenesis.[65] The anti-inflammatory effects of resveratrol have also been related to the inhibition of lipopolysaccharide-induced activation of NF-κB in Caco-2 and SW480 human colon cancer cells.[66] Here resveratrol was found to inhibit the phosphorylation and degradation of the IκB, which are an essential step in the translocation of NF-κB p65 subunits in the nucleus.[66] The molecular network of resveratrol in the regulation of tumor-promoting inflammation and evading immune destruction is shown in Figure 7.8.

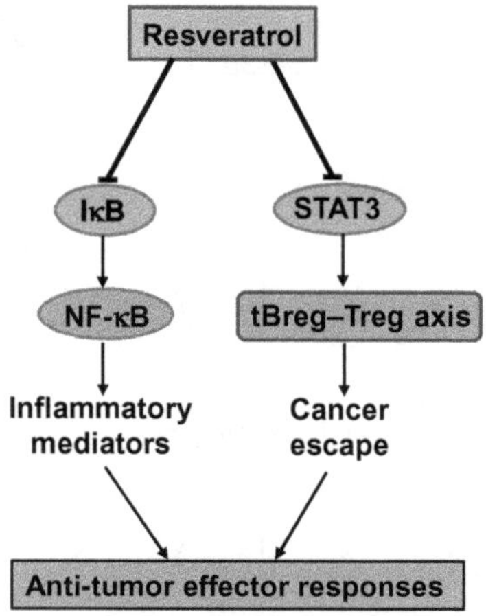

Fig. 7.8. Molecular actions of resveratrol controlling tumor-promoting inflammation and evading immune destruction.

7.3. Concluding remarks

As discussed above, resveratrol has been shown to inhibit key molecular events associated with tumor initiation, promotion, and progression, and to be capable of modulating all the hallmark characteristics of cancer. This is a major reason why resveratrol has been shown to have promise against all the major cancers. This is also illustrated in Figure 7.9 showing the anti-tumor action of resveratrol against major cancers and major mechanisms involved therein. Earlier, we have discussed the challenges associated with the translation of resveratrol in the clinic and also presented the idea that resveratrol chemoprevention capacity can be greatly improved when combining with other agents.[67–70] The data from numerous studies, amassed in last two decades, are very promising and strongly suggest that resveratrol possesses an ability to modulate many pathways involved in cancer development and progression. The molecular actions of resveratrol against cancer described in this chapter may not be a comprehensive and complete description of all the actions of resveratrol. However, it appears that resveratrol is a promising agent and can be useful in cancer management, owing to it's molecular actions, which are extremely relevant to cancer progression.

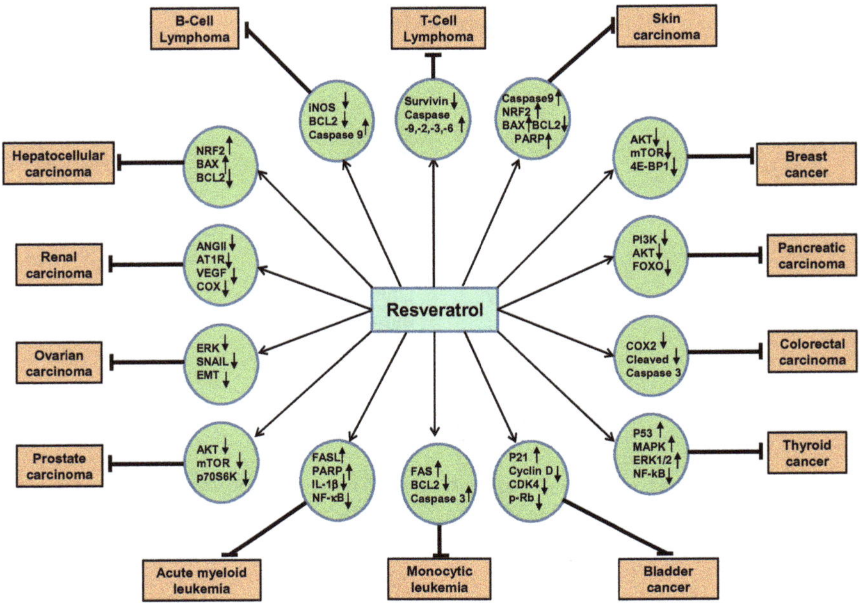

Fig. 7.9. Anti-tumor action of resveratrol against major cancers and mechanisms involved therein.

Acknowledgments

The work on chemoprevention of resveratrol was partially supported by funding from the California Table Grape Commission, as well as the Department of Veterans Affairs (Research Career Scientist Award IK6BX003780 to NA). We also acknowledge the core facilities supported through the UW Carbone Cancer Center Grant P30 CA014520.

References

1. Siegel RL, Miller KD, Jemal A. Cancer statistics, 2018. *CA Cancer J Clin* 2018; **68**: 7–30.
2. Timmers S, Auwerx J, Schrauwen P. The journey of resveratrol from yeast to human. *Aging (Albany NY)* 2012; **4**: 146–58.
3. Jang M, Cai L, Udeani GO, Slowing KV, Thomas CF, Beecher CW, Fong HH, Farnsworth NR, Kinghorn AD, Mehta RG, *et al.* Cancer chemopreventive activity of resveratrol, a natural product derived from grapes. *Science* 1997; **275**: 218–20.

4. Li YR, Li S, Lin CC. Effect of resveratrol and pterostilbene on aging and longevity. *Biofactors* 2017. doi: 10.1002/biof.1400.
5. Bonnefont-Rousselot D. Resveratrol and Cardiovascular Diseases. *Nutrients* 2016; **8**: 250.
6. Bitterman JL, Chung JH. Metabolic effects of resveratrol: addressing the controversies. *Cell Mol Life Sci* 2015; **72**: 1473–88.
7. Howitz KT, Bitterman KJ, Cohen HY, Lamming DW, Lavu S, Wood JG, Zipkin RE, Chung P, Kisielewski A, Zhang LL, *et al.* Small molecule activators of sirtuins extend Saccharomyces cerevisiae lifespan. *Nature* 2003; **425**: 191–6.
8. Hanahan D, Weinberg RA. The hallmarks of cancer. *Cell* 2000; **100**: 57-70.
9. Hanahan D, Weinberg RA. Hallmarks of cancer: the next generation. *Cell* 2011; **144**: 646–74.
10. Campbell SL, Khosravi-Far R, Rossman KL, Clark GJ, Der CJ. Increasing complexity of Ras signaling. *Oncogene* 1998; **17**: 1395–413.
11. Witkiewicz AK, McMillan EA, Balaji U, Baek G, Lin WC, Mansour J, Mollaee M, Wagner KU, Koduru P, Yopp A, *et al.* Whole-exome sequencing of pancreatic cancer defines genetic diversity and therapeutic targets. *Nat Commun* 2015; **6**: 6744.
12. Davies MA, Samuels Y. Analysis of the genome to personalize therapy for melanoma. *Oncogene* 2010; **29**: 5545–55.
13. Saud SM, Li W, Morris NL, Matter MS, Colburn NH, Kim YS, Young MR. Resveratrol prevents tumorigenesis in mouse model of Kras activated sporadic colorectal cancer by suppressing oncogenic Kras expression. *Carcinogenesis* 2014; **35**: 2778–86.
14. Shankar S, Nall D, Tang SN, Meeker D, Passarini J, Sharma J, Srivastava RK. Resveratrol inhibits pancreatic cancer stem cell characteristics in human and KrasG12D transgenic mice by inhibiting pluripotency maintaining factors and epithelial-mesenchymal transition. *PLoS One* 2011; **6**: e16530.
15. Luo H, Umebayashi M, Doi K, Morisaki T, Shirasawa S, Tsunoda T. Resveratrol Overcomes Cellular Resistance to Vemurafenib Through Dephosphorylation of AKT in BRAF-mutated Melanoma Cells. *Anticancer Res* 2016; **36**: 3585–9.
16. Niles RM, McFarland M, Weimer MB, Redkar A, Fu YM, Meadows GG. Resveratrol is a potent inducer of apoptosis in human melanoma cells. *Cancer Lett* 2003; **190**: 157–63.
17. Yuan SX, Wang DX, Wu QX, Ren CM, Li Y, Chen QZ, Zeng YH, Shao Y, Yang JQ, Bai Y, *et al.* BMP9/p38 MAPK is essential for the antiproliferative effect of resveratrol on human colon cancer. *Oncol Rep* 2016; **35**: 939–47.
18. Aggarwal BB, Bhardwaj A, Aggarwal RS, Seeram NP, Shishodia S, Takada Y. Role of resveratrol in prevention and therapy of cancer: preclinical and clinical studies. *Anticancer Res* 2004; **24**: 2783–840.
19. Roccaro AM, Leleu X, Sacco A, Moreau AS, Hatjiharissi E, Jia X, Xu L, Ciccarelli B, Patterson CJ, Ngo HT, *et al.* Resveratrol exerts antiproliferative activity and induces apoptosis in Waldenstrom's macroglobulinemia. *Clin Cancer Res* 2008; **14**: 1849–58.

20. Fung YK, Murphree AL, T'Ang A, Qian J, Hinrichs SH, Benedict WF. Structural evidence for the authenticity of the human retinoblastoma gene. *Science* 1987; **236**: 1657–61.
21. Moses HL, Yang EY, Pietenpol JA. TGF-beta stimulation and inhibition of cell proliferation: new mechanistic insights. *Cell* 1990; **63**: 245–7.
22. Adhami VM, Afaq F, Ahmad N. Involvement of the retinoblastoma (pRb)-E2F/DP pathway during antiproliferative effects of resveratrol in human epidermoid carcinoma (A431) cells. *Biochem Biophys Res Commun* 2001; **288**: 579–85.
23. Ahmad N, Adhami VM, Afaq F, Feyes DK, Mukhtar H. Resveratrol causes WAF-1/p21-mediated G(1)-phase arrest of cell cycle and induction of apoptosis in human epidermoid carcinoma A431 cells. *Clin Cancer Res* 2001; **7**: 1466–73.
24. Kim KH, Back JH, Zhu Y, Arbesman J, Athar M, Kopelovich L, Kim AL, Bickers DR. Resveratrol targets transforming growth factor-beta2 signaling to block UV-induced tumor progression. *J Invest Dermatol* 2011; **131**: 195–202.
25. Ferraz da Costa DC, Casanova FA, Quarti J, Malheiros MS, Sanches D, Dos Santos PS, Fialho E, Silva JL. Transient transfection of a wild-type p53 gene triggers resveratrol-induced apoptosis in cancer cells. *PLoS One* 2012; **7**: e48746.
26. Chow SE, Wang JS, Chuang SF, Chang YL, Chu WK, Chen WS, Chen YW. Resveratrol-induced p53-independent apoptosis of human nasopharyngeal carcinoma cells is correlated with the downregulation of DeltaNp63. *Cancer Gene Ther* 2010; **17**: 872–82.
27. Lin HY, Shih A, Davis FB, Tang HY, Martino LJ, Bennett JA, Davis PJ. Resveratrol induced serine phosphorylation of p53 causes apoptosis in a mutant p53 prostate cancer cell line. *J Urol* 2002; **168**: 748–55.
28. Tang HY, Shih A, Cao HJ, Davis FB, Davis PJ, Lin HY. Resveratrol-induced cyclooxygenase-2 facilitates p53-dependent apoptosis in human breast cancer cells. *Mol Cancer Ther* 2006; **5**: 2034–42.
29. Lin HY, Tang HY, Davis FB, Davis PJ. Resveratrol and apoptosis. *Ann N Y Acad Sci* 2011; **1215**: 79–88.
30. Jiang H, Shang X, Wu H, Gautam SC, Al-Holou S, Li C, Kuo J, Zhang L, Chopp M. Resveratrol downregulates PI3K/Akt/mTOR signaling pathways in human U251 glioma cells. *J Exp Ther Oncol* 2009; **8**: 25–33.
31. Greenberg RA, Chin L, Femino A, Lee KH, Gottlieb GJ, Singer RH, Greider CW, DePinho RA. Short dysfunctional telomeres impair tumorigenesis in the INK4a(delta2/3) cancer-prone mouse. *Cell* 1999; **97**: 515–25.
32. Ferraz da Costa DC, Fialho E, Silva JL. Cancer Chemoprevention by Resveratrol: The p53 Tumor Suppressor Protein as a Promising Molecular Target. *Molecules* 2017; **22**: 1014.
33. Zhai XX, Ding JC, Tang ZM, Li JG, Li YC, Yan YH, Sun JC, Zhang CX. Effects of resveratrol on the proliferation, apoptosis and telomerase ability of human A431 epidermoid carcinoma cells. *Oncol Lett* 2016; **11**: 3015–8.

34. Mirzazadeh A, Kheirollahi M, Farashahi E, Sadeghian-Nodoushan F, Sheikhha MH, Aflatoonian B. Assessment Effects of Resveratrol on Human Telomerase Reverse Transcriptase Messenger Ribonucleic Acid Transcript in Human Glioblastoma. *Adv Biomed Res* 2017; **6**: 73.
35. Kala R, Shah HN, Martin SL, Tollefsbol TO. Epigenetic-based combinatorial resveratrol and pterostilbene alters DNA damage response by affecting SIRT1 and DNMT enzyme expression, including SIRT1-dependent gamma-H2AX and telomerase regulation in triple-negative breast cancer. *BMC Cancer* 2015; **15**: 672.
36. Bashir R, Vale PR, Isner JM, Losordo DW. Angiogenic gene therapy: pre-clinical studies and phase I clinical data. *Kidney Int* 2002; **61**: S110–4.
37. Chen JC, Chen Y, Lin JH, Wu JM, Tseng SH. Resveratrol suppresses angiogenesis in gliomas: evaluation by color Doppler ultrasound. *Anticancer Res* 2006; **26**: 1237–45.
38. Hu Y, Sun CY, Huang J, Hong L, Zhang L, Chu ZB. Antimyeloma effects of resveratrol through inhibition of angiogenesis. *Chin Med J (Engl)* 2007; **120**: 1672–7.
39. Garvin S, Ollinger K, Dabrosin C. Resveratrol induces apoptosis and inhibits angiogenesis in human breast cancer xenografts in vivo. *Cancer Lett* 2006; **231**: 113–22.
40. Cao Z, Fang J, Xia C, Shi X, Jiang BH. trans-3,4,5'-Trihydroxystibene inhibits hypoxia-inducible factor 1alpha and vascular endothelial growth factor expression in human ovarian cancer cells. *Clin Cancer Res* 2004; **10**: 5253–63.
41. Park SY, Jeong KJ, Lee J, Yoon DS, Choi WS, Kim YK, Han JW, Kim YM, Kim BK, Lee HY. Hypoxia enhances LPA-induced HIF-1alpha and VEGF expression: their inhibition by resveratrol. *Cancer Lett* 2007; **258**: 63–9.
42. Wu H, Liang X, Fang Y, Qin X, Zhang Y, Liu J. Resveratrol inhibits hypoxia-induced metastasis potential enhancement by restricting hypoxia-induced factor-1 alpha expression in colon carcinoma cells. *Biomed Pharmacother* 2008; **62**: 613–21.
43. Trapp V, Parmakhtiar B, Papazian V, Willmott L, Fruehauf JP. Anti-angiogenic effects of resveratrol mediated by decreased VEGF and increased TSP1 expression in melanoma-endothelial cell co-culture. *Angiogenesis* 2010; **13**: 305–15.
44. Grant SW, Kyshtoobayeva AS, Kurosaki T, Jakowatz J, Fruehauf JP. Mutant p53 correlates with reduced expression of thrombospondin-1, increased angiogenesis, and metastatic progression in melanoma. *Cancer Detect Prev* 1998; **22**: 185–94.
45. Li W, Ma J, Ma Q, Li B, Han L, Liu J, Xu Q, Duan W, Yu S, Wang F, *et al.* Resveratrol inhibits the epithelial-mesenchymal transition of pancreatic cancer cells via suppression of the PI-3K/Akt/NF-kappaB pathway. *Curr Med Chem* 2013; **20**: 4185–94.
46. Wang H, Zhang H, Tang L, Chen H, Wu C, Zhao M, Yang Y, Chen X, Liu G. Resveratrol inhibits TGF-beta1-induced epithelial-to-mesenchymal

transition and suppresses lung cancer invasion and metastasis. *Toxicology* 2013; **303**: 139–46.
47. Vergara D, Valente CM, Tinelli A, Siciliano C, Lorusso V, Acierno R, Giovinazzo G, Santino A, Storelli C, Maffia M. Resveratrol inhibits the epidermal growth factor-induced epithelial mesenchymal transition in MCF-7 cells. *Cancer Lett* 2011; **310**: 1–8.
48. Buhrmann C, Shayan P, Goel A, Shakibaei M. Resveratrol Regulates Colorectal Cancer Cell Invasion by Modulation of Focal Adhesion Molecules. *Nutrients* 2017; **9**: 1073.
49. Kim SE, Shin SH, Lee JY, Kim CH, Chung IK, Kang HM, Park HR, Park BS, Kim IR. Resveratrol Induces Mitochondrial Apoptosis and Inhibits Epithelial-Mesenchymal Transition in Oral Squamous Cell Carcinoma Cells. *Nutr Cancer* 2018; **70**: 125–35.
50. Li W, Cao L, Chen X, Lei J, Ma Q. Resveratrol inhibits hypoxia-driven ROS-induced invasive and migratory ability of pancreatic cancer cells via suppression of the Hedgehog signaling pathway. *Oncol Rep* 2016; **35**: 1718–26.
51. Gao Q, Yuan Y, Gan HZ, Peng Q. Resveratrol inhibits the hedgehog signaling pathway and epithelial-mesenchymal transition and suppresses gastric cancer invasion and metastasis. *Oncol Lett* 2015; **9**: 2381–7.
52. Warburg O. On the origin of cancer cells. *Science* 1956; **123**: 309–14.
53. Warburg O. On respiratory impairment in cancer cells. *Science* 1956; **124**: 269–70.
54. Hsu PP, Sabatini DM. Cancer cell metabolism: Warburg and beyond. *Cell* 2008; **134**: 703–7.
55. Faber AC, Dufort FJ, Blair D, Wagner D, Roberts MF, Chiles TC. Inhibition of phosphatidylinositol 3-kinase-mediated glucose metabolism coincides with resveratrol-induced cell cycle arrest in human diffuse large B-cell lymphomas. *Biochem Pharmacol* 2006; **72**: 1246–56.
56. Jung KH, Lee JH, Thien Quach CH, Paik JY, Oh H, Park JW, Lee EJ, Moon SH, Lee KH. Resveratrol suppresses cancer cell glucose uptake by targeting reactive oxygen species-mediated hypoxia-inducible factor-1alpha activation. *J Nucl Med* 2013; **54**: 2161–7.
57. Dhar S, Kumar A, Rimando AM, Zhang X, Levenson AS. Resveratrol and pterostilbene epigenetically restore PTEN expression by targeting oncomiRs of the miR-17 family in prostate cancer. *Oncotarget* 2015; **6**: 27214–26.
58. Li W, Ma X, Li N, Liu H, Dong Q, Zhang J, Yang C, Liu Y, Liang Q, Zhang S, *et al.* Resveratrol inhibits Hexokinases II mediated glycolysis in non-small cell lung cancer via targeting Akt signaling pathway. *Exp Cell Res* 2016; **349**: 320–7.
59. Dai W, Wang F, Lu J, Xia Y, He L, Chen K, Li J, Li S, Liu T, Zheng Y, *et al.* By reducing hexokinase 2, resveratrol induces apoptosis in HCC cells addicted to aerobic glycolysis and inhibits tumor growth in mice. *Oncotarget* 2015; **6**: 13703–17.
60. Saunier E, Antonio S, Regazzetti A, Auzeil N, Laprevote O, Shay JW, Coumoul X, Barouki R, Benelli C, Huc L, *et al.* Resveratrol reverses the

Warburg effect by targeting the pyruvate dehydrogenase complex in colon cancer cells. *Sci Rep* 2017; **7**: 6945.
61. Yang L, Pang Y, Moses HL. TGF-beta and immune cells: an important regulatory axis in the tumor microenvironment and progression. *Trends Immunol* 2010; **31**: 220–7.
62. Silva AM, Oliveira MI, Sette L, Almeida CR, Oliveira MJ, Barbosa MA, Santos SG. Resveratrol as a natural anti-tumor necrosis factor-alpha molecule: implications to dendritic cells and their crosstalk with mesenchymal stromal cells. *PLoS One* 2014; **9**: e91406.
63. Lee-Chang C, Bodogai M, Martin-Montalvo A, Wejksza K, Sanghvi M, Moaddel R, de Cabo R, Biragyn A. Inhibition of breast cancer metastasis by resveratrol-mediated inactivation of tumor-evoked regulatory B cells. *J Immunol* 2013; **191**: 4141–51.
64. Olkhanud PB, Damdinsuren B, Bodogai M, Gress RE, Sen R, Wejksza K, Malchinkhuu E, Wersto RP, Biragyn A. Tumor-evoked regulatory B cells promote breast cancer metastasis by converting resting CD4(+) T cells to T-regulatory cells. *Cancer Res* 2011; **71**: 3505–15.
65. Duan J, Yue W, E J, Malhotra J, Lu SE, Gu J, Xu F, Tan XL. In vitro comparative studies of resveratrol and triacetylresveratrol on cell proliferation, apoptosis, and STAT3 and NFkappaB signaling in pancreatic cancer cells. *Sci Rep* 2016; **6**: 31672.
66. Panaro MA, Carofiglio V, Acquafredda A, Cavallo P, Cianciulli A. Anti-inflammatory effects of resveratrol occur via inhibition of lipopolysaccharide-induced NF-kappaB activation in Caco-2 and SW480 human colon cancer cells. *Br J Nutr* 2012; **108**: 1623–32.
67. Singh CK, Ndiaye MA, Ahmad N. Resveratrol and cancer: Challenges for clinical translation. *Biochim Biophys Acta* 2015; **1852**: 1178–85.
68. Singh CK, George J, Ahmad N. Resveratrol-based combinatorial strategies for cancer management. *Ann N Y Acad Sci* 2013; **1290**: 113–21.
69. Singh CK, Liu X, Ahmad N. Resveratrol, in its natural combination in whole grape, for health promotion and disease management. *Ann N Y Acad Sci* 2015; **1348**: 150–60.
70. Singh CK, Siddiqui IA, El-Abd S, Mukhtar H, Ahmad N. Combination chemoprevention with grape antioxidants. *Mol Nutr Food Res* 2016; **60**: 1406–15.

Chapter 8

A Link Between Immunity and Inflammation: Actionable Targets of Resveratrol

Dominique Delmas[*,†,‡,||], Emeric Limagne [*,†,§], Anne Vejux [*,¶], François Ghiringhelli[*,†,‡] and Virginie Aires[*,†,‡]

[*] *University of Bourgogne-Franche Comté, F-21000 Dijon, France*

[†] *INSERM U1231 "Lipids, Nutrition, Cancer", F-21000 Dijon, France*

[‡] *Research team CADIR "Cancer and adaptive immune response"*

[§] *Department of medical Oncology, Georges François Leclerc Centre, F-21000 Dijon, France*

[¶] *Research Team 'Biochemistry of the Peroxisome, Inflammation and Lipid Metabolism' EA 7270, F-21000 Dijon, France*

There are compelling evidences that resveratrol act on immune systems by affecting the major type of immune cells and pro-inflammatory-cytokines associated lymphocytes. Resveratrol is able to interfere on the adaptative and innate immune responses trough a modulation of cytokine-associated immune cells, of the lymphocyte differentiation process and the ratio of the different subtypes of T lymphocytes. In this chapter, we discuss the resveratrol action on the different subtypes of immune cells and subsequently on the interleukin-associated production in various contexts of diseases (auto-immune diseases, cancers). We highlight in each case the molecular mechanism by which resveratrol modulates the immune and inflammatory response, in particular by

[||] E-mail: dominique.delmas@u-bourgogne.fr

modulating signaling pathways such as the NFκB pathway or MAPK pathways. By these immunomodulatory properties, resveratrol could provide a novel strategy to enhance the efficacy of immunotherapy in various human diseases.

8.1. Introduction

Inflammation has been described as an initiator event of major diseases with significant impact in terms of public health such as cardiovascular diseases, autoimmune diseases, eye diseases, age-related diseases and more particularly neurodegenerative diseases, or the occurrence and development of cancers which represent the second cause of death in the world causing nearly seven million of death among the eleven million of people diagnosed and projections for 2030 predict that these data will double. In a cancerous context, inflammation and infiltrated immune cells play an essential role to contribute to carcinogenesis process. Furthermore, chemotherapies could have an impact on inflammation and infiltrated immune cells. In this way, we have recently shown that 5-fluoro-uracil and gemcitabine have a direct cytotoxic effect on cancer cells and could also specifically kill tumor infiltrating myeloid derived suppressor cells without affecting other immune cells.[1] Thus, chemotherapy efficiency could be dependent on its effects on tumor infiltrating cells notably by its capacity to eliminate immunosuppressive cells. This elimination activates anti-tumor immune response and destroys the tumor promoting inflammatory microenvironment. In the tumor microenvironment, there is a delicate balance between antitumor immunity and tumor-originated pro-inflammatory activity, which weaken antitumor immunity.[2,3]

Subsequently, to proliferate and form a clinically detectable tumor, the tumor cells must surpass another control system: the immunosurveillance. This surveillance involves the recognition of tumor cells by the system which perceives them as "non-self" and which ensures their elimination.

For many decades, various investigations have searched to identify new compounds which are able to kill cancer cells without deleterious effects on the normal cells. Among these compounds, dietary polyphenols appear to be of great interest in anticancer strategies notably by their pleiotropic properties on cancer cells and immune cells. The impetus sparking of this scientific inquiry was the result of many epidemiologic studies. Indeed, these studies reveal that phytoconstituents or micronutrients can protect against cancers.[4] Especially, Levi *et al.* show an inverse relation between resveratrol (a non-flavonoid polyphenol) and breast cancer risk.[5] Another cohort study in Finland[6] shows a link between flavonol consumption and reduced risk of lung cancer. These findings lead to the suggestion that polyphenols and flavonoids have beneficial health effects.

The present chapter, after having summarizing the different concepts of immuno-surveillance, immuno-editing and elimination, will first focus on describing the dialogue that may exist between innate and adaptive immunity and the key role played by T lymphocytes. Then, we will discuss how one of the most emblematic polyphenols, resveratrol, can modulate the immune response through action on key targets of immune cells and the process of lymphocyte differentiation.

8.2. Immune system and cancer

"The roots of cancer are in our genes". This famous phrase by Robert A. Weinberg, professor at the Massachusetts Institute of Technology in Boston, USA, summarizes the original process of tumor occurrence: a normal cell is cancerous after a series of mutations which alters his genome. During this process, called oncogenesis, tumor cells acquired abnormal properties. The cell becomes capable of producing survival signals, ignoring growth inhibition signals and escaping cell death. It also adapts its metabolism to survive the conditions of energy and environmental stress in order to replicate indefinitely. The cancer cell is also able to modify its environment to promote its own growth. For this, it promotes angiogenesis and degrades the extracellular matrix to invade other organs (metastases) by the general circulation and/or the destruction of the basement membrane. All of these characteristics constitute the hallmarks of cancer described by Hanahan and Weinberg.[7] Nevertheless, to proliferate and form a clinically detectable tumor, the tumor cells must surpass another control system: the immunosurveillance. This surveillance involves the recognition of tumor cells by the system which perceives them as "non-self" and which ensures their elimination.

8.2.1. *The immunosurveillance concept*

It was at the beginning of the 20th century that Paul Ehrlich for the first time put forward the hypothesis that the immune system could protect organism from the onset of cancer. He even considered the existence of tumor antigens, in other words the "modified self" antigens, but for lack of knowledge in fundamental immunology and without appropriate models, this hypothesis remained unexplored for a long time. Nevertheless, the discovery of Peter Medawar's transplant rejection phenomenon in the 1950s cast doubt on the real existence of a tumor-specific immune response. At the time, it was thought that the immune response induced by tumor cell injection derived from an "anti-graft" reaction, rather than a response against tumor antigens. It was not until the use of syngeneic mouse models that tumor cells had immunological properties different from normal cells. It was in 1964, that Lloyd John Old and Edward Boyse showed that the occurrence of chemically induced cancer could be prevented by vaccinating the mice with a serum from inbred mice already carrying a tumor. These results indicated the presence in the serum of "anti-tumor cell" immunoglobulin, in other words "anti-tumor antigens". This demonstration of a possible immunization against a syngeneic tumor cell transplant is the basis of the theory of immunosurveillance proposed by Mac Farlane Burnett and Lewis Thomas a few years ago.[8]

This discovery suggested the existence of T lymphocytes specific for tumor antigens capable of controlling tumor progression.[9] This work sparked the enthusiasm of the scientific community and research on cancer immunosurveillance accelerated. To confirm and better understand the mechanisms of this process, many research groups then set up immunodeficient syngeneic mouse models. Immunosuppression was induced by injection of a heterologous "anti-lymphocyte" serum, by thymectomy, or by pharmacological agents.[10,11] Other research teams, notably that of Osias Stutman, set up a model of athymic mouse "Nude CBA/H", which was in theory completely devoid of lymphocytes.[12] But the enthusiasm of the researchers quickly gave way to disappointment. These models showed no differences in sensitivity to chemical carcinogenesis compared to immunocompetent mice.[13] The concept of immunosurveillance was then widely criticized and many researchers even thought that the immune system had a rather detrimental effect in promoting inflammation.[14] It took several years to understand the contradiction between these results and those obtained by the team of Mac Farlane Burnett. In fact, mice that were thymectomized or treated with "anti-lymphocyte" serum were extremely sensitive to viral infections, which favored the appearance of viral-induced cancers and lymphomas induced by bacterial and viral hyperstimulation. Furthermore, Nude CBA/H athymic mice have been shown to express the most active isoform of aryl hydroxylase, making them more sensitive to methylcholanthrene than wild-type mice. NK ("Natural Killer") lymphocytes mainly come from the bone marrow and a small proportion of them pass through the thymus for their development.[15] The Nude CBA/H mice were therefore not completely devoid of lymphocytes.[16,17] The lack of knowledge and tools in immunology therefore did not allow at the time to establish experimentally the existence of immunosurveillance.

It was not until the 1990s that the first human tumor antigen, the MZ2-E antigen in melanoma, was able to induce a cytotoxic T lymphocyte reaction via HLA-A1.[18] This historical discovery proved for the first time the ability of "self-transformed" cells to induce a specific immune response of the antigen. Advances in the field of genetic engineering made available immunodeficient transgenic murine models, as well as monoclonal antibodies capable of neutralizing the biological function of cytokines or lymphocyte populations. Thanks to these new technological means, the chemically-induced carcinogenesis experiments could be carried out again. These manipulations have this time solidly proven that immunodeficient mice develop tumors faster than immunocompetent mice.

The concept of immunosurveillance in humans has been proven by epidemiological studies of immunodeficient patients. Indeed, these studies revealed that the immunodeficient patient had an increased risk of developing cancer. As in genetically modified murine models, patients with primary immune deficiency have a greater susceptibility to developing cancer. These patients are more particularly susceptible to viral-induced cancers: non-Hodgkin's lymphoma (induced by the Epstein-Barr virus, EBV), Kaposi's sarcoma (induced by the human herpesvirus 8, HHV8) and uro-genital cancers (induced by human papillomavirus, HPV).[19] Patients with secondary immune deficiency are also likely to develop cancer faster than immunocompetent patients. It is now also known that treatments targeting tumor necrosis factor alpha (TNFα) and methotrexate, used for their immunosuppressive effect to fight against chronic inflammatory diseases (rheumatoid arthritis, inflammatory bowel diseases, systemic lupus erythematosus) or transplant rejection, sensitize the organism to the oncogenesis,[20] as well as patients with human immunodeficiency virus (HIV) who T in small will lose the protection of T-CD4^{+} cells.

8.2.2. *The immunoediting concept*

The diagnosis of cancer in an immunocompetent individual (majority of cases at the time of screening in humans) indicates that some diseased cells may develop despite ongoing monitoring of the immune system. This observation therefore indicates that the action of the immune system can be surpassed and/or limited in time.

The immune system thus plays a double game. It protects the host but in return, it exerts a selection pressure on tumor cells that forces them to find a way (the loss of their immunogenicity) to escape monitoring. This new notion, that the immune system not only protects the host but also "sculpts" the immunogenicity of the tumor is the basis of the theory of "immunoediting".[21] The concept of immunosurveillance has therefore been redesigned, giving way to that of "immuno-editing", which defines the relationships between tumor cells and the immune system over time. This theory "cuts" this process into three phases: elimination, equilibrium, and escape.

The first one, the elimination phase, consists on the detection and the elimination of cancer cells by immune system. This phase could be complete only if all cancer cells are cleared. In the case of incomplete elimination of cancer cells the second phase namely equilibrium phase will be engaged. During this phase the immune system exerts a selective pressure by eliminating some tumor clones. The third step of the immuno-editing, the escape phase, arises when some tumor cells escape from immuno-editing and become ignored by the immune system or induce active process leading to tumor growth.[22]

By its action on the different immune cells, Resveratrol could modulate adaptive and innate immune response and consequently contribute to enhance immuno-editing.

8.3. Resveratrol and immune cells

8.3.1. *Resveratrol and lymphocytes T*

T lymphocytes or T cells play a central role in cell-mediated immunity. They can be distinguished from other lymphocyte types, such as B cells and NK cells by the presence of a special receptor on their cell surface called T cell receptors (TCR). It is well accepted that T lymphocytes play crucial role in the immune response of the host to cancer.[23] Modern cancer immunotherapies are therefore designed to induce or enhance T cell reactivity against tumor antigens.[24] Evidence indicates that many tumors induce T cell apoptosis as a mechanism of anti-tumor activity inhibition.[25] Among the T lymphocytes, we distinguish: (i) regulatory T cells (Treg cells) known as suppressor T cells and they are critical in maintaining immune tolerance and suppressing autoimmunity; (ii) cytotoxic T cells (CTL) that are able to induce the death of infected somatic cells or tumor cells; (iii) T helper (Th cells) that assist other white blood cells in immunologic processes, including maturation of B cells into plasma cells and B memory cells and activation of CTL cells and macrophages. It appears that resveratrol can modulate them to improve immune response or to counteract immune escape which is a characteristic of cancer progression or other diseases.

8.3.1.1. *Resveratrol and regulatory T cells*

The major role of regulatory T cells (Treg cells) is to shut down T cell-mediated immunity toward the end of an immune reaction and to suppress auto-reactive T cells that escaped the process of negative selection in the thymus.

Two major classes of $CD4^+$ regulatory T cells have been described, including the naturally occurring Treg cells and the adaptive Treg cells. Naturally occurring Treg cells (also known as $CD4^+CD25^+$ Treg cells) arise in the thymus and have been linked to interactions between developing T cells with both myeloid ($CD11c^+$) and plasmacytoid ($CD123^+$) dendritic cells, whereas the adaptive Treg cells (also known as Tr1 cells or Th3 cells) may originate during a normal immune response.

The ineffectiveness of numerous immunotherapy strategies may be related, in part, to their negative effects on Treg cells. These last participate to escape immune surveillance and consequently to cancer immunotherapy in tumor-bearing host, inhibiting cell populations such as $CD4^+$, $CD8^+$ T cells, dendritic cells (DC), B cells, macrophages, osteoblasts, mast cells, NK cells then preventing autoimmune, immunopathological, or allergic diseases and help to establish immunological tolerance to organ grafts.[26] In contrast these cells seems deleterious in cancer context by inhibiting anti-tumor immune response.[27] Indeed, for example in breast cancer, Treg cells were increased in the peripheral blood and greatly increased in the tumor microenvironment[28]; the presence of Treg cells is likely to correlate with cancer progression,[29] and specific depletion of Treg cells markedly inhibits tumor growth and maintains a strong and persistent antitumor immune response.[30] One of the most potent and well-studied tumor-induced immunosuppressive phenotypes found in the tumor microenvironment is the regulatory subpopulation cells ($CD4^+CD25^+FoxP3^+$ Treg cells). Recently, a transcriptional factor forkhead box P3 (FoxP3) has been described as crucial for naturally occurring $CD4^+CD25^+$ Treg development. FoxP3 expression is epigenetically regulated by DNA methylation and acetylation. DNA methyltransferase inhibitors can induce FoxP3 expression in naive $CD4^+$ T cells. Moreover, TGFβ, an immunosuppressive cytokine, is a costimulator of FoxP3 expression, which promotes the differentiation of $CD4^+CD25^+$ Treg cells from peripheral $CD4^+CD25^+$ progeny.[31]

Resveratrol and its analogues are able to decrease the number of Treg *in vitro* and *in vivo*.[32–35] Indeed, resveratrol treatment suppresses Treg population in EG7-bearing C57BL/6 mice since a single intraperitoneal administration of 4 mg/kg resveratrol inhibits the $CD4^+CD25^+$ population among $CD4^+$ cells.[32] Moreover, both $CD4^+CD25^+FoxP3^+$ to $CD4^+CD25^+$ cell ratio and $CD4^+CD25^+$ to $CD4^+$ cell ratio are decreased simultaneously by resveratrol in a dose-dependent manner.[32] In a same manner, a resveratrol analogue, the 4-(6-hydroxy-2-naphthyl)-1,3-benzenediol (HS-1793), particularly overcomes the resistance conferred by Bcl-2 by inducing apoptosis and free from the metabolic instability, can diminish the accumulation of Treg population and consequently could restore the antitumor immunity. Indeed, $CD4^+CD25^+$ Treg cells from FM3A breast tumor-bearing mice decreased after a treatment with HS-1793, while the $CD4^+$ T cell population remained unchanged.[35] In contrast, the $CD8^+$ T cell population as well as the interferon IFN-γ-expressing $CD8^+$ T cell population and IFN-γ secretion of splenocytes from tumor-bearing mice were significantly upregulated by HS-1793 treatment.

The ability of resveratrol or its analogues to modulate $CD4^+CD25^+$ $FoxP3^+$ Treg cells down could reveal the ability of resveratrol to be used in combination immunotherapy. Indeed, iimmunotherapy with high-dose interleukin-2 (HDIL-2) is an effective treatment for patients with metastatic melanoma and renal cell carcinoma. Nagarkatti's team has shown that a combination of HDIL-2/resveratrol was more effective in inhibiting tumor metastasis and growth than HDIL-2 treatment alone.[36] These effects were associated with a suppression of HDIL-2-induced expansion of Treg cells.

Biochemical analysis reveals that TGFβ and FoxP3 are involved in the polyphenol action, TGFβ production and $FoxP3^+$ expressing cells among $CD4^+CD25^+$ population are down-regulated by resveratrol leading to the inhibition of Treg cells activity.[32] Interestingly, resveratrol administration (4 mg/kg, i.p.) to lymphoma tumor-bearing C57BL/6 mice resveratrol suppressed the $CD4^+CD25^+$ cell population among $CD4^+$ cells and down-regulated secretion of TGFβ, an immunosuppressive cytokine, measured from the spleens of tumor-bearing mice.[32] Furthermore, resveratrol enhanced IFN-γ expression in $CD8^+$ T cells both *ex vivo* and *in vivo*, leading to immune stimulation, suggesting that resveratrol has a suppressive role on $CD4^+CD25^+$ cell population.

In a very surprisingly manner, as observed for its pro and anti-apoptotic properties, resveratrol can also promote the production of Treg cells in other context of disease where Treg are considered to be protective. It is the case in ischemic stroke where Treg are cerebroprotective immunomodulator.[37] In this context, a pretreatment of resveratrol (50 mg/kg, i.p. 7 days prior to ischemia onset) in cerebral ischemia/reperfusion injury in adult male Sprague-Dawley rats, increased Treg in the spleens and ischemic hemisphere. This observations were associated with a significant increased levels of interleukin-10 (IL-10) in the plasma and ischemic hemisphere, and with a significantly decreased levels of TNFα and IL-6 in the plasma and ischemic hemisphere at 24 hours after ischemia onset.[37] Subsequently, resveratrol pretreatment ameliorated neurological defects and reduced cerebral infarct volume.

Enhancement of Treg production is also benefit in other pathologies such as ulcerative colitis which is an inflammatory bowel disease (IBD). More specifically, Treg cells can suppress the inflammatory response cascade and maintain the balance of the intestinal immune response by secreting and regulating anti-inflammatory cytokines, e.g., interleukin (IL)-10 and transforming growth factor-β (TGF-β).[38] Moreover, IBD seems closely associated with Treg/Th17 imbalance.[39,40] It appear that resveratrol regulates Treg/Th17 balance in favor of Treg production and resveratrol inhibits the expression of HIF-1α, mTOR, and STAT3, suggesting that resveratrol regulates Treg/Th17 balance mainly through inhibiting hypoxia-mTOR-HIF-1α-Th17 and IL-6-STAT3-HIF-1α-Th17 pathways.[41]

A third example is illustrated in a model of diabetic nephropathy is one of the most serious complications of diabetes. Indeed, in

Sprague-Dawley rats were intraperitoneally injected with a single dose of streptozotocin to induce diabetes mellitus, a pretreatment of resveratrol (50 mg/kg, gavage) strongly reduced Th17-related pro-inflammatory cytokines (IL-17A, IL-25) and enhanced Treg-related anti-inflammatory cytokines (IL-35, IL-10) in serum of diabetic rats to compared to the control rats.[42] Again, resveratrol is able to modulate the balance of Th17/Treg-related cytokines in favor of the Treg in order to exercise a beneficial action of the latter.

As many other actions of resveratrol, this polyphenols exert a biphasic action on Treg cells which is dependent of the context.

8.3.1.2. *Resveratrol and cytotoxic T cells*

Cytotoxic T cells (also known as CTL, cytolytic T cell, $CD8^+$ T cells) belong to a sub-group of T lymphocytes that are able to induce the death of infected somatic cells or tumor cells. $CD8^+$ T cells have the capacity to recognize the antigens presented by the peptide/MHC-I complexes through the expression of an $\alpha\beta$-type TCR and the CD8 molecule. The activation and differentiation of $CD8^+$ cells is an essential step for the establishment of a response against intracellular pathogens and against tumor cells. The activation of naive $CD8^+$ cells causes both the development of effector cells with a short-life, extremely functional (memory effector, terminal memory effector), and cells predisposed to survive, little functional, but whose role is to become memory cells. The differentiation of $CD8^+$ cells might be a little different from that of $CD4^+$ cells.

The role of CD8 cells in a cancer context has been fairly widely studied. It is clear that the ability of these cells to produce cytokines and cytostatic and cytotoxic enzymes (IFN-γ, TNFα, TRAIL, granzyme, perforin) allows them to play a central role in the elimination of tumor cells. Studies in mice and humans clearly indicate that the presence of effector T-$CD8^+$ cells is associated with a good prognosis when present in the tumor.[43] In addition, some studies also show that the presence of central memory or stem memory T cells is an important reserve that makes it possible to generate effector cells capable of effectively eliminating tumor cells for a prolonged period. Indeed, in the one hand, the expressions of PD-1, Tim-3 molecules by the effector cells, sensitize them to the death induced by the tumor cells or the presenting cells or the antigens that express their ligand (PD-L1, PD-L2, galectin-9). On the other hand, as for $CD4^+$ cells, the polarization of T-$CD8^+$ cells can affect their ability to induce tumor cells elimination.

In a tumor context, at the inverse of CD4+, resveratrol led to an increase in the number of cytotoxic T lymphocytes associated protein 4 (CTLA-4; also known as CD152)-positive cells and in the gene expression levels of CTLA-4, FoxP3, IL-10 and TGF-β these two last TGF-β and IL-10 being anti-inflammatory cytokines and controlling pathogenic, as well as aging-associated inflammatory conditions.[44] They are selectively produced by peripheral Treg subsets as distant mediators of immunosuppression. These effects are comforted in an ICR mice model.[44] Furthermore, with a Mixed lymphocyte reaction (MLR) assay using $CD8^+$ T cells of OT-1 TCR transgenic mice, which express a TCR specific for the MHC class I-restricted OVA peptide 257–264 antigen ($OVA_{257-264}$) in dendritic cells (DCs), Noh *et al.* demonstrate that resveratrol-mediated IFN-γ-induced modulation of indoleamine 2,3-dioxygenase (IDO) is crucial for $CD8^+$ T cell proliferation by bone marrow-derived dendritic cells (BMDCs) pulsed with $OVA_{257-264}$.[45] Molecular mechanism seems involve a Janus Kinase (JAK)/STAT1- and protein kinase C (PKC) δ-dependent pathway where resveratrol regulates the DC-mediated immune response. In addition, resveratrol not only exerts anti-tumor effect through an IDO-mediated immune modulation in immune cells, but also exerts IDO-mediated tolerance in tumor cells in the tumor environment.[45] In a same manner, resveratrol enhances IFN-γ expression in $CD8^+$ T cells both *ex vivo* and *in vivo*, leading to immune stimulation.[32]

Furthermore, resveratrol metabolites, as previously, can also act on CTL, such as HS-1793, at the inverse of its action on Treg. Indeed, the $CD8^+$ T cell population as well as the IFN-γ-expressing $CD8^+$ T cell population and IFN-γ secretion of splenocytes from tumor-bearing mice were significantly upregulated by HS-1793 treatment.[35]

8.3.1.3. *Resveratrol and T helper*

T helper (Th cells) are also known as $CD4^+$ T cells because they express the CD4 protein on their surface and $CD4^+$ T cells are generally considered to have a pre-defined role as helper T cells within the immune system. Th cells become activated when they are presented with peptide antigens by MHC class II molecules that are expressed on the surface of antigen presenting cells (APCs). Once activated, they divide rapidly and secrete small proteins called cytokines that regulate or assist the active immune response. These cells can differentiate into one of several subtypes (e.g. Th1, Th2, Th3, Th17), which secrete cytokines to facilitate a different type of immune response. The mechanism by which T cells are directed into a particular subtype is poorly understood, though signalling patterns from the APC are thought to play an important role.

8.3.1.3.1. T helper type 1 (Th1) and resveratrol actions

Lymphocytes Th1 are induced by the presence of IL-12 and IFN-γ.[46] IL-12 produced by antigen-presenting cells allows activation of the transcription factor STAT4 *via* the kinase JAK2. STAT4 migrates to the nucleus and binds to the promoter of genes encoding IFN-γ, lymphotoxin-α (LTα), TNF-α and -β and IL-12 receptors (IL-12RB1, IL-12RB2). This process allows the expression of cytokines (IL-12 and IFN-γ) capable of acting autocrine to induce the expression of the transcription factor STAT1, then Tbet, which is the key transcription factor of polarization and maintaining the Th1 subtype. They have an anti-tumor role very widely described, as well as their involvement in inflammation and auto-immune diseases.

Indeed, in the way of inflammatory diseases, IBD, as previously seen in 8.3.1.2. part, is a chronic intestinal inflammation caused by hyperactivated effector immune cells that produce pro-inflammatory cytokines.. it appears, that resveratrol treatment decreases the percentage of the specific marker of Th1, CXCR3 in the spleen, in the mesenteric lymph nodes and in the intestinal lamina propria.

This decrease was associated with a decrease of pro-inflammatory cytokines, i.e. TNFα, IL-6, IL-12 and IL-1β.[47]

Resveratrol seems also act on various auto-immune diseases through a modulation of Th1 levels and associated cytokines. For example, in a model of disease such as periodontal disease, resveratrol treated-rats present a lower Th1/Th2 levels compared with placebo rats and subsequently resveratrol inhibits experimental periodontitis and cigarette smoke inhalation-induced supporting alveolar bone loss and has a beneficial effect on osteo-immuno-inflammatory markers.[48,49] In an other model of multisystemic auto-immune disease, such as stemic lupus erythematosus (SLE), resveratrol is able to suppressed CD69 and CD71 expression on $CD4^+$ T cells as well as $CD4^+$ T cell proliferation, induced $CD4^+$ T cell apoptosis, and decreased CD4 IFN-γ^+ Th1 cells and the ratio of Th1/Th2 cells *in vitro*.[50] In vitro antibody production and proliferation of B cells were also inhibited.

Furthermore, resveratrol could also used in combination with other drug to modify the ratio of Th1 which play an essential role in an autoimmune disease such as multiple sclerosis. Indeed, to mimc this pathology, experimental autoimmune encephalitis (EAE) is an inflammatory demyelinating disease, which served as a useful model providing considerable insights into the pathogenesis of multiple sclerosis. Combination treatment of resveratrol with mouse bone marrow mesenchymal stem cells (mBM-MSC) shown to have neuroprotection capabilities in EAE, promotes a shift in balance from Th1 to Th2 cytokine in EAE mice.[51] A significant decrease in IFN-γ and TNFα Th1 cytokines, and increase in IL-4 and IL-10 Th2 cytokines in the mBM-MSCs or resveratrol treatment groups were observed. In addition, a more prominent decrease in IFN-γ/TNFα Th1 cytokines and increase in IL-4/IL-10 Th2 cytokines in the combination treatment group were observed.[51]

Interestingly, resveratrol metabolite (i.e. piceatannol) significantly inhibits differentiation of $CD4^+$ $CD25^+$ $CD62L^+$ naïve CD4 T cells into Th1, Th2 and Th17 cells through isolated murine splenic T cells from C57BL/6 mice.[52]

8.3.1.3.2. T helper type 2 (Th2) and resveratrol actions

Th2 lymphocytes are induced in the presence of IL-4. This cytokine allows the activation of STAT6 and the expression of the transcription factor GATA-3, the key transcription factor of the Th2 polarization. They can be identified by the expression of CCR4 and CRTH2 receptors and by the absence of expression of CCR6 and CXCR3. The production of IL-4, IL-5 and IL-13 by Th2 antagonizes the differentiation of Th1, which consequently limits the activation of cytotoxic lymphocytes. Then, the ratio of Th2 with other lymphocyte T helper is essential in the control of various diseases. By this way, the hydroxystilbene resveratrol could prevent or diminished the progression of various pathologies linked to this ratio.

If we take again the example of the periodontitis for which we have seen previously that the resveratrol could decrease the ratio Th1/Th2 but also the ratio Th17/Th2.[49] In similar manner, in EAE, resveratrol-mBM-MSC combination treatment enhanced the immunomodulatry effects by increasing anti-inflammatory cytokines IL-4 and IL-10 secreted by Th2.[51]

At the inverse, atopic dermatitis (AD) is a chronic relapsing skin disease that is manifested by type 2 helper T cells (Th2) dominant hyperactive immune disorder. High mobility group box (HMGB1), a non-histone chromatin-associated protein, is implicated as a mediator of both infectious and non-infectious inflammatory conditions. It is secreted actively by immune cells such as monocytes, macrophages, and dendritic cells,[53] and has been proposed to have a crucial role in the pathogenesis of various chronic inflammatory and autoimmune diseases. NFκB is the center of interest, since it promotes the transcription of Th2 cytokines and adhesion molecules.[54] Resveratrol treatment (20 mg/kg/day, p.o. during 2 weeks) of AD mouse model, inhibited the elevation of Th1 and Th2-related cytokines (IL-4 and IFN-γ) in serum as compared to that of the AD group.[55] Moreover resveratrol treatment seemed to down-regulate some key proteins i.e. HMGB1, RAGE, p-NFκB, p-PI3K, p-ERK1/2, COX2, TNFα, IL-1β, IL-2Rα.

Resveratrol was able to inhibit Th2-cytokines (IL-3, IL-4 and IL-13 as well as TNFα) and paxillin-linked allergic responses dependent upon MAPK signaling.[56] Furthermore, resveratrol is able to protect against pristane-induced lupus animal model and when SMCs isolated from pristane-induced lupus mice were activated with ConA with or without resveratrol, the authors observed a decreased of CD69 and CD71 expression on $CD4^+$ T cells and reduced the percentage of Th1 and subsequently decreased the Th1/Th2 ratio where CD4 $IL\text{-}4^+$Th2 were no affected.[50]

In similar manner of Th1, piceatannol, inhibits differentiation of $CD4^+$ $CD25^+$ $CD62L^+$ naïve CD4 T cells into Th1, Th2 and Th17 cells through isolated murine splenic T cells from C57BL/6 mice.[52]

8.3.1.3.3. T helper type 17 (Th17) and resveratrol actions

The establishment of an adaptive immune response of humoral or cytotoxic type is intimately related to the differentiation of $CD4^+$ T cells. Th17s have inflammatory properties derived from their ability to produce various interleukins such as IL-17A, IL-17F, IL-6, IL-21, IL-22 and TNFα. These cytokines play a major role in the recruitment and activation of neutrophils, but they also play an important role in the activation of NK cells, of CD8 T cells and B cell differentiation.[57,58] The pro-inflammatory properties of Th17 are, however, dependent on the cellular and cytokine environment in which they are located. From a molecular point of view, Th17 polarization requires the expression and activation of a large number of interdependent transcription factors expressed sequentially. The initiation of the program begins with the signals induced by TGF-β and IL-6 after binding to their respective receptors (TGF-β-RI and -RII and IL-6R). IL-6 and IL-21, by binding to their receptor, activate signal STAT3 (Signal Transducer and Activator of Transcription 3). This increases the expression of transcription factors RORγt (Retinoid-related Orphan Receptor gamma t) and RORα. Activation of these factors promotes the expression of IL-17A, IL-17F, IL-21 and IL-22. IL-23, in turn, stabilizes the Th17 pathway.

Currently, the therapeutic strategies envisaged for targeting Th17 are mainly based on the neutralization of the cytokines produced by these cells. Clinical studies in autoimmune diseases such as psoriasis or rheumatoid arthritis have shown that targeting by antibodies to IL-17A (Secukinumab), TNFα (Adalimumab), IL-23 (Tidrakizumab) or the IL-17A/F receptor (IL-17RA, Brodalumab) are effective strategies.

In a same manner, resveratrol could act on various auto-immune diseases or cancers through a blockage of interleukins-associated Th17 and through a disruption of Th17 polarization. Indeed, resveratrol modulates inflammatory arthritis in rodents by reducing serum levels of pro-inflammatory cytokines (IL-17, IFNγ, TNFα, IL-1β, IL-6) in collagen-induced arthritis mice and suppresses Th17cell polarization by aryl hydrocarbon receptor *in vitro*.[59]

As the modulation of Th1/Th2 ratio, resveratrol is able to modulate the levels of Th17/Th2 cells as well as mRNA level for RANKL in gingival tissues of rats with experimental periodontitis and subjected to cigarette smoke inhalation.[49] In an other model, resveratrol treatment (50 and 100 mg/kg) was able to regulate the imbalance between Th17 and Treg cells trough reducing the number of Th17 cells and up-regulating the number of Treg cells in dextran sulphate sodium (DSS) murine model of ulcerative colitis.[41] These ratio alterations were associated with a decrease of Il-6 and IL-17 pro-inflammatory cytokines and an increase of IL-10 and TGF-β anti-inflammatory cytokines in the plasma and colonic tissues of mice. Moreover, nuclear HIF-1α and cytoplasmic mTOR and STAT3 protein expression were significantly reduced in colitis mice treated with resveratrol.[41] In this way of reduction of Th17-related cytokine, the polyphenol could also reduce the levels of TGF-β1, IL-17 and IL-25 in diabetic rats.[42]

In an inflammatory and cancerous context, tumoral cells produce chemokines leading to migration of Th17, specific immune cells from the differentiation of naive T cells. Migration and stimulation of these immune Th17cells then trigger the secretion of interleukin IL-17, responsible, firstly, for chronic and acute inflammation which will strongly contribute to tumoral growth. Secondly, this interleukin

IL-17 will promote tumor angiogenesis by stimulating the production of vascular endothelial growth factor, VEGF, by both tumoral cells and fibroblasts. We have very recently shown that resveratrol was able to decrease IL-17-associated pro-inflammatory Th17.[60] Indeed, naïve T cells isolated from the lymph node and mouse spleen, were differentiated *in vitro* in Th17 cells in the presence of resveratrol (10 μM, 3 days). We observed both reduction of IL-17 secretion but also a strong decrease of mRNA levels of gene encoding IL-17. very interestingly, when we treat naïve T cells with resveatrol, it appears that resveratrol alters the expression of RORγt, a transcriptional nuclear factor essential in the process of lymphocyte differentiation.[60] The antitumoral effect of resveratrol involving Th17 cells has been demonstrated *in vivo* by using IL-17 knock down mices where the antitumoral resveratrol activity observed in normal mice was abolished in IL-17 KO mice.[60] Furthermore, resveratrol effect on the IL-17 production contributes to reduce VEGF levels and subsequently contributes to strongly decreases the number of pulmonary metastasis in tumorigenesis-induced mice.[60]

8.3.2. *Resveratrol and lymphocytes B*

B cells are lymphocytes that play a large role in the humoral immune response. The principal functions of B cells are the production of antibodies against antigens, the role of APCs and eventually their development into memory B cells after activation by antigen interaction. B cells are an essential component of the adaptive immune system.

Resveratrol can modulate the activity of B cells as evidenced by a significant inhibition of proliferation, antibody production and lymphokin secretion.[61] Moreover, CD19+ B cells from splenic mononuclear cells of mice and cultured with LPS (10 μg/ml), were decreased by a game range of resveratrol (10 to 40 μM).[50] More specifically, resveratrol strongly decreased the different Ig isotype (IgG1, IgG2a, IgG3, IgGM and IgA).

In a model of multiple sclerosis (MS) for which it is well established that B cells play an essential role in the pathophysiology of MS, a resveratrol treatment normalized production of lymphotoxin

(LT) and TNFα.[62] This mechanism seems to involve a novel microRNA-132-sirtuib-1 axis. Indeed, the authors of this study show that miR-132 enhances LT and TNFα production in association with SIRT1 suppression in B cells. By inhibition of sirtuin-1 in normal B cells, LT and TNFα production were induced.[62]

Very interestingly, resveratrol prevents the Epstein-Barr virus (EBV)-immortalization of primary human B cells without toxicity to normal B cells.[63] Moreover, resveratrol does not block the entry of EBV into B cells but the polyphenol induces apoptosis of EBV-infected B cells. This observation was associated with a decreased of various cytokines such as IL-13, IL-6, IL-10 and with an inhibition of NFκB and STAT3.[63]

8.3.3. *Resveratrol and immunosuppressive cells*

8.3.3.1. *Resveratrol and dendritic cells and myeloid-derived suppressor cells*

8.3.3.1.1. Dendritic cells (DCs) and resveratrol actions

Dendritic cells (DCs) that are key players linking the innate and adaptive immune responses, represent a terminally differentiated stage of monocytes[64] and are the key antigen-presenting cells of the immune system. DCs play a main role as immune sentinels in the initiation of T-cell responses to microbial pathogens, tumors, and inflammation.[65,66]

Resveratrol can exert an effect on DCs during their differentiation and maturation. Indeed, resveratrol treatment down-regulates the expression of various costimulatory molecules (CD11c, CD40, CD80, CD83 and CD86), as well as the expression of major histocompatibility complex (MHC) class I and II molecules in DCs or in mouse bone marrow-derived dendritic cells (BM-DC).[67,68] More specifically, resveratrol up-regulates inhibitory immunoglobulin-like transcript -3 and -4 (ILT3, ILT4) molecules that are among the few well-established markers of TDCs. Interestingly, when resveratrol is used to treat DC cultures, there is a dose-dependent decrease in the ability of the treated DCs to stimulate allogeneic whole $CD4^+$ T cells compared with controls. Both resveratrol-treated DCs, and

BM-DCs lose the ability to produce proinflammatory intracellular IL-12p40/p70 and secreted IL-12p70 after activation, but had an increased ability to produce IL-10 upon stimulation with LPS.[67,68] In the same way, when immature DCs are stimulated by TNFα or LPS, resveratrol treatment reduces DC activation, DC maturation and alters DC cytokine production (IL-12, IL-23).[69] Moreover, resveratrol inhibits T cell proliferation in response to TNF-α stimulated DC.

Biochemical analysis seems to reveal a common mechanism involving kinases, the cyclo-oxygenase (COX), the transcription factors STAT and NF-κB.[69] Indeed, polyphenols inhibit the COX-2 expression, the PGE_2production,[70] the activation and DNA binding of STAT1 to the interferon regulatory factor 1 (IRF-1) promoter in response to IFNγ.[70,71] Recently a role for a miRNA, miR-34, was highlight in chronic obstructive pulmonary disease (COPD) patients. Indeed, resveratrol was able to inhibit upregulation of CD80 and CD86 and secretion of IFN-α.[72] Furthermore, resveratrol induces a marked increase of miR-34 expression in DCs from healthy smoker subjects and COPD patients comparatively to healthy non-smoker subjects. Interestingly, miR-34 inhibited the expression of CD80 and CD86 and also secretion of IFN-α in Dcs, suggesting that the polyphenol could inhibit dysfunction of DCs through promoting miR-34.[72] Another molecular mechanism seems involve advanced glycation end products (AGEs) and their receptor RAGE that play a role in the pathogenesis of auto-immune reactions through DC activation. It appears that resveratrol prevents phenotypical maturation and RAGE upregulation on DC surface in response to AGE-albumin.[73] It also inhibited pro-inflammatory cytokine expression, allostimulatory ability upregulation, mitogen-activated protein (MAP) kinases, and NFκB activation in AGE-albumin-stimulated oDCs.

Very recently, in order to increase the resveratrol efficiency, resveratrol has been associated with nanostructured lipid carriers (NLC) as delivery systems. Interestingly, stimulating DC with TNF-α in the presence of resveratrol-NLC revealed that these construction significantly decrease surface expression of activation marker CD83 and MHC II molecule human leukocyte antigen — antigen D related, both upregulated in response to TNF-α stimulation.[74] Moreover, resveratrol-NLC is able to inhibit NFκB phosphorylation and significantly decrease the level of interleukin-12/23, both upregulated in response to TNF-α.

Together these data show that polyphenols limit maturation of DCs and their capacity to produce proinflammatory cytokines. These events may promote a tolerogenic state and limit inflammatory process.

8.3.3.1.2. Myeloid derived suppressor cells (MDSCs) and resveratrol actions

Myeloid derived suppressor cells (MDSCs) are a heterogeneous population characterized by the co-expression of $CD11b^+$ and $Gr\text{-}1^+$ and have long been known for their immunosuppressive function. Recent studies show that MDSC cells are only found in the "trace" state in the blood or secondary lymphoid organs of healthy mice. It is the same in humans who do not have any inflammatory or cancerous pathologies. These cells are nevertheless found in large quantities in the blood and secondary lymphoid organs in case of cancer.

Treatment with 100 mg/kg of resveratrol reversed the impact of DSS exposure $Apc^{Min/+}$ mice on immune T cells ($CD4^+$ and $CD8^+$), B cells ($CD19^+$), MDSCs, and NK cells in mesenteric lymph nodes (MLNs).[75] Interestingly, in IL-$10^{(-/-)}$ mice, the percentage and absolute numbers of $CD11b^+$ and $Gr\text{-}1^+$ cells in the lamina propria (LP) and spleen were increased after resveratrol treatment as compared with the vehicle treatment.[47] Moreover, co-culture of resveratrol-induced $CD11b^+$ $Gr\text{-}1^+$ cells with T cells, attenuated T cell proliferation, and most importantly reduced IFN-γ and GM-CSF production by LP derived T cells from vehicle treated IL-$10^{(-/-)}$ mice with chronic colitis.[47] In another model of acute inflammation such as lug inflammation, resveratrol treatment upregulated $CD11b^+$ and $Gr\text{-}1^+$ MDSCs that inhibited staphylococcal enterotoxin B (SEB)-mediated T cell activation *in vitro*.[76]

Resveratrol induced expansion and suppressive function of MDSC which inhibited the development of vascular leak syndrome (VLS) after adoptive transfer.[36] It is good point since MDSC is the dominant suppressor cell than regulatory T cell in the development of VLS.[36]

Due to these properties, resveratrol could be used in combination with immunotherapy such as high-dose interleukin-2 (HDIL-2) which an effective treatment for patients with metastatic melanoma and renal cell carcinoma.[77,78] Indeed, resveratrol alone inhibits Treg expansion and simultaneously induces expansion and suppressive functionality of MDSCs.[36] These results were increased by a cotreatment of resveratrol-HDIL-2.

8.3.3.2. *Resveratrol and natural killer (NK) cells*

Natural killer (NK) cells have been described to participate to early response against virus infection and in tumor immunosurveillance. NK cells were first shown to induce directly tumor cell death through perforin and granzyme B production, but also secrete different cytokines (mainly IFN-γ) that affect DCs and macrophage functions.[79]

Resveratrol present a biphasic effect on NK cells where at low concentrations (0.075 to 1.25 μg/mL) the stilbene enhances CTL and NK cell cytotoxic activity and at high concentrations (5 to 20 μg/mL) induces an inhibition of their activities.[80–82] The molecular mechanism contributing to NK cells activation by resveratrol involves different signaling pathways. Indeed, resveratrol lead to (i) an activation of c-Jun-N-terminal kinase (JNK) and extracellular-regulated kinase (ERK)[80,83]; (ii) an enhancement of NKG2D expression, which is an upstream signaling molecule of the MAP kinase pathway playing an important role in NK cell-mediated tumor cell recognition[83]; (iii) an increase in perforin gene and protein expression.[83] These findings are supported by the fact that resveratrol effects were abolished by inhibition of JNK and ERK by pharmacological inhibitors (SP600125 and PD98059, respectively) and/or by siRNA against JNK-1 and ERK-2.[80,83]

Thus polyphenols could be potential activators of NK cells, but could also inhibit their cytotoxic anti-tumor activity depending on the experimental context.

8.4. Conclusion and perspectives

There are compelling evidences that resveratrol act on immune systems by affecting the major type of immune cells and pro-inflammatory-cytokines associated lymphocytes. Resveratrol is able to interfere on the adaptative and innate immune responses trough a modulation of cytokine-associated immune cells, of the lymphocyte differentiation process and the ratio of the different subtypes of lymphocytes T. This modulation is very dependent of the cellular environment and of the disease considered. Indeed, in some case, resveratrol can increase a subtype of immune cells and in other cases it can decrease it. The key molecular targets such as NFκB, STAT3/Sirtuin-1 pathways contribute to the immunomodulatory actions of resveratrol. Other targets such as microRNAs will be of immense interest to both academics and industry and will lead to the discovery of novel therapeutic pathways in inflammatory and autoimmune diseases. Furthermore, resveratrol does not present any cytotoxicity in animal models and in normal cells[84–87] and could provide a novel strategy to enhance the efficacy of immunotherapy in various human diseases.

References

1. Vincent, J., Mignot, G., Chalmin, F., Ladoire, S., *et al.*, 5-Fluorouracil selectively kills tumor-associated myeloid-derived suppressor cells resulting in enhanced T cell-dependent antitumor immunity. *Cancer Res* 2010; **70**: 3052–3061.
2. Ben-Baruch, A., Inflammation-associated immune suppression in cancer: the roles played by cytokines, chemokines and additional mediators. *Semin. Cancer Biol.* 2006; **16**: 38–52.
3. Kim, R., Emi, M., Tanabe, K., Arihiro, K., Tumor-driven evolution of immunosuppressive networks during malignant progression. *Cancer Res* 2006; **66**: 5527–5536.
4. Renaud, S. C., Gueguen, R., Schenker, J., d'Houtaud, A., Alcohol and mortality in middle-aged men from eastern France. *Epidemiology* 1998; **9**: 184–188.
5. Levi, F., Pasche, C., Lucchini, F., Ghidoni, R., *et al.*, Resveratrol and breast cancer risk. *Eur J Cancer Prev* 2005; **14**, 139–142.
6. Knekt, P., Jarvinen, R., Seppanen, R., Hellovaara, M., *et al.*, Dietary flavonoids and the risk of lung cancer and other malignant neoplasms. *Am J Epidemiol* 1997; **146**: 223–230.

7. Hanahan, D., Weinberg, R. A., Hallmarks of cancer: The next generation. *Cell* 2011; **144**: 646–674.
8. Burnet, M., Cancer: A biological approach. III. Viruses associated with neoplastic conditions. IV. Practical applications. *Br. Med. J.* 1957; **1**: 841–847.
9. Burnet, M., Cancer; a biological approach. I. The processes of control. *Br. Med. J.* 1957; **1**: 779–786.
10. Grant, G. A., Miller, J. F., Effect of neonatal thymectomy on the induction of sarcomata in C57 BL mice. *Nature* 1965; **205**: 1124–1125.
11. Stutman, O., Immunodepression and malignancy. *Adv. Cancer Res.* 1975; **22**: 261–422.
12. Pantelouris, E. M., Absence of thymus in a mouse mutant. *Nature* 1968; **217**, 370–371.
13. Stutman, O., Tumor development after 3-methylcholanthrene in immunologically deficient athymic-nude mice. *Science* 1974; **183**: 534–536.
14. Rygaard, J., Povlsen, C. O., The mouse mutant nude does not develop spontaneous tumours. An argument against immunological surveillance. *Acta Pathol Microbiol Scand B Microbiol Immunol* 1974; **82**: 99–106.
15. Ribeiro, V. S., Hasan, M., Wilson, A., Boucontet, L., *et al.*, Cutting edge: Thymic NK cells develop independently from T cell precursors. *J Immunol* 2010; **185**: 4993–4997.
16. Maleckar, J. R., Sherman, L. A., The composition of the T cell receptor repertoire in nude mice. *J Immunol* 1987; **138**: 3873–3876.
17. Ikehara, S., Good, R. A., Modak, M. J., Pahwa, R. N., Adenosine 5'-triphosphate (ATP)-mediated stimulation and suppression of DNA synthesis in lymphoid cells. II. Suppressive effect of ATP on murine T-cell functions. *Int. J. Immunopharmacol.* 1983; **5**: 567–573.
18. van der Bruggen, P., Traversari, C., Chomez, P., Lurquin, C., *et al.*, A gene encoding an antigen recognized by cytolytic T lymphocytes on a human melanoma. *Science* 1991; **254**: 1643–1647.
19. Pardoll, D., Does the immune system see tumors as foreign or self? *Annu. Rev. Immunol.* 2003; **21**: 807–839.
20. Caso, F., Ramonda, R., Del Puente, A., Darda, M. A., *et al.*, Influenza vaccine with adjuvant on disease activity in psoriatic arthritis patients under anti-TNF-alpha therapy. *Clin. Exp. Rheumatol.* 2016; **34**: 507–512.
21. Dunn, G. P., Bruce, A. T., Ikeda, H., Old, L. J., Schreiber, R. D., Cancer immunoediting: From immunosurveillance to tumor escape. *Nat Immunol* 2002; **3**, 991–998.
22. Swann, J. B., Smyth, M. J., Immune surveillance of tumors. *J Clin Invest* 2007; **117**, 1137–1146.
23. Shiku, H., Importance of CD4+ helper T-cells in antitumor immunity. *Int. J. Hematol.* 2003; **77**: 435–438.
24. Tuting, T., Storkus, W. J., Lotze, M. T., Gene-based strategies for the immunotherapy of cancer. *J Mol Med (Berl)* 1997; **75**: 478–491.
25. Hadden, J. W., Immunodeficiency and cancer: Prospects for correction. *Int Immunopharmacol* 2003; **3**: 1061–1071.

26. Wan, Y. Y., Flavell, R. A., 'Yin-Yang' functions of transforming growth factor-beta and T regulatory cells in immune regulation. *Immunol. Rev.* 2007; **220**: 199–213.
27. Ohkura, N., Hamaguchi, M., Sakaguchi, S., FOXP3+ regulatory T cells: Control of FOXP3 expression by pharmacological agents. *Trends Pharmacol. Sci.* 2011; **32**: 158–166.
28. Liyanage, U. K., Moore, T. T., Joo, H. G., Tanaka, Y., *et al.*, Prevalence of regulatory T cells is increased in peripheral blood and tumor microenvironment of patients with pancreas or breast adenocarcinoma. *J Immunol* 2002; **169**: 2756–2761.
29. Perez, S. A., Karamouzis, M. V., Skarlos, D. V., Ardavanis, A., *et al.*, CD4+CD25+ regulatory T-cell frequency in HER-2/neu (HER)-positive and HER-negative advanced-stage breast cancer patients. *Clin Cancer Res* 2007; **13**: 2714–2721.
30. Knutson, K. L., Dang, Y., Lu, H., Lukas, J., *et al.*, IL-2 immunotoxin therapy modulates tumor-associated regulatory T cells and leads to lasting immune-mediated rejection of breast cancers in neu-transgenic mice. *J Immunol* 2006; **177**: 84–91.
31. Chen, W., Jin, W., Hardegen, N., Lei, K. J., *et al.*, Conversion of peripheral CD4+CD25- naive T cells to CD4+CD25+ regulatory T cells by TGF-beta induction of transcription factor Foxp3. *J Exp Med* 2003; **198**: 1875–1886.
32. Yang, Y., Paik, J. H., Cho, D., Cho, J. A., Kim, C. W., Resveratrol induces the suppression of tumor-derived CD4(+)CD25(+) regulatory T cells. *Int Immunopharmacol* 2008; **8**: 542–547.
33. Bhattacharyya, S., Md Sakib Hossain, D., Mohanty, S., Sankar Sen, G., *et al.*, Curcumin reverses T cell-mediated adaptive immune dysfunctions in tumor-bearing hosts. *Cell Mol Immunol* 2010; **7**: 306–315.
34. Guo, T. L., Chi, R. P., Hernandez, D. M., Auttachoat, W., Zheng, J. F., Decreased 7,12-dimethylbenz[a]anthracene-induced carcinogenesis coincides with the induction of antitumor immunities in adult female B6C3F1 mice pretreated with genistein. *Carcinogenesis* 2007; **28**: 2560–2566.
35. Choi, Y. J., Yang, K. M., Kim, S. D., Yoo, Y. H., *et al.*, Resveratrol analogue HS-1793 induces the modulation of tumor-derived T cells. *Exp Ther Med* 2012; **3**: 592–598.
36. Guan, H., Singh, N. P., Singh, U. P., Nagarkatti, P. S., Nagarkatti, M., Resveratrol prevents endothelial cells injury in high-dose interleukin-2 therapy against melanoma. *PLoS One* 2012; **7**, e35650.
37. Yang, H., Zhang, A., Zhang, Y., Ma, S., Wang, C., Resveratrol Pretreatment Protected against Cerebral Ischemia/Reperfusion Injury in Rats via Expansion of T Regulatory Cells. *Journal of stroke and cerebrovascular diseases: The official journal of National Stroke Association* 2016; **25**: 1914–1921.
38. Unutmaz, D., Pulendran, B., The gut feeling of Treg cells: IL-10 is the silver lining during colitis. *Nat Immunol* 2009; **10**: 1141–1143.
39. Ogino, H., Nakamura, K., Ihara, E., Akiho, H., Takayanagi, R., CD4+CD25+ regulatory T cells suppress Th17-responses in an experimental colitis model. *Dig. Dis. Sci.* 2011; **56**: 376–386.

40. Abraham, C., Cho, J., Interleukin-23/Th17 pathways and inflammatory bowel disease. *Inflamm. Bowel Dis.* 2009; **15**: 1090–1100.
41. Yao, J., Wei, C., Wang, J. Y., Zhang, R., *et al.*, Effect of resveratrol on Treg/Th17 signaling and ulcerative colitis treatment in mice. *World J Gastroenterol* 2015; **21**: 6572–6581.
42. Wenbin, Z., Guojun, G., Resveratrol Ameliorates Diabetes-induced Renal Damage through Regulating the Expression of TGF-beta1, Collagen IV and Th17/Treg-related Cytokines in Rats. *The West Indian medical journal* 2014; **63**: 20–25.
43. Fridman, W. H., Pages, F., Sautes-Fridman, C., Galon, J., The immune contexture in human tumours: impact on clinical outcome. *Nat Rev Cancer* 2012; **12**: 298–306.
44. Weng, B. B., Lin, W. S., Chang, J. C., Chiou, R. Y., The phytogestrogenic stilbenes, arachidin-1 and resveratrol, modulate regulatory T cell functions responsible for successful aging in aged ICR mice. *Int J Mol Med* 2016; **38**: 1895–1904.
45. Noh, K. T., Chae, S. H., Chun, S. H., Jung, I. D., *et al.*, Resveratrol suppresses tumor progression via the regulation of indoleamine 2,3-dioxygenase. *Biochem Biophys Res Commun* 2013; **431**: 348–353.
46. Trinchieri, G., Interleukin-12 and the regulation of innate resistance and adaptive immunity. *Nat Rev Immunol* 2003; **3**: 133–146.
47. Singh, U. P., Singh, N. P., Singh, B., Hofseth, L. J., *et al.*, Role of resveratrol-induced CD11b(+) Gr-1(+) myeloid derived suppressor cells (MDSCs) in the reduction of CXCR3(+) T cells and amelioration of chronic colitis in IL-10(−/−) mice. *Brain Behav Immun* 2012; **26**: 72–82.
48. Bakheet, S. A., Alzahrani, M. Z., Ansari, M. A., Nadeem, A., *et al.*, Resveratrol Ameliorates Dysregulation of Th1, Th2, Th17, and T Regulatory Cell-Related Transcription Factor Signaling in a BTBR T + tf/J Mouse Model of Autism. *Mol Neurobiol* 2017; **54**: 5201–5212.
49. Ribeiro, F. V., Pino, D. S., Franck, F. C., Benatti, B. B., *et al.*, Resveratrol Inhibits Periodontitis-Related Bone Loss in Rats Subjected to Cigarette Smoke Inhalation. *J Periodontol* 2017; **88**, 788–798.
50. Wang, Z. L., Luo, X. F., Li, M. T., Xu, D., *et al.*, Resveratrol possesses protective effects in a pristane-induced lupus mouse model. *PLoS One* 2014; **9**, e114792.
51. Wang, D., Li, S. P., Fu, J. S., Bai, L., Guo, L., Resveratrol augments therapeutic efficiency of mouse bone marrow mesenchymal stem cell-based therapy in experimental autoimmune encephalomyelitis. *International journal of developmental neuroscience: The official journal of the International Society for Developmental Neuroscience* 2016; **49**: 60–66.
52. Kim, D. H., Lee, Y. G., Park, H. J., Lee, J. A., *et al.*, Piceatannol inhibits effector T cell functions by suppressing TcR signaling. *Int Immunopharmacol* 2015; **25**: 285–292.
53. Lotze, M. T., Tracey, K. J., High-mobility group box 1 protein (HMGB1): nuclear weapon in the immune arsenal. *Nat Rev Immunol* 2005; **5**: 331–342.

54. Makarov, S. S., NF-kappaB as a therapeutic target in chronic inflammation: Recent advances. *Mol. Med. Today* 2000; **6**: 441–448.
55. Karuppagounder, V., Arumugam, S., Thandavarayan, R. A., Pitchaimani, V., *et al.*, Resveratrol attenuates HMGB1 signaling and inflammation in house dust mite-induced atopic dermatitis in mice. *Int Immunopharmacol* 2014; **23**: 617–623.
56. Han, S. Y., Choi, Y. J., Kang, M. K., Park, J. H., Kang, Y. H., Resveratrol Suppresses Cytokine Production Linked to FcepsilonRI-MAPK Activation in IgE-Antigen Complex-Exposed Basophilic Mast Cells and Mice. *Am J Chin Med* 2015; **43**: 1605–1623.
57. Korn, T., Bettelli, E., Oukka, M., Kuchroo, V. K., IL-17 and Th17 Cells. *Annu. Rev. Immunol.* 2009; **27**: 485–517.
58. Leonard, W. J., Spolski, R., Interleukin-21: a modulator of lymphoid proliferation, apoptosis and differentiation. *Nat Rev Immunol* 2005; **5**: 688–698.
59. Xuzhu, G., Komai-Koma, M., Leung, B. P., Howe, H. S., *et al.*, Resveratrol modulates murine collagen-induced arthritis by inhibiting Th17 and B-cell function. *Ann Rheum Dis* 2012; **71**, 129-135.
60. Limagne, E., Thibaudin, M., Euvrard, R., Berger, H., *et al.*, Sirtuin-1 Activation Controls Tumor Growth by Impeding Th17 Differentiation via STAT3 Deacetylation. *Cell reports* 2017; **19**, 746-759.
61. Sharma, S., Chopra, K., Kulkarni, S. K., Agrewala, J. N., Resveratrol and curcumin suppress immune response through CD28/CTLA-4 and CD80 co-stimulatory pathway. *Clin Exp Immunol* 2007; **147**: 155–163.
62. Miyazaki, Y., Li, R., Rezk, A., Misirliyan, H., *et al.*, A novel microRNA-132-sirtuin-1 axis underlies aberrant B-cell cytokine regulation in patients with relapsing-remitting multiple sclerosis [corrected]. *PLoS One* 2014; **9**, e105421.
63. Espinoza, J. L., Takami, A., Trung, L. Q., Kato, S., Nakao, S., Resveratrol prevents EBV transformation and inhibits the outgrowth of EBV-immortalized human B cells. *PLoS One* 2012; **7**, e51306.
64. Palucka, K. A., Taquet, N., Sanchez-Chapuis, F., Gluckman, J. C., Dendritic cells as the terminal stage of monocyte differentiation. *J Immunol* 1998; **160**: 4587–4595.
65. Crowley, M., Inaba, K., Steinman, R. M., Dendritic cells are the principal cells in mouse spleen bearing immunogenic fragments of foreign proteins. *J Exp Med* 1990; **172**: 383–386.
66. Steinman, R. M., The dendritic cell system and its role in immunogenicity. *Annu. Rev. Immunol.* 1991; **9**: 271–296.
67. Svajger, U., Obermajer, N., Jeras, M., Dendritic cells treated with resveratrol during differentiation from monocytes gain substantial tolerogenic properties upon activation. *Immunology* 2010; **129**: 525–535.
68. Kim, G. Y., Cho, H., Ahn, S. C., Oh, Y. H., *et al.*, Resveratrol inhibits phenotypic and functional maturation of murine bone marrow-derived dendritic cells. *Int Immunopharmacol* 2004; **4**, 245-253.

69. Silva, A. M., Oliveira, M. I., Sette, L., Almeida, C. R., *et al.*, Resveratrol as a natural anti-tumor necrosis factor-alpha molecule: Implications to dendritic cells and their crosstalk with mesenchymal stromal cells. *PLoS One* 2014; **9**, e91406.
70. Jeong, Y. I., Jung, I. D., Lee, J. S., Lee, C. M., *et al.*, (-)-Epigallocatechin gallate suppresses indoleamine 2,3-dioxygenase expression in murine dendritic cells: Evidences for the COX-2 and STAT1 as potential targets. *Biochem Biophys Res Commun* 2007; **354**: 1004–1009.
71. Jeong, Y. I., Kim, S. W., Jung, I. D., Lee, J. S., *et al.*, Curcumin suppresses the induction of indoleamine 2,3-dioxygenase by blocking the Janus-activated kinase-protein kinase Cdelta-STAT1 signaling pathway in interferon-gamma-stimulated murine dendritic cells. *J Biol Chem* 2009; **284**: 3700–3708.
72. Wang, X., Zhang, C., Huang, G., Han, D., *et al.*, Resveratrol inhibits dysfunction of dendritic cells from chronic obstructive pulmonary disease patients through promoting miR-34. *International journal of clinical and experimental pathology* 2015; **8**: 5145–5152.
73. Buttari, B., Profumo, E., Facchiano, F., Ozturk, E. I., *et al.*, Resveratrol prevents dendritic cell maturation in response to advanced glycation end products. *Oxid Med Cell Longev* 2013; **2013**: 574029.
74. Barbosa, J. P., Neves, A. R., Silva, A. M., Barbosa, M. A., *et al.*, Nanostructured lipid carriers loaded with resveratrol modulate human dendritic cells. *Int J Nanomedicine* 2016; **11**: 3501–3516.
75. Altamemi, I., Murphy, E. A., Catroppo, J. F., Zumbrun, E. E., *et al.*, Role of microRNAs in resveratrol-mediated mitigation of colitis-associated tumorigenesis in Apc(Min/+) mice. *J Pharmacol Exp Ther* 2014; **350**: 99–109.
76. Rieder, S. A., Nagarkatti, P., Nagarkatti, M., Multiple anti-inflammatory pathways triggered by resveratrol lead to amelioration of staphylococcal enterotoxin B-induced lung injury. *Br J Pharmacol* 2012; **167**: 1244–1258.
77. Dutcher, J. P., Current status of interleukin-2 therapy for metastatic renal cell carcinoma and metastatic melanoma. *Oncology (Williston Park)* 2002; **16**: 4–10.
78. Schwartz, R. N., Stover, L., Dutcher, J. P., Managing toxicities of high-dose interleukin-2. *Oncology (Williston Park)* 2002; **16**: 11–20.
79. Vivier, E., Raulet, D. H., Moretta, A., Caligiuri, M. A., *et al.*, Innate or adaptive immunity? The example of natural killer cells. *Science* 2011; **331**: 44–49.
80. Bachleda, P., Vrzal, R., Dvorak, Z., Resveratrol enhances NK cell cytotoxicity: Possible role for aryl hydrocarbon receptor. *J Cell Physiol* 2010; **225**: 289–290.
81. Falchetti, R., Fuggetta, M. P., Lanzilli, G., Tricarico, M., Ravagnan, G., Effects of resveratrol on human immune cell function. *Life Sci* 2001; **70**: 81–96.
82. Li, T., Fan, G. X., Wang, W., Li, T., Yuan, Y. K., Resveratrol induces apoptosis, influences IL-6 and exerts immunomodulatory effect on mouse

lymphocytic leukemia both *in vitro* and *in vivo*. *Int Immunopharmacol* 2007; **7**: 1221–1231.
83. Lu, C. C., Chen, J. K., Resveratrol enhances perforin expression and NK cell cytotoxicity through NKG2D-dependent pathways. *J Cell Physiol* 2010; **223**: 343–351.
84. Billard, C., Izard, J. C., Roman, V., Kern, C., *et al.*, Comparative antiproliferative and apoptotic effects of resveratrol, epsilon-viniferin and vine-shots derived polyphenols (vineatrols) on chronic B lymphocytic leukemia cells and normal human lymphocytes. *Leuk Lymphoma* 2002; **43**: 1991–2002.
85. Clement, M. V., Hirpara, J. L., Chawdhury, S. H., Pervaiz, S., Chemopreventive agent resveratrol, a natural product derived from grapes, triggers CD95 signaling-dependent apoptosis in human tumor cells. *Blood* 1998; **92**: 996–1002.
86. Lu, J., Ho, C. H., Ghai, G., Chen, K. Y., Resveratrol analog, 3,4,5,4'-tetrahydroxystilbene, differentially induces pro-apoptotic p53/Bax gene expression and inhibits the growth of transformed cells but not their normal counterparts. *Carcinogenesis* 2001; **22**: 321–328.
87. Colin, D., Gimazane, A., Lizard, G., Izard, J. C., *et al.*, Effects of resveratrol analogs on cell cycle progression, cell cycle associated proteins and 5fluorouracil sensitivity in human derived colon cancer cells. *Int J Cancer* 2009; **124**: 2780–2788.

Chapter 9

Resveratrol: A Therapeutic Tool Against Age-Related Lung Diseases

Cecilia G. Sanchez

Department of Medicine,
University of Arizona Health Sciences,
Tucson, AZ 85721, USA

This chapter reviews the literature supporting the protective role of resveratrol in age-related respiratory diseases, including lung cancer, Chronic Obstructive Pulmonary Disease, Pulmonary Fibrosis and common pulmonary infections of the elderly. The mechanisms of RSV action relevant to each of those respiratory diseases are discussed as well as alternatives for pulmonary delivery and future directions. The research summarized here will provide a better understanding of the multiple targets involved in the pathogenesis of age-related lung diseases that are known to be therapeutically targeted by resveratrol and to allow a rational approach for developing safe and effective interventions against age-related lung diseases.

9.1. Introduction

The dysfunction and reduction of anti-aging molecules are associated with the development of systemic inflammation and age-related lung diseases.[1–4] Resveratrol (RSV) can modulate inflammation after acute lung injury.[5,6] In addition, RSV reduces chronic inflammation in murine models of Chronic Obstructive Pulmonary Disease (COPD) and asthma.[5]

Moreover, RSV has been shown to regulate the mechanistic target of rapamycin complex 1 (mTORC1) signaling pathway. The mTOR pathway can influence longevity and aging. This pathway is hyper-activated in hyper-proliferative disorders, a variety of cancers and tumor syndrome, as well as hypertrophic fibroblasts relevant to fibrotic disorders.[7–10]

There is some evidence suggesting that RSV can bind directly to DNA and RNA.[11] RSV can also act as a specific modulator of alternative splicing.[12,13] As a key process in the regulation of gene expression, alternative splicing is able to influence virtually all cellular processes, including apoptosis, cell cycling, DNA repair and glucose metabolism.[14] Both gene expression and alternative splicing undergo dramatic changes during aging and age-related lung diseases including COPD and Idiopathic Pulmonary Fibrosis.[15–22]

Resveratrol-induced autophagy seems to be critical in mediating many beneficial effects of resveratrol. It is known that autophagy diminishes with aging. Additionally, accelerated aging can be attributed to reduced autophagy. Specifically, in age-related lung disorders such as idiopathic pulmonary fibrosis, impaired autophagy may contribute to its pathogenesis.[23] In other diseases such as lung cancer, autophagy may be critical for tumor cell survival as well as therapeutic efficacy.[24–26]

Recent studies demonstrate that tobacco-smoke, e-cigarette vapor and/or inhaled-nicotine exposure induces proteostasis/autophagy-impairment, leading to perinuclear accumulation of polyubiquitinated proteins.[27–30] In consequence, it is anticipated that an intervention with autophagy-inducing drugs could reduce alveolar damage, tissue remodeling and lung function decline, resulting in a decrease in the current mortality rates by age-related lung diseases. Early studies in lung aging demonstrated that autophagic capacity in the aging lung is reduced. RSV can overcome the effect of TGF-β1 and the impaired autophagic response in human lungs fibroblasts, relevant to the pathogenesis of pulmonary fibrosis.[23]

The mechanisms of action described above are only few from the large array of biological targets supporting the therapeutic use of RSV in the fight against age-related lung diseases.

9.2. RSV in lung cancer

Lung cancer is a leading cause of cancer-related death in most developed countries. Age is a risk factor for cancer, contributing to the increase in the number of lung cancer patients over age of 65 years old. The median age for the diagnosis for lung cancer patients is between 64 and 70 years old. At advanced age, the presence of comorbid conditions influences effective treatment decisions.[31,32] Interventions aiming to promote healthy aging could contribute to reduce the incidence of lung cancer and to prevent multi-morbidities in the aging population.

An epidemiologic study reported that high intake of beer or spirits is correlated with increased relative risk of lung cancer, whereas consumption of red wine, which contains RSV (trans-3,4′5-trihydroxystilbene), is correlated with a reduced risk.[33]

RSV can inhibit chemical induced lung tumorigenesis.[34] Mechanistically, RSV was found to prevent tumor growth and lung metastasis in an animal model of Lewis lung carcinoma via inhibition of DNA synthesis and angiogenesis.[35] In addition, induction of pro-apoptotic p53/Bax expression by RSV promotes cancer cell death while preserving normal cells.[36] Other studies determined that RSV induces cell cycle arrest and stress-induced senescence, evident by measured changes in expression of BRCA1, cyclin B1, pRb and p21 as well as telomere instability.[37] Using microarray analysis in adenocarcinoma cells, it was demonstrated that RSV induces gene and protein expression involved in multiple biological processes. One example is the up-regulation of p53 and p21 and the activation of the caspases.[38] A549 cells treated with the RSV analogue 3,4,4′-trihydroxy-trans-stilbene showed a dose-dependent induction of apoptosis and autophagy concomitant with down-regulation of Bcl-2, inhibition of the mTOR pathway and elevated ROS levels.[39] In other study, A549 cells exposed to oxidative damage, via hydrogen peroxide, were treated with RSV-loaded nanoparticles. In this model, RSV reduced significantly the H_2O_2 induced ROS levels by activating Nrf2-Keap1 signaling leading to the accumulation of Nrf2.[40] Nrf2 is critical in regulating the antioxidant response that protect against oxidative damage triggered by injury and inflammation.

The upstream effect of RSV in lung cancer has been attributed in part to the activation of SIRT1 deacetylase (SIRT1), a deacetylate of histones and non-histone proteins involved in a variety of functions including the regulation of metabolism, stress response, survival, senescence, inflammation, endothelial functions, and circadian rhythms.[37,41] Decreased SIRT1 was implicated in more-invasive types of lung cancers in humans, by contrast SIRT1 activation by RSV antagonized migration of lung cancer cells by suppressing EMT in vitro.[42] Others attributed the effect of RSV to the induction of adenosine-monophosphate-activated protein kinase (AMPK).[43] AMPK is a key player in energy homeostasis. Both, Sirt1 and AMPK are energy sensors that share many target molecules.

Until today the molecular mechanisms of action for RSV in cancer cells remain controversial. Probably multiple mechanisms contribute to its anti-cancer potential. A recent study using human cancer cells from different origins including lung, liver and breast cancer revealed that the sensitivity to RSV is cell dependent and that RSV-induced DNA fragmentation is only detected in leukemic cells. In those cells, the activation of SIRT1 and adenosine-monophosphate-activated protein kinase was not responsible for the effects of RSV.[44]

9.2.1. *TGF-β1 signaling in lung cancer and metastasis*

TGF-β1 is known to induce Epithelial-to-mesenchymal transition (EMT), a cellular process in which epithelial polarized cells become motile mesenchymal-appearing cells, to promote lung adenocarcinoma invasion and metastasis. RSV inhibits TGF-β1 mediated EMT. Adenocarcinoma cells treated with 20 μM RSV express higher levels of the epithelial phenotype markers such as E-cadherin and reduced expression of the mesenchymal phenotype markers, like Fibronectin and Vimentin, during the initiation of TGF-β1-induced EMT.[45] The inhibition of TGF-β1 mediated EMT by RSV was confirmed in cancer cells from diverse origins.[46] Microarray analysis confirmed that RSV exerts a direct impact on the TGF-β pathway, specifically the Smad proteins.[38] Furthermore, it was found that RSV activates SIRT7 deacetylase activity, another member of the sirtuins family, contributing to the inhibition of SMAD4-mediated breast cancer lung metastasis.[47] Interestingly, in studies using immortalized mammary gland cells, RSV counteracts the TGF-β1 EMT phenotype and affects the alternative exon usage of pre-mRNAs that encode crucial factors in adhesion and migration -including CD44, ENAH, and FGFR2.[12]

9.2.2. *Therapeutic effects of RSV in animal models of lung cancer*

Contrary to the extensive in vitro studies, few reports are available involving the role of RSV in lung cancer *in vivo*. Studies with female mice injected with a RSV analog, 4′-chloro-3,5-dihydroxystilbene DHS 50 mg/kg, intraperitoneally for seven days, showed a decrease in tumor growth.[48] DHS induced a rapid and persistent increase in the intracellular reactive oxygen species in the cells.[48] In fact, administration of DHS in drinking water can also decrease cell proliferation, tumor angiogenesis and metastasis of other tumors including LLC tumors.[49,50] RSV reduced M2 polarization of human monocyte derived macrophage induced by lung cancer cells.[50] Oral administration, of 1 or 3 g/kg/day RSV for 28 days, also reduces significantly the tumor growth in female Balb/c mice injected subcutaneously with human non-small cell lung carcinoma cells (SPC-A-1) cells.[51] Finally, RSV reduces tumor growth and metastasis in nude mice inoculated with A549 cells.[52]

9.3. RSV in COPD

Chronic obstructive pulmonary disease (COPD) is a chronic, progressive, inflammatory disease and the fourth leading cause of death worldwide. Aging hallmarks are features of COPD, a disease that is characterized by airways and alveolar walls inflammation due to exposure to various stimuli. Cellular senescence is believed to induce inflammation or immuno-modulation in COPD.

Smoking is a risk factor for COPD, probably due to cigarette smoke-induced endoplasmic reticulum stress, cellular senescence and apoptosis. RSV was found to have a therapeutic effect in our rat COPD model, concomitant with inhibition of oxidative stress and inflammatory response.[5] RSV treatment decreased serum levels of IL-6 and IL-8. The effect of RSV has been attributed in part to the induction of the SIRT1/proliferator-activated receptor-γ coactivator-1α (PGC-1α) signaling pathways, a master regulator of mitochondrial biogenesis.[5]

9.3.1. *Cellular senescence and inflammation*

Cellular senescence is found in lung epithelial cells, endothelial cells, fibroblasts and progenitor or stem cells from patients with COPD.[53–56]

Cellular senescence is believed to induce inflammation or immunomodulation in COPD. Sirtuin-1 (SIRT1) is an important regulator of genomic stability and cellular senescence. SIRT1 is significantly reduced in the lungs of patients with COPD as well as animal models of COPD. A decrease in SIRT1 expression results in an increase of pro-inflammatory cytokines due to NF-κB activation, and in the acceleration of cellular senescence mediated by a decrease of FOXO3 mediated anti-senescent activity.[57] By contrast, induction of SIRT1 can protect against cigarette smoke-induced cellular senescence and apoptosis.[57,58]

Cellular senescence can be suppressed in a SIRT1 transgenic mouse model through a FOXO3 pathway that is deteriorated in SIRT1 knockout mice.[57,58] Recent studies revealed that RSV can induce upregulation of both SIRT1 as well as oxygen-regulated protein 150 (ORP150) mRNA and protein expression levels, which are associated with the decrease in apoptosis in human bronchial epithelial cell.[59] ORF150 is a member of the 70kDa heat shock protein family critical for cellular survival. Deficiency of ORP150 promotes endoplasmic reticulum (ER)-stress and apoptosis.[60,61] Alveolar cell senescence is also seen in COPD patients, co-expressing p16 (INK4a) and phosphorylated NF-κB, which could exacerbates pulmonary inflammation.[54]

An important study used prophylactic RSV inhalational treatments of terc−/− mice aged 2–4 months to test the effect of RSV to preserve youthful whole lung function and epithelial type 2 cells (AEC2) survival. The authors found that RSV treatments were effective in slowing aging-related degeneration of whole lung and AEC2.[62]

Lymphocytes from lungs of patients with COPD also have signs of accelerated aging.[53] To compensate for the decline in the adaptive immune function there is a paradoxical up-regulation of innate immune system resulting in a pro-inflammatory state.[63] The consequence is the recruitment and retention of neutrophils, macrophages, and CD4(+) and CD8(+) T cells in the lungs of smokers with COPD followed by a self-perpetuating cascade of inflammation and lung parenchymal damage.[64] Dietary intake of RSV has been shown to counteract immunosenescence, promotes rejuvenation with a significant increase in T helper cells (CD4(+)), in middle-aged and old Wistar male rats, and re-establishes age-related changes in the humoral response to antigen challenge, leading to rejuvenation.[65]

Gammadelta T cells ($\gamma\delta$ T cells) are T cells that express a unique T-cell receptor (TCR) composed of one γ-chain and one δ-chain. Gammadelta T cells are of low abundance in the body, are found in the gut mucosa, skin, lungs and uterus. A study in healthy Japanese population found that the absolute numbers of $\gamma\delta$ T cells decreases with aging.[66] $\gamma\delta$ T cells are involved in the initiation and propagation of immune responses. $\gamma\delta$ T cells have a unique capacity to present antigens to both CD8 and CD4 cells. A recent study assessed the effects of repeated doses of RSV for 28 days on circulating immune cells in healthy Japanese individuals. RSV promotes the increases in the numbers of circulating $\gamma\delta$ T cells and regulatory T cells and resulted in the decreases in the plasma levels of the pro-inflammatory cytokines TNF-α and MCP-1 and a significant increase in the plasma antioxidant activity compared to control individuals.[67,68] Unfortunately, there are a limited number of studies on pulmonary $\gamma\delta$ T-cells in chronic obstructive inflammatory lung diseases and their contribution to the pathogenesis still unclear.

9.4. RSV in pulmonary fibrosis

Idiopathic pulmonary fibrosis (IPF) is a severe and progressive fibrotic interstitial lung disease, which ultimately results in respiratory failure and death. The median age at diagnosis is 66 years, and the incidence increases with age.[67] Our studies as well as studies from other laboratories demonstrates that RSV inhibits the process of fibroblast-myofibroblast differentiation, inhibiting the TGF-β-induced α-smooth muscle actin (α-SMA) expression and collagen production, as well as cell proliferation of lung fibroblasts from control and patients with idiopathic pulmonary fibrosis (IPF).[23,69] Our studies revealed that RSV promotes effective autophagy as well as expression of Sirtuin3 and mitochondrial homeostasis, key players in the maintenance of the fibroblast phenotype.[23,70]

In vivo, RSV has anti-inflammatory effects in bleomycin-induced lung damage, a common animal model of experimental pulmonary fibrosis.[71] Other studies investigated the role of RSV during the phase of fibrogenesis in the bleomycin animal model, by given the RSV 14 days after bleomycin and continued until sacrifice. The studies demonstrate a significant reduction in the fibrotic tissue compared to controls without RSV.[71a]

We can expect that induction of Sirtuin1 by RSV would be potentially favorable against pulmonary fibrosis. A recent study revealed that Sirt1 as an important regulator of lung fibrosis and a potential novel therapeutic strategy for IPF. Sirt1 expression was found to be up-regulated of Sirt1 in BLM-induced lung fibrosis, as well as in the lungs of IPF patients. Nevertheless, it was the activation or overexpression of Sirt1 that attenuates TGF-β-mediated lung fibroblast differentiation and diminishes the severity of experimental lung fibrosis in mice. By contrast, deficiency of Sirt1 promoted the pro-fibrogenic activity of TGF-β 1 in lung fibroblasts.[72] Epithelial senescence as well as aberrant splicing have been described as contributors to the pathogenesis of idiopathic pulmonary fibrosis.[21,73] As we discussed previously, RSV can prevent cellular senescence in alveolar epithelial cells. Nevertheless, the impact of RSV in the correction of aberrant splicing in IPF needs to be investigated.

RSV also has a preventive potential for bleomycin-induced pulmonary fibrosis, an important consideration for future prophylactic use in the elderly.

9.5. The use of RSV against major pulmonary infections in the elderly

Influenza virus infection causes high morbidity and mortality rates in the aging population. The elderly tend to have a delayed onset of many immune responses as well as prolonged inflammatory responses, leading to an overall weakened response to infection. RSV shows anti-influenza effects. The effect was attributed to the membrane damage caused by the drug and the hindered transfer of polyomavirus from the endoplasmic reticulum to the cytoplasm. Those studies support the use of RSV for the clinical control of proliferative and viral pathologies.[74] In a model of alveolar epithelial cells A549 infected with an influenza virus, RSV was found to attenuate influenza viral replication and to inhibit H1N1 replication via Toll-like receptor 9-induced interferon beta production; IFN-β likely acts synergistically with RES.[75] RSV also significantly improved survival and decreased pulmonary viral titers in influenza virus-infected mice, suggesting that RSV can be a valuable anti-influenza drug.[76]

Influenza is not the only major respiratory infection in the aging population. Bacterial lung infection is a leading cause of death for people over 65 years old. Those patients often require intensive care and mechanical ventilation.

Administration of antibiotics is the standard of care for bacterial pneumonia, but its overuse has led to the emergence of resistant organisms. In consequence, there is a need for new therapeutic strategies to help minimize the effects of bacterial pneumonia in the aging population.

BPIFA1/SPLUNC1 (short palate, lung, and nasal epithelium clone 1) is an abundantly secreted innate immune protein in the mammalian respiratory tract that exerts bacteriostatic and anti-biofilm effects, binds to lipopolysaccharide, and acts as a fluid-spreading surfactant. SPLUNC1 is important in the initial phase of infection because it inhibits the binding and entry of virus and bacteria into airway epithelial cells. Importantly, RSV has been shown to upregulate SPLUNC1 expression in models of streptococcus pneumoniae (SP) infection, and enhance cell protection in a concentration- and time-dependent manner.[77]

9.6. Pulmonary delivery of RSV

Bioavailability seems to be a limitation in the clinical utility of RSV, such that it is difficult to maintain a therapeutically relevant level in the bloodstream and much less in the lung. To overcome this problem, the direct delivery into the lung could contribute to bring the right concentrations of RSV in the right place at the right time. Pharmaceutical companies and researchers are trying to combat this by either re-deriving or reformulating RSV. Larrosa and colleagues have reformulated RSV to devise a set of RSV-prodrug forms that resist absorption in the upper gastrointestinal tract, allowing it to become active through normal metabolic processes at a later time.[78] Others stabilized the trans-RSV into protein microparticles using sodium caseinate (SC); SC-based microencapsulation increased the stability of trans-RSV to UVA light exposure and simulated digestion conditions.[79] We know now that RSV can be potentially developed as a dry powder for inhalation for the treatment of inflammatory lung diseases like COPD. A recent study demonstrates that RSV can be manufactured with a spherical morphology and particle diameter less than 5 mum, allowing for a spray-dried RSV to be used in an inhalable formulation. The cytotoxicity results of RSV on Calu-3 revealed that the cells could tolerate high concentration of RSV and effective against the effects of TGF-β1.[80]

A multidrug inhalable formulation with anti-oxidant and anti-inflammatory activities was proposed using inhalable micro-particles containing budesonide (BD) and RSV for treatment of chronic lung diseases. The co-spray dried microparticle formulations of RSV and BD have morphology and aerosol properties suitable for inhalation drug delivery, with negligible *in vitro* toxicity and enhanced efficacy to control inflammation and oxidative stress in LPS-induced alveolar macrophages.[81] Nevertheless, there is a critical need to better understand the metabolism of RSV in the lung for the development of safe and effective targeted therapy.[82,83]

9.7. Conclusion

A significant amount of scientific evidence points out the beneficial effects of resveratrol in pulmonary health and the inhibition of cellular damage, inflammation, cellular differentiation and cellular senescence in a cell type-dependent manner — providing the biological relevance and therapeutic potential for a variety of pulmonary disorders discussed in this chapter. Nevertheless, the molecular mechanisms behind resveratrol's effects in the pulmonary system remain unclear. We expect that resveratrol has impacts on multiple cellular pathways. Long-term studies are necessary to evaluate the effects of resveratrol on the prevention and control of chronic lung diseases, as well as the effect of resveratrol in the pulmonary system using longitudinal studies in non-human primates. The development of resveratrol in a safe, inhalable formulation will lead to a new therapy for pro-inflammatory diseases of the pulmonary system, including COPD.

References

1. Li L, Zhang M, Zhang L, Cheng Y, Tu X, Lu Z. Klotho Regulates Cigarette Smoke-Induced Autophagy: Implication in Pathogenesis of COPD. *Lung* 2017; **195**(3): 295–301.
2. Pako J, Barta I, Balogh Z, Kerti M, Drozdovszky O, Bikov A, *et al.* Assessment of the Anti-Aging Klotho Protein in Patients with COPD Undergoing Pulmonary Rehabilitation. *Copd* 2017; **14**(2): 176–180.
3. Ito K, Mercado N. STOP accelerating lung aging for the treatment of COPD. *Experimental Gerontology* 2014; **59**: 21–27.
4. MacNee W. Accelerated lung aging: a novel pathogenic mechanism of chronic obstructive pulmonary disease (COPD). *Biochemical Society Transactions* 2009; **37**(Pt 4): 819–823.
5. Wang XL, Li T, Li JH, Miao SY, Xiao XZ. The Effects of RSV on Inflammation and Oxidative Stress in a Rat Model of Chronic Obstructive Pulmonary Disease. *Molecules* 2017; **22**(9).

6. Torun AC, Tutuncu S, Ustun B, Akdemir HU. A Study of the Therapeutic Effects of RSV on Blunt Chest Trauma-Induced Acute Lung Injury in Rats and the Potential Role of Endocan as a Biomarker of Inflammation. *Inflammation* 2017; **40**(5): 1803–1810.
7. Tang ZM, Zhai XX, Ding JC. Expression of mTOR/70S6K signaling pathway in pathological scar fibroblasts and the effects of RSV intervention. *Molecular Medicine Reports* 2017; **15**(5): 2546–2550.
8. Wang L, Wang C, Jia Y, Liu Z, Shu X, Liu K. RSV Increases Anti-Proliferative Activity of Bestatin Through Downregulating P-Glycoprotein Expression Via Inhibiting PI3K/Akt/mTOR Pathway in K562/ADR Cells. *Journal of Cellular Biochemistry* 2016; **117**(5): 1233–1239.
9. Alayev A, Berger SM, Holz MK. RSV as a novel treatment for diseases with mTOR pathway hyperactivation. *Ann N Y Acad Sci* 2015; **1348**(1): 116–123.
10. Widlund AL, Baur JA, Vang O. mTOR: more targets of RSV? *Expert Reviews in Molecular Medicine* 2013; **15**: e10.
11. Usha S, Johnson IM, Malathi R. Interaction of RSV and genistein with nucleic acids. *Journal of Biochemistry and Molecular Biology* 2005; **38**(2): 198–205.
12. Moshiri A, Puppo M, Rossi M, Gherzi R, Briata P. RSV limits epithelial to mesenchymal transition through modulation of KHSRP/hnRNPA1-dependent alternative splicing in mammary gland cells. *Biochimica et Biophysica Acta* 2017; **1860**(3): 291–298.
13. Markus MA, Marques FZ, Morris BJ. RSV, by modulating RNA processing factor levels, can influence the alternative splicing of pre-mRNAs. *PLoS One* 2011; **6**(12): e28926.
14. Johnson JM, Castle J, Garrett-Engele P, Kan Z, Loerch PM, Armour CD, *et al.* Genome-wide survey of human alternative pre-mRNA splicing with exon junction microarrays. *Science* 2003; **302**(5653): 2141–2144.
15. de Vries M, Faiz A, Woldhuis RR, Postma DS, de Jong TV, Sin DD, *et al.* Lung tissue gene-expression signature for the ageing lung in COPD. *Thorax* 2017.
16. Corley M, Solem A, Phillips G, Lackey L, Ziehr B, Vincent HA, *et al.* An RNA structure-mediated, posttranscriptional model of human alpha-1-antitrypsin expression. *Proceedings of the National Academy of Sciences of the United States of America* 2017; **114**(47): E10244–E10253.
17. Lackey L, McArthur E, Laederach A. Increased Transcript Complexity in Genes Associated with Chronic Obstructive Pulmonary Disease. *PLoS One* 2015; **10**(10): e0140885.
18. Barber R, Baillie GS, Bergmann R, Shepherd MC, Sepper R, Houslay MD, *et al.* Differential expression of PDE4 cAMP phosphodiesterase isoforms in inflammatory cells of smokers with COPD, smokers without COPD, and nonsmokers. *American Journal of Physiology Lung Cellular and Molecular Physiology* 2004; **287**(2): L332–343.
19. Barratt SL, Blythe T, Jarrett C, Ourradi K, Shelley-Fraser G, Day MJ, *et al.* Differential Expression of VEGF-Axxx Isoforms Is Critical for

Development of Pulmonary Fibrosis. *Am J Respir Crit Care Med* 2017; **196**(4): 479–493.
20. Nance T, Smith KS, Anaya V, Richardson R, Ho L, Pala M, *et al.* Transcriptome analysis reveals differential splicing events in IPF lung tissue. *PLoS One* 2014; **9**(3): e92111.
21. Deng N, Sanchez CG, Lasky JA, Zhu D. Detecting splicing variants in idiopathic pulmonary fibrosis from non-differentially expressed genes. *PLoS One* 2013; **8**(7): e68352.
22. Mitchell SM, Vaxillaire M, Thomas H, Parrizas M, Benmezroua Y, Costa A, *et al.* Rare variants identified in the HNF-4 alpha beta-cell-specific promoter and alternative exon 1 lack biological significance in maturity onset diabetes of the young and young onset Type II diabetes. *Diabetologia* 2002; **45**(9): 1344–1348.
23. Sosulski ML, Gongora R, Danchuk S, Dong C, Luo F, Sanchez CG. Deregulation of selective autophagy during aging and pulmonary fibrosis: the role of TGFbeta1. *Aging Cell* 2015; **14**(5): 774–783.
24. Kulkarni YM, Kaushik V, Azad N, Wright C, Rojanasakul Y, O'Doherty G, *et al.* Autophagy-Induced Apoptosis in Lung Cancer Cells by a Novel Digitoxin Analog. *J Cell Physiol* 2016; **231**(4): 817–828.
25. Chen YK, Wang HC, Ho CT, Chen HY, Li S, Chan HL, *et al.* 5-demethylnobiletin promotes the formation of polymerized tubulin, leads to G2/M phase arrest and induces autophagy via JNK activation in human lung cancer cells. *The Journal of Nutritional Biochemistry* 2015; **26**(5): 484–504.
26. Zhang J, Chiu J, Zhang H, Qi T, Tang Q, Ma K, *et al.* Autophagic cell death induced by RSV depends on the Ca(2+)/AMPK/mTOR pathway in A549 cells. *Biochemical Pharmacology* 2013; **86**(2): 317–328.
27. Bai X, Stitzel JA, Bai A, Zambrano CA, Phillips M, Marrack P, *et al.* Nicotine Impairs Macrophage Control of Mycobacterium tuberculosis. *Am J Respir Cell Mol Biol* 2017; **57**(3): 324–333.
28. Patel N, Trumph CD, Bodas M, Vij N. Role of second-hand smoke (SHS)-induced proteostasis/autophagy impairment in pediatric lung diseases. *Molecular and Cellular Pediatrics* 2017; **4**(1): 3.
29. Bodas M, Patel N, Silverberg D, Walworth K, Vij N. Master Autophagy Regulator Transcription Factor EB Regulates Cigarette Smoke-Induced Autophagy Impairment and Chronic Obstructive Pulmonary Disease-Emphysema Pathogenesis. *Antioxidants & Redox Signaling* 2017; **27**(3): 150–167.
30. Vij N, Chandramani-Shivalingappa P, Van Westphal C, Hole R, Bodas M. Cigarette smoke-induced autophagy impairment accelerates lung aging, COPD-emphysema exacerbations and pathogenesis. *American Journal of Physiology Cell Physiology* 2018; **314**(1): C73–C87.
31. Blanco JA, Toste IS, Alvarez RF, Cuadrado GR, Gonzalvez AM, Martin IJ. Age, comorbidity, treatment decision and prognosis in lung cancer. *Age and Ageing* 2008; **37**(6): 715–718.

32. Ramsey SD, Howlader N, Etzioni RD, Donato B. Chemotherapy use, outcomes, and costs for older persons with advanced non-small-cell lung cancer: evidence from surveillance, epidemiology and end results-Medicare. *Journal of Clinical Oncology: Official Journal of the American Society of Clinical Oncology* 2004; **22**(24): 4971–4978.
33. Benedetti A, Parent ME, Siemiatycki J. Consumption of alcoholic beverages and risk of lung cancer: results from two case-control studies in Montreal, Canada. *Cancer Causes & Control: CCC* 2006; **17**(4): 469–480.
34. Hecht SS, Kenney PM, Wang M, Trushin N, Agarwal S, Rao AV, *et al.* Evaluation of butylated hydroxyanisole, myo-inositol, curcumin, esculetin, RSV and lycopene as inhibitors of benzo[a]pyrene plus 4-(methylnitrosamino)-1-(3-pyridyl)-1-butanone-induced lung tumorigenesis in A/J mice. *Cancer Letters* 1999; **137**(2): 123–130.
35. Kimura Y, Okuda H. Effects of naturally occurring stilbene glucosides from medicinal plants and wine, on tumour growth and lung metastasis in Lewis lung carcinoma-bearing mice. *The Journal of Pharmacy and Pharmacology* 2000; **52**(10): 1287–1295.
36. Lu J, Ho CH, Ghai G, Chen KY. RSV analog, 3,4,5,4′-tetrahydroxystilbene, differentially induces pro-apoptotic p53/Bax gene expression and inhibits the growth of transformed cells but not their normal counterparts. *Carcinogenesis* 2001; **22**(2): 321–328.
37. Rusin M, Zajkowicz A, Butkiewicz D. RSV induces senescence-like growth inhibition of U-2 OS cells associated with the instability of telomeric DNA and upregulation of BRCA1. *Mechanisms of Ageing and Development* 2009; **130**(8): 528–537.
38. Whyte L, Huang YY, Torres K, Mehta RG. Molecular mechanisms of RSV action in lung cancer cells using dual protein and microarray analyses. *Cancer Research* 2007; **67**(24): 12007–12017.
39. Zhang L, Dai F, Sheng PL, Chen ZQ, Xu QP, Guo YQ. RSV analogue 3,4,4′-trihydroxy-trans-stilbene induces apoptosis and autophagy in human non-small-cell lung cancer cells in vitro. *Acta Pharmacologica Sinica* 2015; **36**(10): 1256–1265.
40. Kim JH, Park EY, Ha HK, Jo CM, Lee WJ, Lee SS, *et al.* RSV-loaded Nanoparticles Induce Antioxidant Activity against Oxidative Stress. *Asian-Australasian Journal of Animal Sciences* 2016; **29**(2): 288–298.
41. Pulliero A, Wu Y, Fenoglio D, Parodi A, Romani M, Soares CP, *et al.* Nanoparticles increase the efficacy of cancer chemopreventive agents in cells exposed to cigarette smoke condensate. *Carcinogenesis* 2015; **36**(3): 368–377.
42. Sun L, Li H, Chen J, Dehennaut V, Zhao Y, Yang Y, *et al.* A SUMOylation-dependent pathway regulates SIRT1 transcription and lung cancer metastasis. *Journal of the National Cancer Institute* 2013; **105**(12): 887–898.
43. Fan XX, Yao XJ, Xu SW, Wong VK, He JX, Ding J, *et al.* (Z)3,4,5,4′-trans-tetramethoxystilbene, a new analogue of RSV, inhibits gefitinb-resistant non-small cell lung cancer via selectively elevating intracellular calcium level. *Scientific Reports* 2015; **5**: 16348.

44. Takashina M, Inoue S, Tomihara K, Tomita K, Hattori K, Zhao QL, *et al.* Different effect of RSV to induction of apoptosis depending on the type of human cancer cells. *International Journal of Oncology* 2017; **50**(3): 787–797.
45. Wang H, Zhang H, Tang L, Chen H, Wu C, Zhao M, *et al.* RSV inhibits TGF-beta1-induced epithelial-to-mesenchymal transition and suppresses lung cancer invasion and metastasis. *Toxicology* 2013; **303**: 139–146.
46. Ji Q, Liu X, Han Z, Zhou L, Sui H, Yan L, *et al.* RSV suppresses epithelial-to-mesenchymal transition in colorectal cancer through TGF-beta1/Smads signaling pathway mediated Snail/E-cadherin expression. *BMC cancer* 2015; **15**: 97.
47. Tang X, Shi L, Xie N, Liu Z, Qian M, Meng F, *et al.* SIRT7 antagonizes TGF-beta signaling and inhibits breast cancer metastasis. *Nature Communications* 2017; **8**(1): 318.
48. Wu JY, Tsai KW, Shee JJ, Li YZ, Chen CH, Chuang JJ, *et al.* 4′-Chloro-3,5-dihydroxystilbene, a RSV derivative, induces lung cancer cell death. *Acta Pharmacologica Sinica* 2010; **31**(1): 81–92.
49. Savio M, Ferraro D, Maccario C, Vaccarone R, Jensen LD, Corana F, *et al.* RSV analogue 4,4′-dihydroxy-trans-stilbene potently inhibits cancer invasion and metastasis. *Scientific Reports* 2016; **6**: 19973.
50. Sun L, Chen B, Jiang R, Li J, Wang B. RSV inhibits lung cancer growth by suppressing M2-like polarization of tumor associated macrophages. *Cellular Immunology* 2017; **311**: 86–93.
51. Zhao W, Bao P, Qi H, You H. RSV down-regulates survivin and induces apoptosis in human multidrug-resistant SPC-A-1/CDDP cells. *Oncology Reports* 2010; **23**(1): 279–286.
52. Yin HT, Tian QZ, Guan L, Zhou Y, Huang XE, Zhang H. In vitro and in vivo evaluation of the antitumor efficiency of RSV against lung cancer. *Asian Pacific Journal of Cancer Prevention: APJCP* 2013; **14**(3): 1703–1706.
53. Aoshiba K, Nagai A. Senescence hypothesis for the pathogenetic mechanism of chronic obstructive pulmonary disease. *Proceedings of the American Thoracic Society* 2009; **6**(7): 596–601.
54. Tsuji T, Aoshiba K, Nagai A. Alveolar cell senescence in patients with pulmonary emphysema. *Am J Respir Crit Care Med* 2006; **174**(8): 886–893.
55. Hashimoto Y, Sugiura H, Togo S, Koarai A, Abe K, Yamada M, *et al.* 27-Hydroxycholesterol accelerates cellular senescence in human lung resident cells. *American Journal of Physiology Lung Cellular and Molecular Physiology* 2016; **310**(11): L1028–1041.
56. Paschalaki KE, Starke RD, Hu Y, Mercado N, Margariti A, Gorgoulis VG, *et al.* Dysfunction of endothelial progenitor cells from smokers and chronic obstructive pulmonary disease patients due to increased DNA damage and senescence. *Stem Cells* 2013; **31**(12): 2813–2826.
57. Yao H, Chung S, Hwang JW, Rajendrasozhan S, Sundar IK, Dean DA, *et al.* SIRT1 protects against emphysema via FOXO3-mediated reduction

of premature senescence in mice. *The Journal of Clinical Investigation* 2012; **122**(6): 2032–2045.
58. Kusaczuk M, Cechowska-Pasko M. Molecular chaperone ORP150 in ER stress-related diseases. *Current Pharmaceutical Design* 2013; **19**(15): 2807–2818.
59. Zhang L, Guo X, Xie W, Li Y, Ma M, Yuan T, *et al.* RSV exerts an anti-apoptotic effect on human bronchial epithelial cells undergoing cigarette smoke exposure. *Molecular Medicine Reports* 2015; **11**(3): 1752–1758.
60. Tagawa Y, Hiramatsu N, Kasai A, Hayakawa K, Okamura M, Yao J, *et al.* Induction of apoptosis by cigarette smoke via ROS-dependent endoplasmic reticulum stress and CCAAT/enhancer-binding protein-homologous protein (CHOP). *Free Radical Biology & Medicine* 2008; **45**(1): 50–59.
61. Wu YB, Li HQ, Ren MS, Li WT, Lv XY, Wang L. CHOP/ORP150 ratio in endoplasmic reticulum stress: a new mechanism for diabetic peripheral neuropathy. *Cellular Physiology and Biochemistry: International Journal of Experimental Cellular Physiology, Biochemistry, and Pharmacology* 2013; **32**(2): 367–379.
62. Navarro S, Reddy R, Lee J, Warburton D, Driscoll B. Inhaled RSV treatments slow ageing-related degenerative changes in mouse lung. *Thorax* 2017; **72**(5): 451–459.
63. Ito K, Colley T, Mercado N. Geroprotectors as a novel therapeutic strategy for COPD, an accelerating aging disease. *International Journal of Chronic Obstructive Pulmonary Disease* 2012; **7**: 641–652.
64. Sharma G, Hanania NA, Shim YM. The aging immune system and its relationship to the development of chronic obstructive pulmonary disease. *Proceedings of the American Thoracic Society* 2009; **6**(7): 573–580.
65. Yuan J, Lu L, Zhang Z, Zhang S. Dietary intake of RSV enhances the adaptive immunity of aged rats. *Rejuvenation Research* 2012; **15**(5): 507–515.
66. Michishita Y, Hirokawa M, Guo YM, Abe Y, Liu J, Ubukawa K, *et al.* Age-associated alteration of gammadelta T-cell repertoire and different profiles of activation-induced death of Vdelta1 and Vdelta2 T cells. *International Journal of Hematology* 2011; **94**(3): 230–240.
67. Jo HE, Randhawa S, Corte TJ, Moodley Y. Idiopathic Pulmonary Fibrosis and the Elderly: Diagnosis and Management Considerations. *Drugs & Aging* 2016; **33**(5): 321–334.
68. Espinoza JL, Trung LQ, Inaoka PT, Yamada K, An DT, Mizuno S, *et al.* The Repeated Administration of RSV Has Measurable Effects on Circulating T-Cell Subsets in Humans. *Oxidative Medicine and Cellular Longevity* 2017; **2017**: 6781872.
69. Fagone E, Conte E, Gili E, Fruciano M, Pistorio MP, Lo Furno D, *et al.* RSV inhibits transforming growth factor-beta-induced proliferation and differentiation of ex vivo human lung fibroblasts into myofibroblasts through ERK/Akt inhibition and PTEN restoration. *Experimental Lung Research* 2011; **37**(3): 162–174.

70. Sosulski ML, Gongora R, Feghali-Bostwick C, Lasky JA, Sanchez CG. Sirtuin 3 Deregulation Promotes Pulmonary Fibrosis. *The Journals of Gerontology Series A, Biological Sciences and Medical Sciences* 2017; **72**(5): 595–602.
71. Impellizzeri D, Talero E, Siracusa R, Alcaide A, Cordaro M, Maria Zubelia J, *et al.* Protective effect of polyphenols in an inflammatory process associated with experimental pulmonary fibrosis in mice. *The British Journal of Nutrition* 2015; **114**(6): 853–865.
71a. Akgedik R, Akgedik Ş, Karamanli H, *et al.* Effect of resveratrol on treatment of bleomycin-induced pulmonary fibrosis in rats. *Inflammation* 2012; **35**: 1732.
72. Zeng Z, Cheng S, Chen H, Li Q, Hu Y, Wang Q, *et al.* Activation and overexpression of Sirt1 attenuates lung fibrosis via P300. *Biochemical and Biophysical Research Communications* 2017; **486**(4): 1021–1026.
73. Nance T, Smith KS, Anaya V, Richardson R, Ho L, Pala M, *et al.* Transcriptome analysis reveals differential splicing events in IPF lung tissue. *PLoS One* 2014; **9**(5): e97550.
74. Berardi V, Ricci F, Castelli M, Galati G, Risuleo G. RSV exhibits a strong cytotoxic activity in cultured cells and has an antiviral action against polyomavirus: potential clinical use. *Journal of Experimental & Clinical Cancer Research: CR* 2009; **28**: 96.
75. Lin CJ, Lin HJ, Chen TH, Hsu YA, Liu CS, Hwang GY, *et al.* Correction: Polygonum cuspidatum and its active components inhibit replication of the influenza virus through toll-like receptor 9-induced interferon beta expression. *PLoS One* 2015; **10**(4): e0125288.
76. Palamara AT, Nencioni L, Aquilano K, De Chiara G, Hernandez L, Cozzolino F, *et al.* Inhibition of influenza A virus replication by RSV. *The Journal of Infectious Diseases* 2005; **191**(10): 1719–1729.
77. Shang YP, Lin L, Li CC. [Streptococcus pneumoniae induces SPLUNC1 and the regulatory effects of RSV]. *Zhongguo dang dai er ke za zhi = Chinese Journal of Contemporary Pediatrics* 2017; **19**(1): 111–116.
78. Larrosa M, Tome-Carneiro J, Yanez-Gascon MJ, Alcantara D, Selma MV, Beltran D, *et al.* Preventive oral treatment with RSV pro-prodrugs drastically reduce colon inflammation in rodents. *Journal of Medicinal Chemistry* 2010; **53**(20): 7365–7376.
79. Koga CC, Andrade JE, Ferruzzi MG, Lee Y. Stability of Trans-RSV Encapsulated in a Protein Matrix Produced Using Spray Drying to UV Light Stress and Simulated Gastro-Intestinal Digestion. *Journal of Food Science* 2016; **81**(2): C292–300.
80. Trotta V, Lee WH, Loo CY, Haghi M, Young PM, Scalia S, *et al.* In vitro biological activity of RSV using a novel inhalable RSV spray-dried formulation. *International Journal of Pharmaceutics* 2015; **491**(1–2): 190–197.
81. Trotta V, Lee WH, Loo CY, Young PM, Traini D, Scalia S. Co-spray dried RSV and budesonide inhalation formulation for reducing inflammation

and oxidative stress in rat alveolar macrophages. *European Journal of Pharmaceutical Sciences: Official Journal of the European Federation for Pharmaceutical Sciences* 2016; **86**: 20–28.
82. Foti RS, Tyndale RF, Garcia KL, Sweet DH, Nagar S, Sharan S, *et al.* "Target-Site" Drug Metabolism and Transport. *Drug Metabolism and Disposition: The Biological Fate of Chemicals* 2015; **43**(8): 1156–1168.
83. Sharan S, Nagar S. Pulmonary metabolism of RSV: in vitro and in vivo evidence. *Drug Metabolism and Disposition: The Biological Fate of Chemicals* 2013; **41**(5): 1163–1169.

Chapter 10

Resveratrol and Aging

Sergio D. Paredes*, Lisa Rancan†, M. Cruz García†,
Elena Vara† and Jesús A. F. Tresguerres*

*Department of Physiology, School of Medicine,
Complutense University of Madrid,
Avda. Complutense, s/n, 28040, Madrid, Spain

†Department of Biochemistry and Molecular Biology,
Complutense University of Madrid,
Avda. Complutense, s/n, 28040, Madrid, Spain

With the aging of the population, the number of patients with chronic diseases associated with adverse cardiovascular diseases, including cardio-metabolic syndrome, diabetes mellitus and chronic kidney disease have increased over the past several decades. Cognitive decline has also emerged as one of the greatest health threats of old age, with nearly 50% of adults over the age of 85 afflicted with Alzheimer's disease. Strategies to combat the unwanted effects of aging encompass nutraceutical approaches to manage the cardiovascular and neurological dysfunctions observed in the aged organisms. Resveratrol, a compound found in appreciable amounts in a variety of plant foodstuffs, such as grapes and other berries, is currently a widely investigated molecule for its potentially beneficial effects to counteract and/or mitigate age-related conditions, at least in preclinical trials. Specifically, resveratrol appears as a potentially beneficial agent that may exert protective effects against heart and brain damage related to aging. Thus, it has been shown to play an important role in relieving and reversing cardiovascular alterations in various models of aging. Recent research also suggests that resveratrol supplementation may help in maintaining neurological health and protect against aging-associated neurodegeneration. Furthermore, it has been reported that this polyphenolic molecule can extend lifespan in some organisms, such as the budding yeast *Saccharomyces cerevisiae* and small metazoans, involving a similar mechanism to that of calorie restriction. More detailed clinical trials with well-defined protocols are needed to reach definitive statements regarding the therapeutic potential of resveratrol for human aging.

10.1. The aging process

Aging is a physiological process of multifactorial origin. This process provokes a series of biological changes including genetic, molecular, cellular, tissue and organ alterations that take place over the years. These changes and alterations are universal and occur continuously and progressively, resulting in metabolic, endocrine, and behavioral deficits, together with loss of function.[1] Aging is also a heterogeneous process, i.e., it may occur at different rates across diverse organisms, and even organisms of the same species can age at variable rates. It is also a process of heterochronic nature meaning that cells and tissues within a single organism age asynchronically.[2] At the biological level, aging is characterized by the accumulation of molecular and cellular damage, which leads to structural and functional aberrancies in cells and tissues, such as loss of mitochondrial homeostasis, impaired intercellular communication, senescence (cell arrest that hampers growth and division), and decreased regenerative capacity.[3]

As organisms' progress in age, a gradual loss of physiological functions is accompanied by the development of age-related diseases. The ability of organisms to overcome stress and respond to external environmental challenges/insults is blunted within aged individuals when compared to younger counterparts.[4,5] Particularly, old age is a risk factor for the physiology of heart and blood vessels, as evidenced from an exponential increase in the incidence of cardiovascular disease in the elderly population aged 65 or older. In fact, cardiovascular diseases are considered to be the most common cause of death in the global population. The annual socioeconomic cost for stroke, hypertension, diabetes, ischemic heart disease, and hyperlipidemia, which are categorized as cardiovascular diseases, has become a tremendous burden worldwide.[6] Despite numerous medical and scientific achievements aimed at overcoming the adverse effects of cardiovascular disease, new preventive or therapeutic measures are still needed in both developed and developing countries. Cardiac aging per se increases the risk of cardiovascular disease.

Furthermore, old age prolongs the exposure to several cardiovascular risk factors including hypertension, diabetes, hypercholesterolemia, and smoking.[7]

Aging is also accompanied by stereotypical structural and neurophysiological changes in the brain and variable degrees of cognitive decline. Cognitive frailty is emerging as one of the greatest health threats of the twenty-first century. As the life expectancy of the population has increased, so too has the prevalence of cognitive decline and dementia, largely in the form of Alzheimer's disease (AD). The greatest risk factor for cognitive decline and AD in older adults is age itself. Therefore, the development of these pathologies must be understood in the context of the molecular biology of the ageing process.[8]

Aging mechanisms and how to elude them have always fascinated the human being. The finding that mortality at advanced ages can be postponed, and indeed is being postponed, resolved a millennia-old debate about whether survival could be extended among the elderly.[9] Unless radical breakthroughs are achieved, perhaps as a result of research on other species, humans will continue to suffer senescence, although currently the process is not intractable. Mortality has been postponed considerably, as a result not of revolutionary advances in slowing the process of ageing but of ongoing progress in improving health. The debility that often characterizes the last years of life also seems to have been delayed, although the evidence is mixed, in part because deterioration is difficult to define and measure.[10]

Nutrient and stress sensors mediate lifespan extensions that occur in response to many different environmental and physiological signals. The best known of these signals is dietary restriction, which extends lifespan in many species, from yeast to primates,[11] although, at least in *Drosophila*, the relative amounts of protein and carbohydrate in the diet, rather than an overall reduction in calories per se, may be behind this effect.[12] In rodents, calorie restriction can extend lifespan up to 50%, while suppressing diverse age-associated conditions such as cancer, autoimmune disease, type 2 diabetes, neurodegeneration, and many others.[13] Several physiological pathways have been proposed to regulate calorie restriction-dependent lifespan extension and health span maintenance, including insulin/insulin-like growth factor-1 (IGF-1) signaling, mTOR signaling, AMP-activated kinase (AMPK) signaling, and sirtuins.[14] Given that sirtuin catalytic activity is NAD^+-dependent, and NAD^+ levels rise upon nutrient stress in certain tissues, up-regulation of specific sirtuin functions has been proposed to represent a means to reap some beneficial effects of calorie restriction.[15]

The protective action of calorie restriction against aging has led to a growing interest in searching for calorie restriction mimetics, i.e., small-molecule therapeutics or other chemicals/interventions that can recapitulate the rejuvenation effects attributed to calorie restriction. Examples of some of these compounds include resveratrol. Resveratrol is a natural dietary phenolic compound which is extensively present in many edible fruits, including grape skin, blueberries, cranberries, bilberries, lingonberries, partridgeberries, pomegranates, mulberries, peanuts, Japanese knotweed (*Polygonum cuspidatum*, a perennial shrub), and strawberries.[16] Its occurrence has also been reported in other plants such as *Eucalyptus* or spruce. Interestingly, this naturally-occurring polyphenol has been identified as a small molecule sirtuin-activating compound. In this regard and similar to caloric restriction, resveratrol treatment decreases age-dependent cognitive decline and AD-like pathologies in animal models,[17] as well as exhibiting antioxidant, anti-platelet, anti-inflammatory, anti-cancer, and anti-mutagenic activity, inhibition of lipid peroxidation, vasorelaxation, and protection from atherosclerotic disease and cardioprotective degeneration,[18] thereby improving general health in mammals.[19] The beneficial actions of resveratrol not only involve these capacities but also the regulation of the activities and expression levels of enzymes and proteins associated with the survival signals, regulation of ion channels, and antioxidative actions.[20] Preclinical studies with resveratrol suggest that it mimics the beneficial effects of calorie restriction via pharmacologic activation of sirtuin 1.[21] These data support the notion that compounds able to regulate the aging molecular mechanisms, as seems to be the case with resveratrol, may be exploited pharmacologically to delay or prevent diseases of aging, prolonging health span.

A plethora of protective properties against aging have been attributed to resveratrol, including resistance to age-related bone loss,[22] inhibition of skin photoaging,[23] enhancement of the expression of the anti-aging Klotho gene in the kidney,[24] or attenuation of inflammatory and oxidative damage in the aged immune system,[25] to name a few. Resveratrol administration has also been able to increase sirtuin 1 mRNA expression, as well as decreasing the expression of the transcription factor NFκB and reducing the proinflammatory and prooxidant status associated with age in the pancreas of senescence-accelerated prone mice.[26] Here, the focus has been to review and summarize the possible functions of this intriguing molecule in relation to cardiovascular and neural aging, as well as longevity, with the aim to provide gerontologists, geriatricians, and researchers on aging in general with a state-of-the-art view, in the hope of stimulating studies that will clarify the exact physiological roles that resveratrol may play in the heart and brain aging processes, and lifespan.

10.2. Effects of resveratrol in cardiovascular aging

A large body of evidence suggests that aging produces disorders in the anatomy and physiology of the heart.[7] The aged vasculature is characterized by increased arterial thickness, stiffness, and endothelial dysfunction. Such structural and functional changes result in hypertension, arteriosclerosis, stroke, and poor tissue perfusion, among other pathological conditions.[27] This has been shown to cause an increased afterload on the left ventricle, an increase in systolic blood pressure, and left ventricular hypertrophy, as well as other changes in the left ventricular wall that seem to prolong relaxation of the left ventricle in diastole. A dropout of atrial pacemaker cells resulting in a decrease in intrinsic heart rate has been reported.[28] Loss of myocytes with subsequent hypertrophy of the remaining cells has usually been observed. Calcification involving the conduction and valvular apparatus has been seen in most elderly individuals and might predispose to the common arrhythmias in old age.[29] Finally, the aging process is accompanied by a series of changes in the autonomic control of the cardiovascular system, including higher cardiac spillover and reduced plasma clearance rate of catecholamines, favoring heightened cardiac sympathetic tone with parasympathetic withdrawal and blunted cardiovagal baroreflex sensitivity.[30] These changes set the stage for isolated systolic hypertension, diastolic dysfunction and heart failure, atrioventricular conduction defects, and aortic valve calcification, all diseases seen in the elderly.[31]

Cells undergoing senescence are associated with hyperactivity of NF-κB and overexpression of inflammatory cytokines such as TNF-α, IL-6, and IFN-γ in circulating macrophages. Indeed, low-grade inflammation that is associated with age increases the incidence of both stroke and coronary artery disease in aged subjects. Increased NAD(P)H oxidase activity and overproduction of reactive oxygen species seem to underlie the vascular oxidative stress associated with age and promote inflammation and endothelial damage.[32] Resveratrol has been shown to affect several chemical compounds involved in macrophage lipid metabolism, including cyclooxygenase (COX)-2, peroxisome proliferator-activated receptor gamma (PPAR-c), liver X receptors (LXRs), ATP-binding cassette (ABC) transporters A1 and G1, and nitric oxide (NO).[33] In fact, resveratrol restricts atherosclerosis-associated inflammation via transcriptional regulation of COX-2 activity, ultimately suppressing prostaglandin E2 production, a key player in inflammation catalyzed by COX-2.[34] Moreover, it has been shown to reverse oxidative damage by reducing the expression of NADPH oxidase while increasing the expression of sirtuin 1 in the aorta of aged rats.[35]

Cardiovascular aging is also characterized by a reduced bioavailability of NO, increased production of collagen by vascular smooth muscle cells, and changes in the expression of proteins that regulate calcium handling.[36] Endothelium-derived NO functions to protect the cardiovascular system during aging, as evidenced by the observation that mice deficient in the endothelial nitric oxide synthase (eNOS) gene display premature cardiac aging concomitant with early mortality.[37] Oxidative stress is a major cause that affects NO synthesis in the aged vasculature. Oxidative stress reduces the bioavailability of NO, which has been associated with hypertension, arteriosclerosis, and a reduced vasodilatory response. High levels of free radicals and the low bioavailability of NO lead to a positive feedback loop of further oxidative stress, organelle damage, poor repair, and endothelial dysfunction. The mechanisms by which resveratrol confers vascular protection involve a higher biosynthesis of NO, increases in the activity of proteins that regulate oxidative metabolism, increases in the activity of several antioxidant enzymes, and the stimulation of cellular self-repair processes (collectively known as autophagy) within the endothelial cell.[38] Consistent with these positive effects, it should be noted that resveratrol upregulates eNOS, and increases NO bioavailability, therefore improving myocardial damage during ischemia-reperfusion.[39]

Cardiac damage is a major contributor to the morbidity and mortality particularly associated with coronary artery disease. Moreover, it is also related to some metabolic diseases such as diabetes and to some side effects of drug treatments. Resveratrol have been proven to provide cardioprotection against cardiac insults not only in aged animals but also in young individuals. Specifically, resveratrol has been found to protect cardiomyocytes against oxidative stress, autophagy, apoptosis, and cardiac fibrosis.[33] In particular, resveratrol selectively inhibits production of reactive oxide species via sirtuin 1 activation. This compound has also displayed antioxidant activity in the heart of old mice, as well as in liver and muscle, which was accompanied by a decrease in the levels of lipid peroxidation.[40]

Taking together, the afore-mentioned results suggest that compromised populations such as the elderly would benefit from treatment with resveratrol.[41] Nonetheless, other authors reveal no benefit from resveratrol supplementation.[42] For this reason, further research needs to be conducted to validate in humans the positive results that are clearly observed in the preclinical context.

10.3. Effects of resveratrol in neurological aging

The physiological process of aging involves a progressive cognitive loss caused by deteriorating brain function, which includes a decrease in learning and memory skills and slower responses to intellectual stimuli. Neurodegenerative diseases are characterized by inflammatory conditions, along with chronic and progressive pathologies, whereas activation of microglia causes neuronal damage in the central nervous system and increased reactive oxygen species generation. Alzheimer's disease (AD) is one of the most devastating neurodegenerative diseases. The exact cause of the disease is still not known although many scientists believe in the beta amyloid hypothesis which states that the accumulation of the amyloid peptide beta (Aβ) in brain is the initial cause which consequently leads to pathological neuroinflammation. AD is also the most common cause of dementia and an increasingly common cause of morbidity and mortality in the elderly. The prevalence and incidence of AD increases as life expectancies advance around the globe. AD is not only a growing healthcare burden, but a growing social and economic burden as well.

Resveratrol has been shown to cause resistance against neurodegeneration and preserve cognitive functions alone and in combination with flavonoids, as proved by various trials and epidemiological reports.[43] Certainly, it has exhibited effective properties against various neuronal ailments, including Huntington's disease, Parkinson's disease, diabetic neuropathy, amyotrophic lateral sclerosis and AD.[44] For example, in a transgenic model of AD, treatment with resveratrol diminished plaque formation in a region-specific manner, although changes occurred without detectable activation of sirtuin 1 or alterations in amyloid precursor protein processing.[45] Similarly, resveratrol triggered a decrease in hippocampal neuronal deterioration, inhibiting cognitive dysfunction and reducing the acetylation of sirtuin 1 substrates in models for AD and amyotrophic lateral sclerosis.[46] It also ameliorated both motor deficits and pathological changes in mice treated with 1-methyl-4-phenyl-1, 2, 3, 6-tetrahydropyridine (MPTP), a neurotoxin that damages the nigrostriatal dopaminergic pathway, as seen in Parkinson's disease; this effect was via activation of sirtuin 1 and subsequent microtubule-associated protein 1 light chain 3 (LC3) deacetylation-mediated autophagic degradation of α-synuclein.[47]

While sirtuin activation is the presumed target, resveratrol may have many other molecular effects, including anti-inflammatory, antioxidant, and anti-Aβ aggregation.[17] For example, resveratrol-administered rats improved memory and safety from amyloid-triggered neurotoxicity via suppressed levels of lipid peroxidation and inducible nitric oxide synthase (iNOS) and enhanced synthesis of heme oxygenase-1.[48] Other putative mechanisms of action include sirtuin activation with enhanced α-cleavage of transmembrane protein amyloid precursor protein and promotion of autophagy with removal of toxic intraneuronal proteinaceous aggregates.

Senescence-accelerated prone mouse neurons have resulted to be more vulnerable to inhibitors and uncouplers than their senescence-resistant counterparts. This increased vulnerability, indicative of frailty-associated senescence, was protected by resveratrol. The sirtuin 1 inhibitor, sirtinol, demonstrated that this neuroprotection was partially mediated by increased sirtuin 1 expression.[49] In humans, supplementation with resveratrol and quercetin enhanced hippocampal functional connectivity and increased glucose metabolism in the brain of healthy overweight older adults, as well as improving memory performance.[50] Resveratrol treatment also decreased the levels of metalloproteinase 9 in the cerebrospinal fluid of individuals with mild to moderate AD, as well as modulating neuroinflammation, and inducing adaptive immunity, which suggests that sirtuin 1 activation may be a viable target for treatment or prevention of neurodegenerative disorders.[51] However, other authors have found no evidence to state that resveratrol improves memory or cognitive performance as well as mood state in both healthy older adults and individuals with mild cognitive decline.[52] Again, further investigation is needed to determine whether resveratrol consistently exerts beneficial effects on neurological aging, including AD progression, and whether the promise of animal studies will translate into the clinic.

10.4. Effects of resveratrol in longevity and lifespan

The prevalence of obesity presents a serious health and societal issue all over the world. Type-2 diabetes, hypertension and dyslipidemia are various metabolic issues that cause accumulation of fat in the body. As shown above, this further leads to cardiac diseases as well as shortening of life.

Calorie restriction is the most robust intervention demonstrated to extend lifespan and delay the physiological deterioration associated with aging.[53] Because calorie restriction involves a number of overlapping and interconnected signaling pathways, the mechanism(s) underlying its beneficial effects are difficult to identify with certainty. Small molecules that have been identified as activators of the sirtuin pathway, such as the natural polyphenol resveratrol, have been shown to recapitulate calorie restriction-like effects. Specifically, this polyphenol reduces IGF-1 levels, increases AMPK enzymatic activity, and boosts mitochondrial number.[2]

In relation to longevity and lifespan, resveratrol mimics calorie restriction by stimulating Sir2 (whose homolog in mammals is sirtuin 1), DNA stability and extending lifespan by 70% in the budding yeast *Saccharomyces cerevisiae*.[54] In addition, some studies conducted in metazoans *Caenorhabditis elegans* and *Drosophila melanogaster* and the short-lived seasonal fish *Nothobranchius furzeri* have demonstrated that resveratrol extended medium and maximal lifespan by modulating sirtuin 1.[55] In mammals, resveratrol has also been successful in shifting the physiology of mice consuming excess calories towards that of mice on a standard diet, modulating known longevity pathways, and improving health, as indicated by a variety of measures, including survival, motor function, insulin sensitivity, organ pathology, proliferator-activated receptor gamma coactivator 1-alpha (PGC-1α) activity, and mitochondrial number.[19] In this case, however, whether these effects were due to resveratrol working primarily through sirtuin 1 or through a combination of interactions remained to be determined. In fact, a growing body of evidence indicates that other important signaling pathways other than sirtuins convey the antioxidant and anti-inflammatory properties of resveratrol. Such pathways include the afore-mentioned activation of PGC-1α via AMPK, as well as downregulating NF-κB, which signals pro-inflammatory responses in aging.[56]

Results from the study of a wide range of human degenerative diseases, both inherited and acquired, points to limiting telomeres as key pathogenetic elements driving degenerative pathologies, increasing cancer risk and shortening lifespan.[57] In fact, a positive link between telomere length and longevity has been reported in centenarians and their offspring; in particular, those with longer telomeres had an overall improved health profile, with decreased age-associated disease and better cognitive function and lipid profiles relative to controls.[58] Adequate telomerase activity in maintenance of telomere length appears to be essential. Resveratrol has been reported to preserve telomere length and increase telomerase activity in aortic rings from aged rats.[59]

Nonetheless, resveratrol has so far failed to extend lifespan in other murine models. For example, resveratrol had no significant effect on the lifespan of genetically heterogeneous mice[60] or on survival in genetically heterogeneous mice.[61] Similarly, it did not improve lifespan in a mouse model for Werner syndrome, although it ameliorated insulin resistance, hyperglycemia and hepatosteatosis.[62] Still, sirtuin 1 seems to be directly activated by resveratrol and other sirtuin-activating compounds via binding in the N-terminus of sirtuin 1.[63] Specifically, a mutation in this site blocked activation by resveratrol and other sirtuin-activating compounds. Additionally, recent studies have provided strong evidence of beneficial effects of small-molecule activators of sirtuin 1, including resveratrol, in lifespan and health span of mice fed with both standard and high fat diets.[64]

10.5. Concluding remarks

Further complicating matters related to cardiovascular or brain diseases, the pace of aging, at least in mammals, is influenced by external factors, including dietary intake, intrinsic stresses such as reactive oxygen species and telomere erosion, and germline variation in genetic elements governing DNA repair, among other processes.[57] Resveratrol has become an attractive candidate in nutraceutical strategies that aim to improve body functions in the elderly.

Considering the broad range of responses of resveratrol in the aging process and that it has not been found to exert adverse effects in animal models or clinical intervention studies, it is tempting to view this molecule as a panacea. However, it should be born in mind that the effects are often statistically significant but relatively weak, so a definitive response or cure is not a reasonable expectation.[65] In general, from all the mechanisms described for resveratrol in preclinical models, only a few of them have somehow been confirmed in humans, including signaling cascades via protein kinase B (Akt) pathway and regulation of some target genes or proteins in peripheral blood cells or muscle.[55] The highest degree of evidence in humans (although still scarce) deals with the cardioprotective activity of resveratrol through effects related to the improvement of inflammatory markers, atherogenic profile, glucose metabolism and endothelial function.[55] Additional studies with a higher number of participants, longer treatment periods and different dosages of resveratrol will be necessary to confirm the real potential of this stilbene derivative in aging and the effects of long-term supplementation.

Acknowledgments

The authors gratefully acknowledge the support of Red Cooperativa de Envejecimiento y Fragilidad 2013–2016. Instituto de Salud Carlos III, Madrid (Spain).

References

1. Harman D. Aging: overview. *Ann. NY Acad. Sci.* 2001; **928**: 1–21.
2. Carmona JJ, Michan S. Biology of healthy aging and longevity. *Rev. Inves. Clin.* 2016; **68**: 7–16.
3. López-Otín C, Blasco MA, Partridge L, Serrano M, Kroemer G. The hallmarks of aging. *Cell.* 2013; **153**: 1194–1217.
4. Johnson TE, Cypser J, de Castro E, de Castro S, Henderson S, Murakami S, Rikke B, Tedesco P, Link C. Gerontogenes mediate health and longevity in nematodes through increasing resistance to environmental toxins and stressors. *Exp. Gerontol.* 2000; **35**: 687–694.
5. Pérez VI, Buffenstein R, Masamsetti V, Leonard S, Salmon AB, Mele J, Andziak B, Yang T, Edrey Y, Friguet B, Ward W, Richardson A, Chaudhuri A. Protein stability and resistance to oxidative stress are

determinants of longevity in the longest-living rodent, the naked mole-rat. *Proc. Natl. Acad. Sci. U S A.* 2009; **106**: 3059–3064.

6. Moodie DS. The global burden of cardiovascular disease. *Congenit. Heart Dis.* 2016; **11**: 213.
7. Dai DF, Chiao YA, Wessells RJ, Bodmer R, Szeto HH, Rabinovitch PS. *Handbook of the Biology of Aging*, Eighth edition, eds. Kaeberlein M. and Martin, G. M., Chapter 17 "Cardiac Aging" (Academic Press), 2016; pp. 459–494.
8. Bishop NA, Lu T, Yankner BA. Neural mechanisms of ageing and cognitive decline. *Nature* 2010; **464**: 529–535.
9. Vaupel JW. Biodemography of human ageing. *Nature* 2010; **464**: 536–542.
10. Christensen K, Doblhammer G, Rau R, Vaupel JW. Ageing populations: The challenges ahead. *Lancet.* 2009; **374**, 1196–1208.
11. Colman RJ, Beasley TM, Kemnitz JW, Johnson SC, Weindruch R, Anderson RM. Caloric restriction reduces age-related and all-cause mortality in rhesus monkeys. *Nat. Commun.* 2014; **5**: 3557.
12. Tatar M, Post S, Yu K. Nutrient control of Drosophila longevity. *Trends Endocrinol. Metab.* 2014; **25**: 509–517.
13. Speakman JR, Mitchell SE. Caloric restriction. *Mol. Aspects Med.* 2011; **32**: 159–221.
14. Guarente L. Calorie restriction and sirtuins revisited. *Genes Dev.* 2013; **27**: 2072–2085.
15. Giblin W, Lombard DB. *Handbook of the Biology of Aging*, Eighth edition, eds. Kaeberlein M. and Martin, G. M., Chapter 3 "Sirtuins, Healthspan, and Longevity in Mammals" (Academic Press), 2016; pp. 459–494.
16. Gupta C, Prakash D. Phytonutrients as therapeutic agents. *J. Complement. Integr. Med.* 2014; 11: 151–169.
17. Sawda C, Moussa C, Turner RS. Resveratrol for Alzheimer's disease. *Ann. N Y Acad. Sci.* 2017; **1403**: 142–149.
18. Ramis MR, Esteban S, Miralles A, Tan DX, Reiter RJ. Caloric restriction, resveratrol and melatonin: Role of SIRT1 and implications for aging and related-diseases. *Mech Ageing Dev.* 2015; **146–148**: 28–41.
19. Baur JA, Pearson KJ, Price NL, Jamieson HA, Lerin C, Kalra A, Prabhu VV, Allard JS, Lopez-Lluch G, Lewis K, Pistell PJ, Poosala S, Becker KG, Boss O, Gwinn D, Wang M, Ramaswamy S, Fishbein KW, Spencer RG, Lakatta EG, Le Couteur D, Shaw RJ, Navas P, Puigserver P, Ingram DK, de Cabo R, Sinclair DA. Resveratrol improves health and survival of mice on a high-calorie diet. *Nature* 2006; **444**: 337–342.
20. Baur JA, Sinclair DA. Therapeutic potential of resveratrol: The *in vivo* evidence. *Nat. Rev. Drug. Discov.* 2006; **5**: 493–506.
21. Kulkarni SS, Canto C. The molecular targets of resveratrol. *Biochim. Biophys. Acta.* 2015; **1852**: 1114–1123.
22. Tresguerres IF, Tamimi F, Eimar H, Barralet J, Torres J, Blanco L, Tresguerres JA. Resveratrol as anti-aging therapy for age-related bone loss. *Rejuvenation Res.* 2014; **17**: 439–445.

23. Lee TH, Seo JO, Baek SH, Kim SY. Inhibitory effects of resveratrol on melanin synthesis in ultraviolet B-induced pigmentation in Guinea pig skin. *Biomol. Ther. (Seoul)* 2014; **22**: 35–40.
24. Hsu SC, Huang SM, Chen A, Sun CY, Lin SH, Chen JS, Liu ST, Hsu YJ. Resveratrol increases anti-aging Klotho gene expression via the activating transcription factor 3/c-Jun complex-mediated signaling pathway. *Int. J. Biochem. Cell Biol.* 2014; **53**: 361–371.
25. Wong YT, Gruber J, Jenner AM, Tay FE, Ruan R. Chronic resveratrol intake reverses pro-inflammatory cytokine profile and oxidative DNA damage in ageing hybrid mice. *Age (Dordr.)* 2011; **33**: 229–246.
26. Ginés C, Cuesta S, Kireev R, García C, Rancan L, Paredes SD, Vara E, Tresguerres JA. Protective effect of resveratrol against inflammation, oxidative stress and apoptosis in pancreas of aged SAMP8 mice. *Exp Gerontol.* 2017; **90**: 61–70.
27. North BJ, Sinclair DA. The intersection between aging and cardiovascular disease. *Circ. Res.* 2012; **110**: 1097–1108.
28. Cheitlin MD. Cardiovascular physiology-changes with aging. *Am. J. Geriatr. Cardiol.* 2003; **12**: 9–13.
29. Leipsic J, Blanke P. Calcification of the aortic valve and mitral apparatus: location, quantification and implications for device selection. *EuroIntervention.* 2016; **12**(Y): Y16–20.
30. Santulli G, Iaccarino G. Adrenergic signaling in heart failure and cardiovascular aging. *Maturitas.* 2016; **93**: 65–72.
31. Paredes SD, Forman KA, García C, Vara E, Escames G, Tresguerres JA. Protective actions of melatonin and growth hormone on the aged cardiovascular system. *Horm. Mol. Biol. Clin. Investig.* 2014; **18**: 79–88.
32. Donato AJ, Eskurza I, Silver AE, Levy AS, Pierce GL, Gates PE, Seals DR. Direct evidence of endothelial oxidative stress with aging in humans: Relation to impaired endothelium-dependent dilation and upregulation of nuclear factor-kappaB. *Circ Res.* 2007; **100**: 1659–1666.
33. Cho S, Namkoong K, Shin M, Park J, Yang E, Ihm J, Thu VT, Kim HK, Han J. Cardiovascular Protective Effects and Clinical Applications of Resveratrol. *J. Med. Food.* 2017; **20**: 323–334.
34. O'Leary KA, de Pascual-Teresa S, Needs PW, Bao YP, O'Brien NM, Williamson G. Effect of flavonoids and vitamin E on cyclooxygenase-2 (COX-2) transcription. *Mutat Res.* 2004; **551**: 245–254.
35. Tang Y, Xu J, Qu W, Peng X, Xin P, Yang X, Ying C, Sun X, Hao L. Resveratrol reduces vascular cell senescence through attenuation of oxidative stress by SIRT1/NADPH oxidase-dependent mechanisms. *J. Nutr. Biochem.* 2012; **23**: 1410–1416.
36. Strait JB, Lakatta EG. Aging-associated cardiovascular changes and their relationship to heart failure. *Heart Fail. Clin.* 2012; **8**: 143–164.
37. Li W, Mital S, Ojaimi C, Csiszar A, Kaley G, Hintze TH. Premature death and age-related cardiac dysfunction in male eNOS-knockout mice. *J. Mol. Cell Cardiol.* 2004; **37**: 671–680.

38. Diaz M, Degens H, Vanhees L, Austin C, Azzawi M. The effects of resveratrol on aging vessels. *Exp Gerontol.* 2016; **85**: 41–47
39. Bradamante S, Barenghi L, Villa A. Cardiovascular protective effects of resveratrol. *Cardiovasc. Drug Rev.* 2004; **22**: 169–188.
40. Tung BT, Rodriguez-Bies E, Thanh HN, Le-Thi-Thu H, Navas P, Sanchez VM, López-Lluch G. Organ and tissue-dependent effect of resveratrol and exercise on antioxidant defenses of old mice. *Aging Clin. Exp. Res.* 2015; **27**: 775–783.
41. Smoliga JM, Colombo ES, Campen MJ. A healthier approach to clinical trials evaluating resveratrol for primary prevention of age-related diseases in healthy populations. *Aging.* 2013; **5**: 495–506.
42. Olesen J, Gliemann L, Biensø R, Schmidt J, Hellsten Y, Pilegaard H. Exercise training, but not resveratrol, improves metabolic and inflammatory status in skeletal muscle of aged men. *J. Physiol.* 2014; **592**: 1873–1886.
43. Wahab A, Gao K, Jia C, Zhang F, Tian G, Murtaza G, Chen J. Significance of Resveratrol in Clinical Management of Chronic Diseases. *Molecules.* 2017; **22**, pii: E1329.
44. Tellone E, Galtieri A, Russo A, Giardina B, Ficarra S. Resveratrol: A Focus on Several Neurodegenerative Diseases. *Oxid. Med. Cell. Longev.* 2015; **2015**: 392169.
45. Karuppagounder SS, Pinto JT, Xu H, Chen HL, Beal MF, Gibson GE. Dietary supplementation with resveratrol reduces plaque pathology in a transgenic model of Alzheimer's disease. *Neurochem. Int.* 2009; **54**: 111–118.
46. Kim D, Nguyen MD, Dobbin MM, Fischer A, Sananbenesi F, Rodgers JT, Delalle I, Baur JA, Sui G, Armour SM, Puigserver P, Sinclair DA, Tsai LH. SIRT1 deacetylase protects against neurodegeneration in models for Alzheimer's disease and amyotrophic lateral sclerosis. *EMBO J.* 2007; **26**: 3169–3179.
47. Guo YJ, Dong SY, Cui XX, Feng Y, Liu T, Yin M, Kuo SH, Tan EK, Zhao WJ, Wu YC. Resveratrol alleviates MPTP-induced motor impairments and pathological changes by autophagic degradation of α-synuclein via SIRT1-deacetylated LC3. *Mol. Nutr. Food Res.* 2016; **60**: 2161–2175.
48. Huang TC, Lu KT, Wo YY, Wu YJ, Yang YL. Resveratrol protects rats from Abeta-induced neurotoxicity by the reduction of iNOS expression and lipid peroxidation. *PLoS ONE.* 2011; **6**: e29102.
49. Cristòfol R, Porquet D, Corpas R, Coto-Montes A, Serret J, Camins A, Pallàs M, Sanfeliu C. Neurons from senescence-accelerated SAMP8 mice are protected against frailty by the sirtuin 1 promoting agents melatonin and resveratrol. *J. Pineal Res.* 2012; **52**: 271–281.
50. Witte AV, Kerti L, Margulies DS, Flöel A. Effects of resveratrol on memory performance, hippocampal functional connectivity, and glucose metabolism in healthy older adults. *J. Neurosci.* 2014; **34**: 7862–7870.
51. Moussa C, Hebron M, Huang X, Ahn J, Rissman RA, Aisen PS, Turner RS. Resveratrol regulates neuro-inflammation and induces adaptive immunity in Alzheimer's disease. *J. Neuroinflammation.* 2017; **14**: 1.

52. Farzaei MH, Rahimi R, Nikfar S, Abdollahi M. Effect of resveratrol on cognitive and memory performance and mood: A meta-analysis of 225 patients. *Pharmacol Res.* **2017**; pii: S1043-6618(17)30812-5.
53. Park SJ, Ahmad F, Philp A, Baar K, Williams T, Luo H, Ke H, Rehmann H, Taussig R, Brown AL, Kim MK, Beaven MA, Burgin AB, Manganiello V, Chung JH. Resveratrol ameliorates aging-related metabolic phenotypes by inhibiting cAMP phosphodiesterases. *Cell.* 2012; **148**: 421–433.
54. Howitz KT, Bitterman KJ, Cohen HY, Lamming DW, Lavu S, Wood JG, Zipkin RE, Chung P, Kisielewski A, Zhang LL, Scherer B, Sinclair DA. Small molecule activators of sirtuins extend Saccharomyces cerevisiae lifespan. *Nature.* 2003; **425**: 191–196.
55. Tomé-Carneiro J, Larrosa M, González-Sarrías A, Tomás-Barberán FA, García-Conesa MT, Espín JC. Resveratrol and clinical trials: the crossroad from in vitro studies to human evidence. *Curr. Pharm. Des.* 2013; **19**: 6064–6093.
56. Ungvari Z, Labinskyy N, Mukhopadhyay P, Pinto JT, Bagi Z, Ballabh P, Zhang C, Pacher P, Csiszar A. Resveratrol attenuates mitochondrial oxidative stress in coronary arterial endothelial cells. *Am. J. Physiol. Heart Circ. Physiol.* 2009; **297**: H1876–H1881.
57. Sahin E, DePinho RA. Linking functional decline of telomeres, mitochondria and stem cells during ageing. *Nature.* 2010; **464**: 520–528.
58. Atzmon G, Cho M, Cawthon RM, Budagov T, Katz M, Yang X, Siegel G, Bergman A, Huffman DM, Schechter CB, Wright WE, Shay JW, Barzilai N, Govindaraju DR, Suh Y. Evolution in health and medicine Sackler colloquium: Genetic variation in human telomerase is associated with telomere length in Ashkenazi centenarians. *Proc. Natl. Acad. Sci. U S A.* 2010; **107**(Suppl 1): 1710–1717.
59. da Luz PL, Tanaka L, Brum PC, Dourado PMM, Favarato D, Krieger JE, Laurindo FRM. Red wine and equivalent oral pharmacological doses of resveratrol delay vascular aging but do not extend life span in rats. *Atherosclerosis.* 2012; **224**: 136–142.
60. Strong R, Miller RA, Astle CM, Baur JA, de Cabo R, Fernandez E, Guo W, Javors M, Kirkland JL, Nelson JF, Sinclair DA, Teter B, Williams D, Zaveri N, Nadon NL, Harrison DE. Evaluation of resveratrol, green tea extract, curcumin, oxaloacetic acid, and medium-chain triglyceride oil on life span of genetically heterogeneous mice. *J. Gerontol. A Biol. Sci. Med. Sci.* 2013; **68**: 6–16.
61. Miller RA, Harrison DE, Astle CM, Baur JA, Boyd AR, de Cabo R, Fernandez E, Flurkey K, Javors MA, Nelson JF, Orihuela CJ, Pletcher S, Sharp ZD, Sinclair D, Starnes JW, Wilkinson JE, Nadon NL, Strong R. Rapamycin, but not resveratrol or simvastatin, extends life span of genetically heterogeneous mice. *J. Gerontol. A Biol. Sci. Med. Sci.* 2011; **66**: 191–201.
62. Labbé A, Garand C, Cogger VC, Paquet ER, Desbiens M, Le Couteur DG, Lebel M. Resveratrol improves insulin resistance hyperglycemia and hepatosteatosis but not hypertriglyceridemia, inflammation, and life span in

a mouse model for Werner syndrome, *J. Gerontol. A Biol. Sci. Med. Sci.* 2011; **66**: 264–278.

63. Hubbard BP, Gomes AP, Dai H, Li J, Case AW, Considine T, Riera TV, Lee JE, E SY, Lamming DW, Pentelute BL, Schuman ER, Stevens LA, Ling AJ, Armour SM, Michan S, Zhao H, Jiang Y, Sweitzer SM, Blum CA, Disch JS, Ng PY, Howitz KT, Rolo AP, Hamuro Y, Moss J, Perni RB, Ellis JL, Vlasuk GP, Sinclair DA. Evidence for a common mechanism of SIRT1 regulation by allosteric activators. *Science.* 2013; **339**: 1216–1219.
64. Ingram DK, Roth GS. Calorie restriction mimetics: Can you have your cake and eat it, too? *Ageing Res. Rev.* 2015; **20**: 46–62.
65. Park EJ, Pezzuto JM. The pharmacology of resveratrol in animals and humans. *Biochim. Biophys. Acta.* 2015; **1852**: 1071–1113.

Chapter 11

Resveratrol and Advancing Virtual Drug Discovery

Amanda H. Pham, Mark Andrew Phillips
and Zhong-Ru Xie

*Computational Drug Discovery Laboratory,
School of Electrical and Computer Engineering,
College of Engineering, University of Georgia,
Athens, GA, USA*

For many years, researchers have been fascinated by resveratrol and its potential therapeutic effects. The advancement of technology and an increased understanding of disease mechanisms have allowed scientists to further have a critical outlook on the effects of resveratrol. Here, we identify a number of studies that have utilized time-saving and cost effective computational methods to investigate the links between resveratrol or its derivatives and the prevention or treatment of aging and diseases such as Alzheimer's, various cancers, and cardiovascular diseases.

11.1. Introduction

Resveratrol (trans-3,5,4′-trihydroxystilbene) is a natural compound found in many herbal dietary resources such as grapes, peanuts, berries and red wine. Because resveratrol is good for human health and lifespan, scientists have been interested in exploring its potential as a therapeutic agent, identifying resveratrol derivatives that have improved efficacy, revealing its binding and functioning mechanisms, and developing new formulae or therapies with resveratrol or its derivatives.

Because it easily can take 10+ years and costs in the billions of dollars to bring a drug to market, virtual drug discovery has attracted increasing attention. Computational methods including molecular docking/virtual screening, molecular dynamic (MD) simulation, models for virtual drug design or virtual screening such as pharmacophore and quantitative structure-activity relationship (QSAR) models, and drug repurposing have been well developed and successfully applied at different stages of drug discovery. Because resveratrol and its derivatives bind to many proteins of biological significance, scientists have employed many of the methods mentioned above to study the interactions among resveratrol and/or its derivatives and their target proteins as well as the potential for their use in medicine.

11.2. Resveratrol and molecular docking

In silico studies have been fairly popular among drug discovery researchers.[1] Because these studies can highlight the interactions on the protein surface and the catalytic and inhibition sites, they can provide evidence to find promising drugs that interact and provide the same details and effects as other proteins.[1] *In silico* studies have provided insightful results from computational models that reveal interactions between proteins and compounds that would be identified in a laboratory using experimental screening approaches.[1] Docking studies involve the use of a computational program that analyzes protein-ligand interactions.[1] By conducting a protein-ligand analysis, these programs can produce a series of different conformations of a ligand that will bind correctly to a protein receptor site.[1] Further information provided by docking studies using several software packages includes non-covalent interactions such as van der Waals, electrostatic, and hydrophobic interactions.[2] Together, these interactions contribute to the overall binding strength between the interacting ligand and protein.[2] In addition, docking studies can also provide a series of docking scores,[2] calculated using a variety of mathematical or statistical models that can be used to analyze the strength of an interaction.[2] Based on these scores and depending on case by case studies, these scores can be interpreted differently.[2]

Generally, molecular docking is the most commonly used computational tool in drug discovery.[3] Several factors may affect the accuracy and results of molecular docking studies, including the determined or predicted protein structure, the structure resolution, the identification or prediction of the binding site(s), and the procedure, parameters, software, and hardware. Molecular docking studies are used to determine whether a molecule binds to a target protein, the binding location, orientation, conformation, and affinity as well as which amino acid residues are involved in a protein-drug interaction.

Molecular docking has been applied in many studies to the target proteins that bind to resveratrol and its analogs.

11.2.1. *Docking resveratrol on SIRT1*

Failure to maintain homeostasis in the vascular can lead to coronary complications and eventually in heart-related disorders that culminate in cell death.[4] Such type of injury has irreversible effects. There are two known modes of cell death, apoptosis and autophagy.[4] One mechanism that has been found to prevent and reduce cell death involves proteins known as sirtuins.[5] The primary sirtuin in mammals is known as SIRT1 (silent mating type information regulation 2 homolog).[5] Because it is an NAD+-dependent class III histone and a protein deacetylase, it is able to deacetylate various proteins involved in the regulation of gene expression, cell survival and apoptosis.[6] Specifically, in apoptosis, SIRT1 can deacetylate and inactivate a protein known as p53, which is involved in the induction of apoptosis. In addition, SIRT1 is the main regulator that helps control the cellular response to various levels of stress.[6] Structurally, sirtuin proteins have a conserved catalytic domain made up of approximately 275 amino acids between the N and C terminal domains.[7] SIRT1 has a small domain with a zinc binding site connected with four linking loops and a large domain involving the Rossmann-Fold domain.[7,8] The zinc-binding domain is involved in the structural stability of the catalytic region, whereas the Rossmann-Fold domain allows

NAD+ binding.[8] Sirtuins such as human SIRT1 uniquely contain an extended N-terminal domain that facilitates substrate recognition for the regulation of their activity.[9,10]

Specifically, SIRT1 in the heart has been found to help maintain the longevity of cardiomyocytes under stress conditions.[11] Additional studies have shown that in patients diagnosed with diabetic cardiomyopathy, apoptosis in cardiomyocytes was reduced in coordination with upregulation of SIRT1.[11] Therefore, if SIRT1 could be continuously activated and expressed to prevent further damage to the heart, major risks of heart failure and death could potentially be reduced. One compound has been identified as a SIRT1 activator, **resveratrol**, but this has only been observed in *in vitro* studies.[9–11] An allosteric mechanism has been commonly applied to explain SIRT1/resveratrol interaction.[11] The chemical effect of resveratrol on SIRT1 activation and other SIRT1 activators results from a reduction in the SIRT1 peptide Michaelis constant (K_M).[12] A lower peptide Michaelis constant results in a conformational change at the N-terminal domain of SIRT1, increasing the affinity between resveratrol and SIRT1, in turn resulting in activation.[10,12] In addition, an SIRT1 assay using a peptide substrate known as Fluor de Lys-SIRT1 peptide substrate revealed that resveratrol activates SIRT1.[13] Fluor de Lys-SIRT1 is a fluorophore, a compound that can emit fluorescence if excited, which can also be used for labeling.[13]

Resveratrol and its popular analog pterostilbene have been used to study the effects of SIRT1 activation on the reduction of cardiac myocyte apoptosis and death.[14] Resveratrol is less well absorbed compared to pterostilbene; thus, both compounds have been studied.[14] Guo *et al.* (2016) first exposed *in vitro* cardiomyocytes to a hypoxic environment, causing the inhibition of SIRT1 functionality.[14] The cells were then exposed to different concentrations of pterostilbene and resveratrol.[14] Concentrations ranging from 1 μM to 3 μM effectively reduced cell apoptosis, but higher concentrations contributed to nonspecific cell toxicity.[14] Resveratrol alone was required at a concentration of approximately 20 μM to prevent apoptosis without toxicity.[14] However, the mechanism is unclear.

To investigate the specific mechanisms of resveratrol and SIRT interactions, Guo *et al.* (2016) turned to a flexible molecular docking assay.[14] These docking studies were used for virtual screening to explore the activation rather than the strength of these interactions because other studies had already provided evidence for resveratrol/SIRT1 activation.[1] Specifically, the docking software package was DOCK of MOE (molecular operation environment), which analyzed the interaction of resveratrol at the active site of SIRT1.[14] In addition, SIRT1 and pterostilbene were docked together to analyze their mechanistic functions.[14] The catalytic activity of SIRT1 has been documented for the Cys482, Cys485, Ser442, Arg466, Asp272, Asp481, Asn465, Gly480, Gly440, Gly263, Ser275, Ser441, and Phe273 residues. When docked, pterostilbene was found to have embedded itself to the Cys482 and Arg466 portion of the catalytic pocket.[14] Resveratrol also showed similar interactions with Cys482 and Ser422.[14] In addition, both resveratrol and pterostilbene are also able to interact with the SIRT1 catalytic site via aromatic and hydrogen bond interactions.[14] The docking studies led to the conclusion that pterostilbene was able to interact in the SIRT1 active pocket, particularly at Cys482 and Arg466, which could then activate transcription factors for SIRT1 mRNA that could lead to the gene expression of SIRT1 and help to reduce further cell damage to the cardiomyocytes.[14] It should be noted that the resveratrol analog pterostilbene has the same effects as resveratrol, but it is more efficient because a lower concentration is required.[14] Therefore, further studies on additional resveratrol family members should be beneficial to supporting the hypothesis that these polyphenols can have a direct impact on the role of SIRT1 and the overall maintenance of cardiac health.

11.2.2. *Docking resveratrol on COX*

Cyclooxygenases, also known as COX, are enzymes involved in the conversion of arachidonic acid to bioactive lipid products playing an important role in regulation of processes within blood vessels.[15] There are two forms of cyclooxygenases, COX-1 and COX-2, with different and distinct functions.[15–17] COX-1 is involved in synthesis of thromboxane-A2 and eicosanoids that act as a vasoconstrictor and induce platelet aggregation in blood vessels.[15,16] As a result, the inhibition of COX-1 activity could help prevent the onset of cardiovascular-related diseases.[16] On the other hand, COX-2 catalyzes the formation of prostacyclin, which promotes vasodilation and acts as an antiplatelet agent.[15] Therefore, COX-2 inhibition has the potential of increasing the risk of cardiovascular disease.[16]

It has also been found that resveratrol acts as an inhibitor of both COX-1 and COX-2; however, its selectivity is unclear.[15,16] To further study the effects of resveratrol, wines from which resveratrol could be derived were used to study inhibitory effects. Kutil and coworkers used the GOLD docking software to identify the binding interactions of the wine and wine constituents.[15] Similar to other docking software, GOLD allows for the binding site size parameters for docking to be set, as well as for the incorporation of water molecules in the binding site. In addition, the LigandScout option was used to energetically minimize docking poses. The docking results showed significant COX inhibition by a trans-resveratrol structure. The studies confirmed that resveratrol can act as a selective site inhibitor, as found in a similar study by Handler *et al.* (2007).[16]

A study by Kummerle *et al.* (2005) proposed a model for resveratrol's inhibitory effects. Docking studies employing the FlexX software were used to analyze the binding affinity between resveratrol and COX. Affinities of resveratrol and COX-1 at −20.4 kcal/mol and COX-2 at −17.9 kcal/mol were obtained,[17] suggesting that COX-1 and COX-2 do not bind by the same mechanism.[17] COX-2 exhibits specific hydrogen bond interactions between the two resorcinol ring hydroxyl groups of resveratrol.[17] In addition, interactions between the His207 and Gln203 residues helps to stabilize resveratrol, ensuring inhibitory activity.[17] For COX-1, the 3,5-hydroxyl groups attached to the phenyl ring of resveratrol exhibit hydrogen bond interactions with the backbone of the COX-1 heme group.[17] Hydrogen bond interactions between the hydrogen of the 4′-hydroxyphenyl group of resveratrol and the carbonyl oxygen of the COX-1 Leu294 residue help stabilize resveratrol.[17] However, even though the inhibitory activity of resveratrol has been shown, it cannot be concluded that resveratrol alone is responsible for all of the inhibition.

11.2.3. *Docking resveratrol on MAO and SSAO*

The mechanism of antidiabetic or anti-obesity activities of resveratrol and its derivatives have not been clearly elucidated. Hydrogen peroxide, which triggers insulin-like activation of glucose transport and lipogenesis in adipocytes,[18] is generated by oxidases in the mitochondrial respiratory chain. Resveratrol is able to inhibit monoamine oxidase (MAO)[19] and semicarbazide-sensitive amine oxidase (SSAO).[20,21] Les *et al.* docked both trans- and cis-form resveratrol and trans-3,3′,4,5′-tetrahydroxystilbene (piceatannol) to human MAO-A (PDB ID: 2Z5X), MAO-B (PDB ID: 2V61) and SSAO/VAP-1 (vascular adhesion protein-1, PDB ID: 4BTW) using Schroedinger's Glide docking software (Glide 6.8).[22] The results indicated that both stilbenes bound to the substrate pockets close to the FAD forming hydrogen bonds, and pi-pi interactions occurred between the ligands and the amino acid residues, but the docking scores for the MAOs were higher than the scores for SSAO. Combined with the results of *in vitro* experiments, the docking approach confirmed that both stilbenes interact with MAO in human adipose tissue and exert their anti-adipogenic effects by inhibiting MAO. This study is an example showing that molecular docking can not only be used to virtually screen drug candidates but also to assist in the identification of proteins targeted by resveratrol or other drugs.

11.2.4. *Docking resveratrol on P450*

CYP3A4 is one of the major Cytochrome P450 enzymes, which acts as a monooxygenase and is involved in the metabolism of more than 50% of the drugs on the market.[23] CYP3A4 is a major protein component of the human liver and is found in other organs such as the small intestine, prostate, and kidneys. Both *in vitro* and *in vivo* studies have shown possible irreversible inhibitory effects of resveratrol on CYP3A4.[24–26] One study identified stilbene and CYP3A4 interactions, which showed that methoxy-stilbenes interacted better with CYP3A4 than trans-resveratrol. The higher the lipophilicity of a substrate is, the better its interaction with CYP3A4 will be. In addition to lipophilicity, hydrogen bond interactions with CYP3A4 also play an important role in its affinity. Pharmacophore studies on CYP3A4 found that for a substrate to optimally interact with CYP3A4, two hydrogen bond acceptors, a hydrogen bond donor, and a hydrophobic region are required.

Docking studies were performed by Basheer *et al.* to analyze the interactions between resveratrol and CYP3A4.[23] The use of docking software provides a better understanding of the complex inhibitory effects between resveratrol and modified stilbene on CYP3A4 and the interactions that may cause them. Specifically, the docking software CDOCKER was used to show that there were interactions with the A and B ring of resveratrol. Hydrophobic interactions occurred with the resveratrol A ring and the Ala370 residue of CYP3A4. Electrostatic interactions showed alkyl-pi interactions with the B ring of resveratrol and the Ala305 and Leu482 residues of CYP3A4. In addition, there were also resveratrol and pi-pi interactions with the two pyrrole rings of the heme group, cation-pi interactions with the heme group, and anion-pi interactions with the nitrogen of CYP3A4.

The hydrogen atoms of the three hydroxyl groups that are a part of resveratrol were also found to act as hydrogen donors and facilitate the binding between Ala370 and Glu371 in the A ring and Thr309 in the B ring. Additionally, when docking was extended to modified stilbenes including mono-acetoxy resveratrol and di-acetoxy resveratrol, fewer interactions were found. Specifically, both showed interactions with the $5'$ hydroxyl group of the A ring and Arg105 and the heme group of CYP3A4. In addition, electrostatic interactions were found with Glu374 and hydrophobic interactions with Ala370. Mono-acetoxy resveratrol had the highest affinity of −21.741 kcal/mol. Di-acetoxy resveratrol showed an energy value of −25.733 kcal/mol and resveratrol of −25.504 kcal/mol. In addition, tri-acetoxy resveratrol had a value of −29.487 kcal/mol as well as the weakest inhibitory effect on CYP3A4 because it docks in a different orientation than the other modified stilbenes and resveratrol, and as a result, it does not interact with the heme group but does show an interaction between its 4′ acetoxy group and Arg372 of CYP3A4.

Recent *in vitro* and *in silico* studies have also shown that an aldehyde group added to the A ring of resveratrol will decrease the inhibitory effects on CYP3A4. This further supports the above findings about the significance of electrostatic and hydrophobic interactions as well as hydrogen bond interactions that can help define how well the complex works. In addition, acetoxy groups are strong hydrogen bond acceptors compared to hydroxyl groups, and using these acetoxy groups to modify stilbenes can most likely increase their lipophilicity and introduce more H-bond acceptors on the molecule.

11.2.5. *Docking resveratrol on GAPDH*

Glyceraldehyde 3-phosphate dehydrogenase (GAPDH) is a housekeeping protein, which is ubiquitously expressed in almost all cells. GAPDH has multiple functions including its main function of converting 3-phosphate to 1,3-bisphosphoglycerate using the cofactor NAD+,[27,28] and it is also involved in many cellular processes such as DNA repair, transcription activation, nuclear RNA transport, vesicle shuttling, membrane fusion and so on.[29–33] Therefore, GAPDH is a significant drug target for several diseases. Furthermore, GAPDH is a major component of amyloid lesions in some neurodegenerative disorders (e.g., Parkinson's and Alzheimer's diseases).[31,34–38]

Rodacka *et al.* docked resveratrol and trans-3,3′,5,5′-tetrahydroxy-4′-methoxystilbene (THMS) to GAPDH using LigandFit and CDOCKER (Discovery Studio, Accelrys Inc.).[39] The determined structure of GAPDH (PDB ID: 1J0X) used in molecular docking was retrieved from Protein Data Bank (PDB).[40] A total of 48 potential resveratrol and THMS binding pockets on the GAPDH surface were identified by LigandFit, and 11 and 8 of them were identified as highly likely resveratrol and THMS binding sites, with dock-scores ranging from 20.5 to 48.0 and from 36.1 to 47.7, respectively. The analysis shows that the binding of THMS is stronger, as evidenced by more hydrogen bond interactions than resveratrol, which agrees with the measurement of the reduced activity of GAPDH binding to resveratrol and THMS.[39]

11.2.6. *Docking resveratrol on PKC*

Protein kinase C (PKC) is a kinase from a group of serine and threonine kinases that is essential for regulating cell functions such as cell growth, proliferation, apoptosis, cell differentiation and gene expression.[41,42] There are four types of PKCs, α, $\beta 1$, $\beta 2$, and δ. Structurally, the kinase can be thought of as four subunits. Two subunits are involved in the regulatory responses.[42] The third subunit is involved in the catalytic activity, and the fourth subunit binds to ATP, which provides a source of energy for the kinase.[42] In addition, PKC is also modulated by intracellular calcium and is subjected to activation by diacylglycerol, also known as DAG.[42] Activated PKC affects a cascade of downstream effectors subsequently impinging on a multitude of cellular processes.[42]

On the basis that resveratrol inhibits growth and promote apoptosis in gastric and prostate cancer cells as well as inhibit PKC activity in response to exposure to a tumor promoter, relationship between resveratrol and PKC have been intensively studied.[42,43] *In vitro* studies suggest that resveratrol mainly acts on and binds to a regulatory domain in PKC known as C1. Das *et al.* used resveratrol and modified structural analogs to study its effects on PKC.[42] In addition, the effects of PKC and resveratrol were tested in the presence and absence of a phorbol ester known as TPA, which acts as a tumor promoter that can prolong the activation of PKC and is also known to compete with resveratrol at the binding site. Structural modifications of resveratrol included the addition or alkylation of the hydroxyl groups or the addition of a long chain of hydroxyl groups that has been found to activate components such as DAG and TPA.

Binding of resveratrol and analogs with PKC, notably, the main domain for activation and inhibition on PKC is the C1 domain, was queried in molecular docking studies by the SYBYL software using the SurflexDock function.[42,43] The max conformation and rotations were set. Minimizations of the proteins were performed prior and subsequent to docking. The docking scores provided by the software were used, and a higher docking score represents the better docking of a compound at the receptor site. The results showed that increasing the number of hydroxyl groups increased the binding affinity, and scores were higher than with resveratrol alone.[42,43] The binding scores were also higher if resveratrol was modified by the addition of a long chain.[42,43] In addition, the binding scores between resveratrol and TPA provided evidence about the similarity of their binding affinity to PKC and their competition for the receptor site.[42]

In addition to the C1 domain, other subdomains known as C1A and C1B exist. The literature indicates that TPA has a higher binding affinity at the C1B domain.[42] Due to similar affinities, resveratrol and its derivatives were also analyzed for interaction in the C1B cysteine-rich domain of PKC. Together, resveratrol and resveratrol modified by the addition of hydroxyl groups were also found to interact with the C1B domain by optimally forming hydrogen bonds. The results of these binding studies led to the conclusion that resveratrol and modified derivatives were able to increase binding affinity to the PKC domains and improve PKC activity. Resveratrol was also able to reduce the TPA-induced activity onto PKC, indicating that these derivatives are possible regulators of PKC function.[42] Such stable regulation may underlie the suppression of cell proliferation and possible tumor growth, which are characteristics found in cancers.

11.2.7. *Docking resveratrol to cancer-related target proteins*

Previous studies have shown resveratrol and its derivatives exert antiproliferative activities in many cancer cell lines.[44–46] Resveratrol and its derivatives may suppress proliferation by inhibiting estrogen receptor-α (ER-α)-positive breast cancer cells,[47] regulating epigenetic proteins, i.e., DNA-associated histone proteins and histone deacetylases (HDACs),[48] or inducing apoptosis by activating p53.[49]

Breast cancer is one of the most common cancers, which affects more than a million women and causes approximately 40 thousand deaths per year, in the United States.[50–52] The chemical structure of resveratrol resembles mammalian estrogen 17β-estradiol; therefore, resveratrol or its derivatives are considered to interfere with the binding between 17β-estradiol and the estrogen receptor (ER).[47] Siddiqui *et al.* docked resveratrol and its derivatives to the active site of ER-α in addition to *in vitro* experiments.[47] The results show that some derivatives have lower binding affinities, better efficacy and explain how resveratrol and its derivatives can prevent and treat cancers.

Epigenetic modulation, or DNA-associated histone protein modification, has attracted increasing attention as a cancer treatment, especially regarding changes in acetylation status. Thus, two enzyme families, histone acetyltransferases (HATs) and histone deacetylases (HDACs), are significant target proteins for cancer treatment.[53–55] Venturelli *et al.* docked resveratrol onto human HDAC2, 4, 7, and 8 using GOLD (version 4.1.2, the Cambridge Crystallographic Data Center).[48] The crystal structures of HDAC2, 4, 7, and 8 (PDB ID: 3MAX, 2VQM, 3C10, 1T64) used in the molecular docking were downloaded from PDB. The ligand 3D structures were prepared, the ionization state of residue atoms was assigned, and the hydrogens were added back into the protein structures using MOE. The docking results show that resveratrol fits into the binding pocket of HDACs and interacts with the active site residues and the zinc ion. The *in silico* results agreed with the *in vitro* results, which confirmed that resveratrol inhibits the activity of HDACs.[48] Dayangac-Erden *et al.* also docked resveratrol on HDAC8 (PDB ID: 1T64) with Autodock,[56] although they actually proposed to increase the expression of the SMN2 (survival motor neuron) gene for the treatment of proximal spinal muscular atrophy, a neuromuscular disorder, by inhibiting HDAC8. The results show that resveratrol binds to the active site of HDAC8 and interacts with Gln263, Tyr306, and Phe208.

Integrins are transmembrane receptors in signal transduction pathways that regulate the cell cycle. Resveratrol binds to integrin αvβ3 and activates ERK, and p53 is activated after its Ser15 is phosphorylated, which eventually causes the induction of apoptosis.[57] Hsieh *et al.* docked resveratrol and the derivatives, triacetyl-resveratrol and trimethoxy-resveratrol (obtained from the ZINC database), onto an integrin αvβ3 X-ray structure (PDB ID: 1L5G) using DOCK6.3 from UCSF.[49] The docking results show that the three stilbenes bind to the active site, but only resveratrol and triacetyl-resveratrol show similar binding patterns and form three hydrogen bonds with integrin $\alpha v\beta 3$, which supports the hypothesis that the binding of resveratrol and triacetyl-resveratrol produce similar signaling effects that are different from the effect of trimethoxy-resveratrol binding.[49]

Microtubules are known to be a dynamic intracellular structure that are involved in a variety of cellular mechanisms including intracellular transportation, structural stability, and cell division. Microtubules are a desired drug target in cancer treatment because they are necessary for cell division. Microtubules initially develop from long and hollow filaments that are composed of multiple stacked α-β tubulin heterodimers. These dimers are placed end to end in a pattern, forming a protofilament.[58] Because microtubules are a well-studied target protein for cancer treatment, four main drug binding sites have been identified that affect tubulin stability: the laulimalide site (stabilizing), the taxane site (stabilizing), the colchicine site (destabilizing), and the vinca alkaloid site (destabilizing). Ruan *et al.* synthesized and docked 34 resveratrol derivatives that contained a curcumin moiety onto the tubulin-colchicine binding site (PDB ID: 1SA0) using Auto-Dock 4.0.[59] This study is a good example of virtual screening and identified the drug candidate, (1E,4E)-1-(2,4-dimethoxy-6-((E)-4-methoxystyryl)phenyl)-5-(3,5-dimethoxyphenyl) penta-1,4-dien-3-one, which had the best docking score and interactions among the compound and amino acids.

The specificity and bioavailability of resveratrol is the main concern of resveratrol applications. An oligomerized resveratrol such as balanocarpol produced by plant has better selectivity and potency for modulating the metabolism of sphingolipids, a major class of lipids in cells. Both resveratrol and balanocarpol inhibit the activity of sphingosine kinase 1 (SK1) and reduce SK1 expression in prostate cancer cells.[60,61] Lim *et al.* docked resveratrol and balanocarpol onto SK1 (PDB ID: 3VZB) using GOLD 5.2.[62] The results indicated that the hydroxyl groups form hydrogen bonds with the backbone carbonyl groups of Leu268 and Leu299 as well as a side chain oxygen of Asp178, while balanocarpol forms hydrogen bonds with the backbone carbonyl group of Leu268 and Ala262, a side chain oxygen of Thr196, and both side chain oxygens of Asp178. The models explain why balanocarpol has increased potency for the inhibition of SK1 activity.

11.2.8. *Docking resveratrol on albumin and gelatin: Nanotechnology on medicine*

Natural compounds such as resveratrol have chemotherapeutic activities, but instability and poor solubility result in limited bioavailability, which is the main challenge associated with the use of these as dietary compounds. Nanotechnology provides a possibility to deliver drugs with improved pharmacokinetics and bio-distribution using polymer-based nanoparticles.[63–65] Therefore, to identify or optimize a material that comprises a nanoparticle and binds a drug well is a significant challenge. Docking studies have been used to investigate the binding of resveratrol to two protein-based nanoparticles, albumin and gelatin nanoparticles, to determine which is the best carrier.[66]

Metwally *et al.*[66] used GROMACS v4.6.5 to build a system of 48 gelatin molecules and simulate the structure, then conducted docking studies using the MOE software package. Because resveratrol is a poorly soluble hydrophobic compound, it prefers to bind to a hydrophobic carrier such as bovine serum albumin. The docking results agreed with this hypothesis because the binding energy for docking resveratrol onto albumin was lower than that for a gelatin matrix. Thus, albumin can accommodate a greater mass of resveratrol.

11.3. Resveratrol and MD simulation

It is well-known that a major issue and challenge of molecular docking is target protein flexibility. To reduce the computational load, most of the time, the protein structures are considered as rigid in molecular docking studies. The provision of a wider range of conformations for ligand and protein interactions is essential to obtain accurate docked structures and estimate precise binding affinities.[2] Similar to docking studies, another computational method known as molecular dynamic (MD) simulation can provide both dynamic structural flexibility for proteins and help analyze their interactions with other molecules.[67] A simulation involves a system in which a protein complex is placed in and created based on forcefields, pressure, and interactions with other molecules.[67] Using such simulations, conformations are predicted based on a system created over a period of time.[67] These predictions are then computed over a series of snapshots to help explain the behavior of the molecules under consideration.[67] The advantage to using such a system is to investigate the molecular dynamics under specific conditions to provide greater biological relevance, especially for *in silico* drug research.[67]

11.3.1. *MD simulation of SIRT1 binding*

There have been additional computational studies that have raised concern about whether resveratrol has a direct impact on the SIRT1 activation. Aside from a conformational change that activates SIRT1 functionality, resveratrol cannot activate SIRT1 *in vitro* in the absence of a fluorophore in its substrate.[9,13] In addition, resveratrol appears to only activate SIRT1 with substrates with bulky hydrophobic groups.[6,9] Without the appropriate requirements for activation, no conformational changes or activation occurs.[9] As a result, it has been proposed that resveratrol acts more as a protein substrate stabilizer and enhances interactions because activation relies on conformational changes between the N-terminal domain and the substrate binding site.[9] Therefore, resveratrol and its role were analyzed in the presence of three types of p53 molecules, a known SIRT1 substrate — p53 — in its native form, p53 with the bulky hydrophobic group 7-amino-4-methylcoumarin (p53 AMC) and p53 with a tryptophan residue (p53 W).[9]

In order to investigate the effects of resveratrol on SIRT1 with these substrates, Hou *et al.* (2016) used molecular dynamics (MD) simulations with the Amber14 molecular dynamics package instead of docking.[9] To implement the MD simulations, the experiment set an Amber14SB forcefield and a TIP3P model for the water molecules.[9]

After 5 independent 0.2-μ s simulations were performed, resveratrol was able to activate SIRT1 in the presence of p53AMC and p53W but not native p53.[9] However, native p53 binds tightly to SIRT1 compared to p53AMC and p53W, and in the absence of resveratrol, p53AMC and p53W are not able to interact with the SIRT1 N-terminal domain.[9] These results indicated that the interaction energy between SIRT1 and p53 increased when resveratrol was introduced into the system.[9] The MD simulations also informed how resveratrol promotes a tighter binding interaction between the substrates and SIRT1 to activate it.[9] Resveratrol significantly limited the ability of the substrate bulky hydrophobic groups to move, therefore promoting a better interaction.

11.3.2. *MD simulation of the inhibition of amyloid polypeptide aggregation*

The term amyloid polypeptide is generally used to describe β-sheet-rich, misfolded, and insoluble protein aggregates and is associated with many neurodegenerative diseases. Wang *et al.* utilized MD simulation as a means of classifying the mechanism by which resveratrol inhibits human islet amyloid polypeptide (hIAPP, aka, amylin).[68] The researchers used AMBER 10.0 for their MD simulations, employing four different simulation systems to investigate potential binding modes of resveratrol. The four systems included one hIAPP pentamer and three different hIAPP-resveratrol complexes. The root-mean-square-deviation (RMSD) values of the backbone atoms indicated that binding to resveratrol caused significant structural changes in the hIAPP pentamers. One of the three hIAPP-resveratrol complexes did not demonstrate stable interactions, making it an unlikely binding mechanism. This study concluded that resveratrol acted as an inhibitor of amyloid aggregation by changing the overall structure of hIAPP. The main structural changes were a "reduction in order degree and beta sheet contents".[68] This structural analysis led to the discovery of two binding sites (Sites I and II). Site I includes Ser29, Thr30, and Asn31 residues from the chains of five hIAPP pentamers. Resveratrol has van der Waals interactions with the Asn31 side chains and a hydrogen bond between one of its hydroxyl groups and the Ser29 carbonyl. Site II contains Arg11, Leu12, and Ala13 residues from multiple hIAPP chains. The hydrogens from the arginine amino groups can form hydrogen bonds with the resveratrol hydroxyl groups. Free-energy calculations for the two sites showed a stronger binding affinity between resveratrol and Site II than between resveratrol and Site I. This work demonstrates the potential mechanism by which *resveratrol reduces the risk of developing Alzheimer's disease (AD)* and indicates that it may be more than just a correlation. This work could also lead to future drug discovery for AD, using hIAPP as a target. Even more effective molecules could be designed that further change the structure of hIAPP and have a higher binding affinity, increased site specificity, and increased bioavailability. In terms of the amyloid cascade hypothesis, disruption of amyloid aggregation would greatly reduce the progression of AD symptoms. Evidence that indicates symptom reversal has not been found.

11.3.3. *MD simulation of HSA binding*

Human serum albumin (HSA), which is abundant in blood plasma and cerebrospinal fluid, can bind many different substrates including hormones, fatty acids, and drugs, and transport them to target positions in the body. Previous studies have shown that resveratrol can stabilize the structure of HSA, inhibit amyloid fibrillation and reduce the cytotoxicity of the aggregates.[69–71] Stirpe *et al.* conducted molecular docking studies with AutoDock Vina,[72] docking resveratrol onto HSA (PDB ID: 1E7H and 1E7I) and did an MD simulation with GROMACS[73] to simulate the interactions between HSA and resveratrol.[74] Resveratrol was docked to the two potential binding sites, Sudlow's drug site I and fatty acid site 1, and the predicted binding affinities were −6.5 and −8.9 kcal/mol. Fatty acid site 1 is more likely the primary binding site for resveratrol, so that binding pose was used to conduct the MD simulation. The results show that there are 4 residues with aromatic side chains, Phe134, Phe165, Tyr138 and Tyr161, surrounding the bound resveratrol and interchanging continuously while forming stacking interactions with the resveratrol phenol rings.

11.3.4. *MD simulation of NF-κB binding*

NF-κB plays a critical role in the immune response and is also a target of resveratrol and its derivatives. The NF-κB complex consists of three subunits, p50, p65, and IκBα. Banaganapalli *et al.* published two studies on the interactions between NF-κB and resveratrol and its two derivatives.[75,76] They excluded IκBα, the inhibitory protein, from the complex structure (PDB ID: 1IKN or 3GUT) and docked resveratrol and two new derivatives, (E)-1,3-bis(prop-2-ynyloxy)-5-(4-(prop-2-ynyloxy)styryl)benzene (RVS(a)) or (E)-5-(4-acetoxystyryl)-1,3-phenylenediacetate (RESL43) onto the p50/p65 heterodimer with the AutoDock4[77] software package in two separate studies. The results showed that resveratrol and RVS(a) both bind to the IκBα binding site on p65 and p50. Resveratrol interacts with Gln49 and Arg50 (p65) with a binding affinity of −6.12 kcal/mol. RVS(a) interacts with more residues, Gln49, Arg50, and His181 (p65), with a lower binding affinity (−7.89 kcal/mol) than resveratrol.[75] RESL43 forms H-bonds with Arg33, Arg35 (p65), Arg354, Lys572, and Gln606 (p50) with a binding affinity of −10.42 kcal/mol.[76] MD simulation was conducted on the apo-form NF-κB, NF-κB-resveratrol, NF-κB-RVS(a) and the NF-κB-RESL43 complex using GROMACS4.0.[78] The simulation trajectories show that the structures of all three NF-κB-ligand binding complexes are more stable than the apo-form because less RMSD fluctuation occurs, and the ligand binding complexes have lower energy than the apo-form NF-κB, implying that ligand binding may interfere with the binding between NF-κB and DNA. These studies proposed two resveratrol derivatives as drug candidates for cancer treatment because they had good potency.

11.3.5. *MD simulation of AR binding*

The androgen receptor (AR) binds to androgen, translocates into the nucleus and binds DNA, resulting in the activation of its target genes, which play a critical role in the development and progression of prostate cancer. Resveratrol and its natural derivatives pterostilbene (trans-3,5-dimethoxy-4′-hydroxystilbene), piceatannol (trans-3,4,3′,5′-tetrahydroxystilbene), and trimethoxy-resveratrol (trans-3,5,4′-trimethoxystilbene) bind to the AR and inhibit tumor growth. Chakraborty *et al.* docked these four compounds onto the ligand-binding domain at the C-terminal of AR (PDB ID: 1E3G) using AutoDock 4. Thr877 was mutated to Ala and designated as the T877A mutant expressed in LNCaP prostate cancer cells.[79] The complex structures obtained from the docking were employed to perform MD simulation using GROMACS. The results indicated that the endogenous androgen showed the strongest binding affinity, and the AR antagonist flutamide showed the lowest binding affinity. The binding affinities of resveratrol and the three analogs fell between the endogenous androgen and flutamide. The order of the estimated binding affinities agreed well with the experimentally derived ED_{50} values. Among resveratrol and its analogs, pterostilbene has the lowest estimated binding affinity, which is similar to that of flutamide. Resveratrol and the three analogs form more stable hydrogen bonds according to the analysis of the trajectories of the MD simulations, i.e., the percentage of the conformations in the simulation trajectories, which form hydrogen bonds. The binding pattern of pterostilbene is the most similar to flutamide, the known AR antagonist; for example, androgen often interacted with Arg752 (71.6% of conformations), but the interaction was totally absent in the flutamide and pterostilbene trajectories. Resveratrol and the other two analogs formed fewer hydrogen bonds with Arg752 compared with androgen, which shows that pterostilbene can be a drug candidate because it is the most potent AR antagonist among resveratrol and its analogs.

11.3.6. *MD simulation of pterostilbene-mediated Nrf2 activation*

Nuclear factor-erythroid 2-related factor 2 (Nrf2) is a protein that responds to oxidative stress, and it binds to the promoters of antioxidant/electrophilic responsive elements to enhance the expression of detoxifying genes. Kelch-like ECH-associated protein-1 (Keap1) binds to Nrf2 and induces its degradation through the ubiquitin-proteasome pathway.[80,81] The dissociation of the Keap1-Nrf2 complex leads to the transcriptional activation of many cytoprotective genes.[82,83] Resveratrol has been found to bind Nrf2, perturb Keap1-Nrf2 binding, and activate Nrf2.[84]

Bhakkiyalakshmi *et al.* first docked a resveratrol derivative, pterostilbene (PTS), to the Keap1-Nrf2 complex structure (PDB IDs: 2FLU) using Schroedinger Glide and performed molecular dynamic simulations following the docking studies.[85] A PTS was docked to either a cysteine pocket[86] or a kelch pocket[87] on the binding interface between Keap1 and Nrf2. The docking results show that PTS should bind to the kelch pocket because the docked poses suggest stronger interactions among PTS and residues such as Arg415, Arg380, Asn414, Arg483, His436, and Ser431 in the kelch pocket.

Following the molecular docking experiments, Bhakkiyalakshmi *et al.* performed a 6 ns MD simulation with the apo-form Keap1-Nrf2 complex and the complex with docked PTS using Desmond. The RMSDs of all C_α and backbone atoms from the initial structures show that the apo-form structure is more stable (the RMSDs remain within 1.8 Angstrom) than the ligand binding structure (the RMSDs are a large as 3 Angstroms). The analysis of the clustered dynamic structures from the simulation trajectories also demonstrates that the binding affinities calculated based on the apo-form structures are lower than those based on the PTS binding structures. All of the data show that the PTS binding will enhance the dissociation of the Keap1-Nrf2 complex and activate Nrf2.[85] This study is a good example of how to combine molecular docking and MD simulation to reveal the details of the functional mechanisms of resveratrol or its derivatives.

11.4. Other computational studies

11.4.1. *Binding pocket analysis of SIRT1*

Fragment-centric topographical mapping has been used to analyze protein-protein interactions.[88] Functionally, this type of program can map out space where a protein-protein interaction occurs and evaluate the binding pocket complementation based on optimal energies and residues.[88]

Despite a knowledge of the interactions between these proteins and their substrates, it is also helpful to analyze the binding pocket domains in which these reactions occur to further explain the mechanism. This can facilitate a drug discovery analysis to find complementary drugs that can interact efficiently with specific active sites. AlphaSpace was used to analyze the SIRT1 binding sites.[9] The AlphaSpace analysis indicated that the highest interaction energies for substrate binding between the native form of p53 and SIRT are based on the residues that occupied the majority of the pocket. These residues were acetyl-lysine, leucine, methionine, and phenylalanine.[9] Resveratrol could increase the pocket sites for these bulky hydrophobic side chains.[9] Resveratrol is able to help form a new binding pocket that allows for interactions between hydrophobic groups such as AMC, which can restore optimal interaction with the substrate for activation.[9]

Therefore, in summary, resveratrol may not be able to activate SIRT1 alone or in the presence of a native p53 protein. However, in the presence of p53 that contains specific bulky hydrophobic groups, resveratrol can act as a substrate stabilizer, which helps proteins that may not have been able to bind in certain domains. Once it can stabilize and allow an optimal binding interaction, it can indirectly serve as an activator because it ensures tight interactions and absolute SIRT1 activation. Further *in vivo* studies should be performed to determine if resveratrol directly participates in SIRT1 activation.

11.4.2. *3D-QSAR models*

QSAR models are regression or classification models that are widely used in computational drug discovery. They propose a relationship between structural descriptors and the biological activities of compounds and can be employed in sequence or in parallel with docking simulations. Given the activities of a group of drug candidates, a QSAR model can help interpret the results of biological experiments, provide a direction for lead optimization, and identify or predict good drug candidates. Comparative molecular field analysis (CoMFA), which predicts affinities based on steric fields and electrostatic fields, and comparative molecular similarity indices analysis (CoMSIA), which is based on the structure indices, are both generally applied models.

Amyloid beta (Aβ) peptides are small peptides whose aggregates lead to Alzheimer's disease (AD), a fatal neurodegenerative disorder. Current AD drugs on the market such as galantamine, rivastigmine, and donepezil have severe side effects,[89, 90] and none of them are targeted against Aβ aggregation. Therefore, the use of resveratrol to stop the aggregation of Aβ has attracted much attention. Awasthi *et al.* built a CoMFA and a CoMSIA based on 30 resveratrol derivatives proven to be Aβ aggregation inhibitors. The models indicated that bulky, electronegative substituents on the 3,5-hydroxyl groups attached to the phenyl ring of resveratrol and that the bulky, electropositive, and hydrophobic substituents on 4′-hydroxyl group attached to the other phenyl ring of resveratrol can improve the inhibition potency of the resveratrol derivatives.[91]

Resveratrol was also shown to have anti-virus activity. For example, it can bind to influenza neuraminidase (NA), which is one of the two main targets on the surface of influenza virus particles. Li *et al.* derived CoMFA and CoMSIA models for 26 resveratrol derivatives as a training set and predicted the pIC_{50} of another 9 resveratrol derivatives (the test set).[92] The performance of both models was good ($r^2 > 0.95$), which means they could accurately predict the pIC_{50} of new resveratrol derivatives binding to NA. The two models also produced contour maps of the regions on the resveratrol structure that would prefer different substituent chemical groups to increase their activities. In addition, Li *et al.* also docked these resveratrol derivatives onto the Zanamivir binding site of NA (PDB ID: 1A4G) using the CDOCKER protocol in Accelrys Discovery Studio 2.5. The docking results support the results of the CoMSIA model contour map, i.e., that the hydrogen bond acceptor at the ortho-position (5′-) of the benzene ring and hydrogen bond donor at the para-position (4-) of the other benzene ring are energy favorable. The 5′-carbonyl oxygens and 4-hydroxyl group of the most potent compound (R42) form ten hydrogen bonds with key residues including Glu225, Arg115, Arg149, Ser117, Arg291, Tyr408, Asp148, and Arg373 on NA, which is similar to the interactions between known NA and its known drug Zanamivir.[92]

11.4.3. *Additional targets but limited computational method research*

Despite extensive research on resveratrol and its possible targets in cardiovascular-related disorders, there are still not as many related *in silico* studies as compared to the vast majority of *in vivo* and *in vitro* studies. There is also a lack of repeated studies or further *in silico* investigations on the same targets. Resveratrol can act on a variety of targets that are involved or contribute to signaling in the membrane and cells, metabolic cascades and gene expression.[42] Laboratory studies can only provide a limited amount of data. Computational methods such as docking, pocket analysis, QSAR and MD simulations can provide another dataset that can further explain why certain protein interactions occur, their mechanisms, and the types of interactions that cause particular effects. These are necessary answers that should be provided by research.[2, 93]

Human DNA polymerase exemplifies a target for which further investigation would be beneficial. Research has shown that resveratrol has selective inhibitory effects on DNA polymerase α and δ *in vitro* and results in the cessation of DNA replication.[94] However, the results from this research are limited because binding modes and more specific interactions have not been found or defined.[94] Therefore, the Computer Drug Discovery Lab at the University of Georgia performed preliminary docking studies with resveratrol and DNA polymerase α. The RCSB protein data bank was used to find a PDB file for a human DNA polymerase α. The human primosome 5EXR was also found, which consisted a complex of primase subunits and DNA polymerase α subunits. To prepare the protein, the primase subunits were removed to maintain the DNA polymerase α complex. The modified PDB file was subsequently submitted to a predicting binding pocket server known as POCASA 1.1.[95]

When the protein structure was submitted for binding site prediction, all settings were kept at default. The protein was subsequently preprocessed using the Schroedinger Glide program. All preprocessing, optimization and minimization were kept at default, and no errors were reported from the program. Each predicted binding site found from POCASA[95] was then separately mapped into an active site on the protein complex. Unlike in prior studies, the native structure of resveratrol was used and prepped as the ligand using the LigPrep function in Schroedinger. Once the protein and ligand were prepped, the docking process was performed using the GlideDock function. Docking was done with both standard precision and flexible extra precision docking. The results showed the highest docking scores using both standard and extra precision docking primarily at the site identified as the DNA polymerase alpha catalytic subunit. Notable interactions at this binding site are an aromatic interaction with ARG175 and a hydrogen bond interaction between the 5′OH and TYR32. The limited results allowed for the conclusion that resveratrol showed the highest inhibitory activity in the catalytic unit of DNA polymerase α. Further studies involving preparing and docking with a more isolated DNA polymerase α complex as well as other *in silico* models including a molecular dynamic simulation will be used to analyze the protein-ligand interactions and to elucidate the exact mechanism.

11.4.4. *Statistical analysis of amino acid preferences in interaction of target proteins with resveratrol*

On PDB, there are 23 entries containing resveratrol binding to proteins. Using 4 angstrom threshold, we determined which residues interact with resveratrol on these 23 complexes. We counted the number of 20 amino acids interacting with resveratrol. Table 11.1 shows the numbers, the percentage (the number of a specific amino acid interacting with resveratrol/the number of all residues interacting with resveratrol), and the enrichment factors (the percentage of an interacting residue/the natural content of the amino acid in PDB). Our analysis shows that resveratrol prefers binding to residues that

Table 11.1. The interactions between bound resveratrol and 20 amino acids sorted according to their enrichment factors (binding percentages divided by the natural content). The statistical analysis was performed on all 23 entries downloaded from PDB.

Residue	No. of interactions	%	Enrichment factor
PHE	37	11.1	2.871
MET	15	4.5	1.942
TYR	17	5.1	1.754
HIS	13	3.9	1.727
THR	30	9.0	1.629
TRP	6	1.8	1.441
ASN	16	4.8	1.223
GLY	27	8.1	1.153
LEU	36	10.8	1.117
ASP	16	4.8	0.875
CYS	4	1.2	0.87
GLU	18	5.4	0.855
PRO	14	4.2	0.837
ILE	15	4.5	0.82
ARG	13	3.9	0.675
GLN	8	2.4	0.616
LYS	10	3.0	0.579
ALA	16	4.8	0.548
VAL	12	3.6	0.535
SER	10	3.0	0.421

contain aromatic rings such as Phe, Tyr, His, and Trp; hydroxyl groups such as Tyr and Thr; and polar side chains such as Asn. Because resveratrol simply consists of two aromatic rings and three hydroxyl groups, it can fit in and bind to the active sites of a wide variety of enzymes and regulatory proteins. The docked structures of resveratrol or its derivatives and the target proteins mentioned in this chapter generally support our observation.

11.5. Summary

Summarizing the computational studies of resveratrol targeting proteins, we noticed that resveratrol affects human health in many ways by binding target proteins, which play significant roles in regulating gene expression, proliferation, apoptosis, inflammation, lipogenesis, angiogenesis, neurodegeneration, and so on. Based on a previous computational system toxicology study,[96] resveratrol is a herbal compound that widely binds to many proteins.

Molecular docking was mainly used to identify interacting residues, confirm and compare the binding positions/conformations and binding affinities of resveratrol and its analogs using the docked poses and scores. The bound residues identified also provide hints that reveal the binding mechanisms, allowing for the design of new drugs and the planning of new research.

Previously, molecular docking has been applied to confirm and interpret the binding of resveratrol. However, we know that resveratrol can bind to the active sites of many enzymes and significant proteins because of its simple chemical structure and characteristics. Docking studies should be designed and conducted on a large-scale experiment to identify all possible target proteins and estimate their binding affinities. The conventional one-drug-one-target-one-disease paradigm cannot explain how resveratrol affects human health in so many different aspects. To construct a multi-drug multi-target network in the system pharmacology era,[97] we should use molecular docking to actively identify new targets and perform multiscale modeling to reveal the functions of resveratrol on a system scale.

References

1. Kumalo, HM, Bhakat, S, Soliman, ME. Theory and applications of covalent docking in drug discovery: merits and pitfalls. *Molecules* 2015; **20**: 1984–2000.
2. Ferreira, LG, Dos Santos, RN, Oliva, G, Andricopulo, AD. Molecular docking and structure-based drug design strategies. *Molecules* 2015; **20**: 13384–13421.
3. Pagadala, NS, Syed, K, Tuszynski, J. Software for molecular docking: a review. *Biophys Rev* 2017; **9**: 91–102.
4. Buja, LM, Vela, D. Cardiomyocyte death and renewal in the normal and diseased heart. *Cardiovasc Pathol* 2008; **17**: 349–374.
5. Akkafa, F, Halil Altiparmak, I, Erkus, ME, Aksoy, N, Kaya, C, Ozer, A, Sezen, H, Oztuzcu, S, Koyuncu, I, Umurhan, B. Reduced SIRT1 expression

correlates with enhanced oxidative stress in compensated and decompensated heart failure. *Redox Biol* 2015; **6**: 169–173.

6. Xia, N., Forstermann, U. Li, H. Resveratrol and endothelial nitric oxide. *Molecules* 2014; **19**: 16102–16121.
7. Moniot, S, Weyand, M, Steegborn, C. Structures, substrates, and regulators of Mammalian sirtuins — opportunities and challenges for drug development. *Front Pharmacol* 2012; **3**: 16.
8. Yuan, H, Marmorstein, R. Structural basis for sirtuin activity and inhibition. *The Journal of biological chemistry* 2012; **287**: 42428–42435.
9. Hou, X, Rooklin, D, Fang, H, Zhang, Y. Resveratrol serves as a protein-substrate interaction stabilizer in human SIRT1 activation. *Scientific Reports* 2016; **6**: 38186.
10. Cao, D, Wang, M, Qiu, X, Liu, D, Jiang, H, Yang, N, Xu, RM. Structural basis for allosteric, substrate-dependent stimulation of SIRT1 activity by resveratrol. *Genes Dev* 2015; **29**: 1316–1325.
11. Sulaiman, M, Matta, MJ, Sunderesan, NR, Gupta, MP, Periasamy, M, Gupta, M. Resveratrol, an activator of SIRT1, upregulates sarcoplasmic calcium ATPase and improves cardiac function in diabetic cardiomyopathy. *Am J Physiol Heart Circ Physiol* 2010; **298**: H833–843.
12. Hubbard, BP, Gomes, AP, Dai, H, Li, J, Case, AW, Considine, T, Riera, TV, Lee, JE, E, SY, Lamming, DW, *et al.* Evidence for a common mechanism of SIRT1 regulation by allosteric activators. *Science* 2013; **339**: 1216–1219.
13. Beher, D, Wu, J, Cumine, S, Kim, KW, Lu, SC, Atangan, L, Wang, M. Resveratrol is not a direct activator of SIRT1 enzyme activity. *Chemical Biology & Drug Design* 2009; **74**: 619–624.
14. Guo, Y, Zhang, L, Li, F, Hu, CP, Zhang, Z. Corrigendum to "Restoration of sirt1 function by pterostilbene attenuates hypoxia-reoxygenation injury in cardiomyocytes" [*Eur. J. Pharmacol.* 776 (2016) 26–33]. *European Journal of Pharmacology* 2016; **782**: 120.
15. Kutil, Z, Temml, V, Maghradze, D, Pribylova, M, Dvorakova, M, Schuster, D, Vanek, T, Landa, P. Impact of wines and wine constituents on cyclooxygenase-1, cyclooxygenase-2, and 5-lipoxygenase catalytic activity. *Mediators Inflamm* 2014; 178931.
16. Handler, N, Brunhofer, G, Studenik, C, Leisser, K, Jaeger, W, Parth, S, Erker, T. 'Bridged' stilbene derivatives as selective cyclooxygenase-1 inhibitors. *Bioorganic & Medicinal Chemistry* 2007; **15**: 6109–6118.
17. Kummerle, AE, Sperandio da Silva, GM, Sant'Anna, CM, Barreiro, EJ, Fraga, CA. A proposed molecular basis for the selective resveratrol inhibition of the PGHS-1 peroxidase activity. *Bioorganic & Medicinal Chemistry* 2005; **13**: 5981–5985.
18. Wu, X, Zhu, L, Zilbering, A, Mahadev, K, Motoshima, H, Yao, J, Goldstein, BJ. Hyperglycemia potentiates H(2)O(2) production in adipocytes and enhances insulin signal transduction: potential role for oxidative inhibition of thiol-sensitive protein-tyrosine phosphatases. *Antioxid Redox Signal* 2005; **7**: 526–537.

19. Fitzgerald, JC, Ufer, C, De Girolamo, LA, Kuhn, H, Billett, EE. Monoamine oxidase-A modulates apoptotic cell death induced by staurosporine in human neuroblastoma cells. *J Neurochem* 2007; **103**: 2189–2199.
20. Shen, SH, Wertz, DL, Klinman, JP. Implication for functions of the ectopic adipocyte copper amine oxidase (AOC3) from purified enzyme and cell-based kinetic studies. *PloS one* 2012; **7**: e29270.
21. Morin, N, Lizcano, JM, Fontana, E, Marti, L, Smih, F, Rouet, P, Prevot, D, Zorzano, A, Unzeta, M, Carpene, C. Semicarbazide-sensitive amine oxidase substrates stimulate glucose transport and inhibit lipolysis in human adipocytes. *J Pharmacol Exp Ther* 2001; **297**: 563–572.
22. Les, F, Deleruyelle, S, Cassagnes, LE, Boutin, JA, Balogh, B, Arbones-Mainar, JM, Biron, S, Marceau, P, Richard, D, Nepveu, F, *et al.* Piceatannol and resveratrol share inhibitory effects on hydrogen peroxide release, monoamine oxidase and lipogenic activities in adipose tissue, but differ in their antilipolytic properties. *Chemico-Biological Interactions* 2016; **258**: 115–125.
23. Basheer, L, Schultz, K, Kerem, Z. Inhibition of cytochrome P450 3A by acetoxylated analogs of resveratrol in *in vitro* and *in silico* models. *Scientific Reports* 2016; **6**: 31557.
24. Detampel, P. Beck, M. Krahenbuhl, S. Huwyler, J. Drug interaction potential of resveratrol. *Drug Metab Rev* 2012; **44**: 253–265.
25. Regev-Shoshani, G, Shoseyov, O, Kerem, Z. Influence of lipophilicity on the interactions of hydroxy stilbenes with cytochrome P450 3A4. *Biochemical and Biophysical Research Communications* 2004; **323**: 668–673.
26. Chan, WK, Delucchi, AB. Resveratrol, a red wine constituent, is a mechanism-based inactivator of cytochrome P450 3A4. *Life Sci* 2000; **67**: 3103–3112.
27. Huberts, DH, van der Klei, IJ. Moonlighting proteins: an intriguing mode of multitasking. *Biochimica et Biophysica Acta* 2010; **1803**: 520–525.
28. Maughan, DW, Henkin, JA, Vigoreaux, JO. Concentrations of glycolytic enzymes and other cytosolic proteins in the diffusible fraction of a vertebrate muscle proteome. *Molecular & Cellular Proteomics : MCP* 2005; **4**: 1541–1549.
29. Tristan, C, Shahani, N, Sedlak, TW, Sawa, A. The diverse functions of GAPDH: views from different subcellular compartments. *Cell Signal* 2011; **23**: 317–323.
30. Ganapathy-Kanniappan, S, Kunjithapatham, R, Geschwind, JF. Glyceraldehyde-3-phosphate dehydrogenase: a promising target for molecular therapy in hepatocellular carcinoma. *Oncotarget* 2012; **3**: 940–953.
31. Huang, J, Xiong, N, Chen, C, Xiong, J, Jia, M, Zhang, Z, Cao, X, Liang, Z, Sun, S, Lin, Z, *et al.* Glyceraldehyde-3-phosphate dehydrogenase: activity inhibition and protein overexpression in rotenone models for Parkinson's disease. *Neuroscience* 2011; **192**: 598–608.
32. Cumming, RC, Schubert, D. Amyloid-beta induces disulfide bonding and aggregation of GAPDH in Alzheimer's disease. *FASEB J* 2005; **19**: 2060–2062.

33. Butterfield, DA, Hardas, SS, Lange, ML. Oxidatively modified glyceraldehyde-3-phosphate dehydrogenase (GAPDH) and Alzheimer's disease: many pathways to neurodegeneration. *J Alzheimers Dis* 2010; **20**: 369–393.
34. Bae, BI, Hara, MR, Cascio, MB, Wellington, CL, Hayden, MR, Ross, CA, Ha, HC, Li, XJ, Snyder, SH, Sawa, A. Mutant huntingtin: nuclear translocation and cytotoxicity mediated by GAPDH. *Proceedings of the National Academy of Sciences of the United States of America* 2006; **103**: 3405–3409.
35. Naletova, I, Schmalhausen, E, Kharitonov, A, Katrukha, A, Saso, L, Caprioli, A, Muronetz, V. Non-native glyceraldehyde-3-phosphate dehydrogenase can be an intrinsic component of amyloid structures. *Biochimica et Biophysica Acta* 2008; **1784**: 2052–2058.
36. Mazzola, JL, Sirover, MA. Reduction of glyceraldehyde-3-phosphate dehydrogenase activity in Alzheimer's disease and in Huntington's disease fibroblasts. *J Neurochem* 2001; **76**: 442–449.
37. Sultana, R, Boyd-Kimball, D, Cai, J, Pierce, WM, Klein, JB, Merchant, M, Butterfield, DA. Proteomics analysis of the Alzheimer's disease hippocampal proteome. *J Alzheimers Dis 11* 2007; 153–164.
38. Avila, CL, Torres-Bugeau, CM, Barbosa, LR, Sales, EM, Ouidja, MO, Socias, SB, Celej, MS, Raisman-Vozari, R, Papy-Garcia, D, Itri, R, *et al.* Structural characterization of heparin-induced glyceraldehyde-3-phosphate dehydrogenase protofibrils preventing alpha-synuclein oligomeric species toxicity. *The Journal of Biological Chemistry* 2014; **289**: 13838–13850.
39. Rodacka, A, Strumillo, J, Serafin, E, Puchala, M. Analysis of Potential Binding Sites of 3,5,4′-Trihydroxystilbene (Resveratrol) and trans-3,3′,5,5′-Tetrahydroxy-4′-methoxystilbene (THMS) to the GAPDH Molecule Using a Computational Ligand-Docking Method: Structural and Functional Changes in GAPDH Induced by the Examined Polyphenols. *The Journal of Physical Chemistry B* 2015; **119**: 9592–9600.
40. Berman, HM, Westbrook, J, Feng, Z, Gilliland, G, Bhat, TN, Weissig, H, Shindyalov, IN, Bourne, PE. The Protein Data Bank. *Nucleic Acids Research* 2000; **28**: 235–242.
41. Lim, PS, Sutton, CR, Rao, S. Protein kinase C in the immune system: from signalling to chromatin regulation. *Immunology* 2015; **146**: 508–522.
42. Das, J, Pany, S, Majhi, A. Chemical modifications of resveratrol for improved protein kinase C alpha activity. *Bioorganic & Medicinal Chemistry* 2011; **19**: 5321–5333.
43. Pany, S, Majhi, A, Das, J. PKC activation by resveratrol derivatives with unsaturated aliphatic chain. *PloS one* 2012; **7**: e52888.
44. Ahmad, N, Adhami, VM, Afaq, F, Feyes, DK, Mukhtar, H. Resveratrol causes WAF-1/p21-mediated G(1)-phase arrest of cell cycle and induction of apoptosis in human epidermoid carcinoma A431 cells. *Clin Cancer Res* 2001; **7**: 1466–1473.
45. Schneider, Y, Vincent, F, Duranton, B, Badolo, L, Gosse, F, Bergmann, C, Seiler, N, Raul, F. Anti-proliferative effect of resveratrol, a natural component

of grapes and wine, on human colonic cancer cells. *Cancer Lett* 2000; **158**: 85–91.

46. Huang, C, Ma, WY, Goranson, A, Dong, Z. Resveratrol suppresses cell transformation and induces apoptosis through a p53-dependent pathway. *Carcinogenesis* 1999; **20**: 237–242.
47. Siddiqui, A, Dandawate, P, Rub, R, Padhye, S, Aphale, S, Moghe, A, Jagyasi, A, Venkateswara Swamy, K, Singh, B, Chatterjee, A, *et al.* Novel Aza-resveratrol analogs: synthesis, characterization and anticancer activity against breast cancer cell lines. *Bioorganic & Medicinal Chemistry Letters* 2013; **23**: 635–640.
48. Venturelli, S, Berger, A, Bocker, A, Busch, C, Weiland, T, Noor, S, Leischner, C, Schleicher, S, Mayer, M, Weiss, TS, *et al.* Resveratrol as a pan-HDAC inhibitor alters the acetylation status of histone [corrected] proteins in human-derived hepatoblastoma cells. *PloS one* 2013; **8**: e73097.
49. Hsieh, TC, Wong, C, John Bennett, D, Wu, JM. Regulation of p53 and cell proliferation by resveratrol and its derivatives in breast cancer cells: an *in silico* and biochemical approach targeting integrin alphavbeta3. *Int J Cancer* 2011; **129**: 2732–2743.
50. Gluz, O, Liedtke, C, Gottschalk, N, Pusztai, L, Nitz, U, Harbeck, N. Triple-negative breast cancer–current status and future directions. *Ann Oncol* 2009; **20**: 1913–1927.
51. Glass, AG, Lacey, JV, Jr, Carreon, JD, Hoover, RN. Breast cancer incidence, 1980–2006: combined roles of menopausal hormone therapy, screening mammography, and estrogen receptor status. *J Natl Cancer Inst* 2007; **99**: 1152–1161.
52. Siegel, R, Naishadham, D, Jemal, A. Cancer statistics, 2012. *CA Cancer J Clin* 2012; **62**: 10–29.
53. Ellis, L, Atadja, PW, Johnstone, RW. Epigenetics in cancer: targeting chromatin modifications. *Mol Cancer Ther* 2009; **8**: 1409–1420.
54. Marks, PA, Xu, WS. Histone deacetylase inhibitors: potential in cancer therapy. *J Cell Biochem* 2009; **107**: 600–608.
55. Acharya, MR, Sparreboom, A, Venitz, J, Figg, WD. Rational development of histone deacetylase inhibitors as anticancer agents: a review. *Molecular Pharmacology* 2005; **68**: 917–932.
56. Dayangac-Erden, D, Bora, G, Ayhan, P, Kocaefe, C, Dalkara, S, Yelekci, K, Demir, AS, Erdem-Yurter, H. Histone deacetylase inhibition activity and molecular docking of (e)-resveratrol: its therapeutic potential in spinal muscular atrophy. *Chemical Biology & Drug Design* 2009; **73**: 355–364.
57. Lin, HY, Lansing, L, Merillon, JM, Davis, FB, Tang, HY, Shih, A, Vitrac, X, Krisa, S, Keating, T, Cao, HJ, *et al.* Integrin alphaVbeta3 contains a receptor site for resveratrol. *FASEB J* 2006; **20**: 1742–1744.
58. Hawkins, T, Mirigian, M, Selcuk Yasar, M, Ross, JL. Mechanics of microtubules. *J Biomech* 2010; **43**: 23–30.
59. Ruan, BF, Lu, X, Li, TT, Tang, JF, Wei, Y, Wang, XL, Zheng, SL, Yao, RS, Zhu, HL. Synthesis, biological evaluation and molecular docking studies

of resveratrol derivatives possessing curcumin moiety as potent antitubulin agents. *Bioorganic & Medicinal Chemistry* 2012; **20**: 1113–1121.
60. Lim, KG, Gray, AI, Pyne, S, Pyne, NJ. Resveratrol dimers are novel sphingosine kinase 1 inhibitors and affect sphingosine kinase 1 expression and cancer cell growth and survival. *British Journal of Pharmacology* 2012; **166**: 1605–1616.
61. Brizuela, L, Dayon, A, Doumerc, N, Ader, I, Golzio, M, Izard, JC, Hara, Y, Malavaud, B, Cuvillier, O. The sphingosine kinase-1 survival pathway is a molecular target for the tumor-suppressive tea and wine polyphenols in prostate cancer. *FASEB J* 2010; **24**: 3882–3894.
62. Lim, KG, Gray, AI, Anthony, NG, Mackay, SP, Pyne, S, Pyne, NJ. Resveratrol and its oligomers: modulation of sphingolipid metabolism and signaling in disease. *Archives of Toxicology* 2014; **88**: 2213–2232.
63. Ranganathan, R, Madanmohan, S, Kesavan, A, Baskar, G, Krishnamoorthy, YR, Santosham, R, Ponraju, D, Rayala, SK, Venkatraman, G. Nanomedicine: towards development of patient-friendly drug-delivery systems for oncological applications. *Int J Nanomedicine* 2012; **7**: 1043–1060.
64. Walsh, MD, Hanna, SK, Sen, J, Rawal, S, Cabral, CB, Yurkovetskiy, AV, Fram, RJ, Lowinger, TB, Zamboni, WC. Pharmacokinetics and antitumor efficacy of XMT-1001, a novel, polymeric topoisomerase I inhibitor, in mice bearing HT-29 human colon carcinoma xenografts. *Clin Cancer Res* 2012; **18**: 2591–2602.
65. Chu, KS, Hasan, W, Rawal, S, Walsh, MD, Enlow, EM, Luft, JC, Bridges, AS, Kuijer, JL, Napier, ME, Zamboni, WC, *et al.* Plasma, tumor and tissue pharmacokinetics of Docetaxel delivered via nanoparticles of different sizes and shapes in mice bearing SKOV-3 human ovarian carcinoma xenograft. *Nanomedicine* 2013; **9**: 686–693.
66. Metwally, AA, El-Ahmady, SH, Hathout, RM. Selecting optimum protein nano-carriers for natural polyphenols using chemoinformatics tools. *Phytomedicine* 2016; **23**: 1764–1770.
67. Leimkuhler, B, Matthews, C. Introduction. In Molecular Dynamics: With Deterministic and Stochastic Numerical Methods. (Cham %@ 978-3-319-16375-8, Springer International Publishing), 2015; pp. 1–51.
68. Wang, Q, Ning, L, Niu, Y, Liu, H, Yao, X. Molecular mechanism of the inhibition and remodeling of human islet amyloid polypeptide (hIAPP(1-37)) oligomer by resveratrol from molecular dynamics simulation. *The Journal of Physical Chemistry B* 2015; **119**: 15–24.
69. Shariatizi, S, Meratan, AA, Ghasemi, A, Nemat-Gorgani, M. Inhibition of amyloid fibrillation and cytotoxicity of lysozyme fibrillation products by polyphenols. *International Journal of Biological Macromolecules* 2015; **80**: 95–106.
70. Woods, LA, Platt, GW, Hellewell, AL, Hewitt, EW, Homans, SW, Ashcroft, AE, Radford, SE. Ligand binding to distinct states diverts aggregation of an amyloid-forming protein. *Nature Chemical Biology* 2011; **7**: 730–739.

71. Ladiwala, AR, Lin, JC, Bale, SS, Marcelino-Cruz, AM, Bhattacharya, M, Dordick, JS, Tessier, PM. Resveratrol selectively remodels soluble oligomers and fibrils of amyloid Abeta into off-pathway conformers. *The Journal of Biological Chemistry* 2010; **285**: 24228–24237.
72. Trott, O, Olson, AJ. AutoDock Vina: improving the speed and accuracy of docking with a new scoring function, efficient optimization, and multithreading. *Journal of Computational Chemistry* 2010; **31**: 455–461.
73. Pronk, S, Pall, S, Schulz, R, Larsson, P, Bjelkmar, P, Apostolov, R, Shirts, MR, Smith, JC, Kasson, PM, van der Spoel, D, *et al.* GROMACS 4.5: a high-throughput and highly parallel open source molecular simulation toolkit. *Bioinformatics* 2013; **29**: 845–854.
74. Stirpe, A, Pantusa, M, Rizzuti, B, De Santo, MP, Sportelli, L, Bartucci, R, and Guzzi, R. Resveratrol induces thermal stabilization of human serum albumin and modulates the early aggregation stage. *International Journal of Biological Macromolecules* 2016; **92**: 1049–1056.
75. Banaganapalli, B, Mulakayala, C, D, G, Mulakayala, N, Pulaganti, M, Shaik, NA, Cm, A, Rao, RM, Al-Aama, JY, Chitta, SK. Synthesis and biological activity of new resveratrol derivative and molecular docking: dynamics studies on NFkB. *Appl Biochem Biotechnol* 2013; **171**: 1639–1657.
76. Banaganapalli, B, Mulakayala, C, Pulaganti, M, Mulakayala, N, Anuradha, CM, Suresh Kumar, C, Shaik, NA, Yousuf Al-Aama, J, Gudla, D. Experimental and computational studies on newly synthesized resveratrol derivative: a new method for cancer chemoprevention and therapeutics? *Omics : A Journal of Integrative Biology* 2013; 17: 568–583.
77. Morris, GM, Huey, R, Lindstrom, W, Sanner, MF, Belew, RK, Goodsell, DS, and Olson, AJ. AutoDock4 and AutoDockTools4: Automated docking with selective receptor flexibility. *Journal of Computational Chemistry* 2009; **30**: 2785–2791.
78. Hess, B, Kutzner, C, van der Spoel, D, Lindahl, E. GROMACS 4: Algorithms for highly efficient, load-balanced, and scalable molecular simulation. *Journal of Chemical Theory and Computation* 2008; **4**: 435–447.
79. Chakraborty, S, Kumar, A, Butt, NA, Zhang, L, Williams, R, Rimando, AM, Biswas, PK, Levenson, AS. Molecular insight into the differential anti-androgenic activity of resveratrol and its natural analogs: in silico approach to understand biological actions. *Mol Biosyst* 2016; **12**: 1702–1709.
80. Taguchi, K, Motohashi, H, Yamamoto, M. Molecular mechanisms of the Keap1-Nrf2 pathway in stress response and cancer evolution. *Genes Cells* 2011; **16**: 123–140.
81. Finkel, T, Holbrook, NJ. Oxidants, oxidative stress and the biology of ageing. *Nature* 2000; **408**: 239–247.
82. Baird, L, Dinkova-Kostova, AT. The cytoprotective role of the Keap1-Nrf2 pathway. *Archives of Toxicology* 2011; **85**: 241–272.

83. Nguyen, T, Nioi, P, Pickett, CB. The Nrf2-antioxidant response element signaling pathway and its activation by oxidative stress. *The Journal of Biological Chemistry* 2009; **284**: 13291–13295.
84. Ungvari, Z, Bagi, Z, Feher, A, Recchia, FA, Sonntag, WE, Pearson, K, de Cabo, R, and Csiszar, A. Resveratrol confers endothelial protection via activation of the antioxidant transcription factor Nrf2. *Am J Physiol Heart Circ Physiol* 2010; **299**: H18–24.
85. Bhakkiyalakshmi, E, Dineshkumar, K, Karthik, S, Sireesh, D, Hopper, W, Paulmurugan, R, Ramkumar, KM. Pterostilbene-mediated Nrf2 activation: mechanistic insights on Keap1:Nrf2 interface. *Bioorganic & Medicinal Chemistry* 2016; **24**: 3378–3386.
86. Dinkova-Kostova, AT, Holtzclaw, WD, Cole, RN, Itoh, K, Wakabayashi, N, Katoh, Y, Yamamoto, M, Talalay, P. Direct evidence that sulfhydryl groups of Keap1 are the sensors regulating induction of phase 2 enzymes that protect against carcinogens and oxidants. *Proceedings of the National Academy of Sciences of the United States of America* 2002; **99**: 11908–11913.
87. Itoh, K, Wakabayashi, N, Katoh, Y, Ishii, T, O'Connor, T, Yamamoto, M. Keap1 regulates both cytoplasmic-nuclear shuttling and degradation of Nrf2 in response to electrophiles. *Genes Cells* 2003; **8**: 379–391.
88. Rooklin, D, Wang, C, Katigbak, J, Arora, PS, Zhang, Y. AlphaSpace: fragment-Centric topographical mapping to target protein-Protein interaction interfaces. *Journal of Chemical Information and Modeling* 2015; **55**: 1585–1599.
89. Grutzendler, J, Morris, JC. Cholinesterase inhibitors for Alzheimer's disease. *Drugs* 2001; **61**: 41–52.
90. Chopra, K, Misra, S, Kuhad, A. Current perspectives on pharmacotherapy of Alzheimer's disease. *Expert Opin Pharmacother* 2011; **12**: 335–350.
91. Awasthi, M, Singh, S, Pandey, VP, Dwivedi, UN. Alzheimer's disease: An overview of amyloid beta dependent pathogenesis and its therapeutic implications along with *in silico* approaches emphasizing the role of natural products. *J Neurol Sci* 2016; **361**: 256–271.
92. Li, C, Fang, JS, Lian, WW, Pang, XC, Liu, AL, Du, GH. *In vitro* antiviral effects and 3D QSAR study of resveratrol derivatives as potent inhibitors of influenza H1N1 neuraminidase. *Chemical Biology & Drug Design* 2015; **85**: 427–438.
93. Wong, CF. Flexible receptor docking for drug discovery. *Expert Opinion on Drug Discovery* 2015; **10**: 1189–1200.
94. Locatelli, GA, Savio, M, Forti, L, Shevelev, I, Ramadan, K, Stivala, LA, Vannini, V, Hubscher, U, Spadari, S, Maga, G. Inhibition of mammalian DNA polymerases by resveratrol: mechanism and structural determinants. *Biochem J* 2005; **389**: 259–268.
95. Yu, J, Zhou, Y, Tanaka, I, Yao, M. Roll: a new algorithm for the detection of protein pockets and cavities with a rolling probe sphere. *Bioinformatics* 2010; **26**: 46–52.

96. Wang, YY, Li, J, Wu, ZR, Zhang, B, Yang, HB, Wang, Q, Cai, YC, Liu, GX, Li, WH, Tang, Y. Insights into the molecular mechanisms of Polygonum multiflorum Thunb-induced liver injury: a computational systems toxicology approach. *Acta Pharmacologica Sinica* 2017; **38**: 719–732.
97. Xie, L, Ge, X, Tan, H, Xie, L, Zhang, Y, Hart, T, Yang, X, Bourne, PE. Towards structural systems pharmacology to study complex diseases and personalized medicine. *PLoS Computational Biology* 2014; **10**: e1003554.

Chapter 12

Evolution from Gene to Gene Network: Using Bioinformatics to Gain Insights into the Chemopreventive Mechanisms of Resveratrol

Evelien Schaafsma[*,†], John T. Pinto[*], Joseph Garvey[*], Radha Garvey[*], Joseph M. Wu[*] and Tze-chen Hsieh[*]

[*]*Biochemistry and Molecular Biology, New York Medical College, Valhalla, NY 10595, USA*

[†]*Molecular Cell Biology Program, Dartmouth College, Hanover, NH 03755-3842, USA*

Interest in the use of computational-based approaches to address biomedical problems began in the 1970s. Bioinformatics is an approach that scans and weaves seemingly diverse types of data into progressive cycles of hypothesis-driven and testable outputs. Bioinformatics is applicable to intensely researched biologicals and phytochemicals including resveratrol. Following the seminal reports by Pezzuto *et al.*, the anticancer and chemopreventive efficacy of this grape polyphenolic received unprecedented, widespread support by thousands of published studies. Since then, a remarkable number of diverse health promoting, disease prevention, and therapeutic effects of resveratrol have appeared. A number of targets and mechanisms have been identified and characterized. An outstanding challenge to understanding resveratrol's chemotherapeutic properties is how to best prioritize and identify the most promising targeted genes and binding proteins. Here we review a web-based virtual approach for systematically dissecting multi-modal publicly available databases collected from large-scale studies. We believe that the tools reviewed herein may constitute a guide to help other scientists select a gene-to-gene network strategy that best fits their scientific priorities and needs. In addition, mining of web-available data may help unclog conceptual bottlenecks and unleash unconventional ideas forming the basis for novel testable hypotheses. The intent of this approach is to connect resveratrol both productively and cost-effectively to its primary and secondary target genes and proteins. In addition, this approach intends to expand the identification of other biological functions and to serve as a guide for future studies to investigate more targeted health benefits and disease intervention potentials.

12.1. Introduction to bioinformatics

The term "bioinformatics" was first coined in the 1970s by two Dutch scientists, Hogeweg and Hesper, to describe "the study of informatic processes in biotic systems".[1] This concept combined pattern analysis and dynamic modeling to gain insights into processes in biotic systems at multiple levels, e.g. genotype, phenotype, behavioral, and environmental-levels. As the amount of sequencing data expanded rapidly in the 1980s, the term also expanded in its use, veering towards the more modern notion of bioinformatics: the development and use of computational methods for data management and data analysis of sequence data, protein structure determination, homology-based function prediction, and phylogeny.[2]

The use of bioinformatics has grown to become central in the biomedical sciences as the rapid improvement in technologies over the past decades has led to a substantial increase in data acquisition. The term "Omics" is often used in this context and broadly includes genomics, transcriptomics, metabolomics and proteomics, but additional "Omics" formats are continuously added, e.g. lipidomics, glycomics or mutagenomics, and these new types of "Omics" data are likely to grow in the next decade as we continue to improve our understanding and analytic capabilities.

The evolution of the genomics and transcriptomics fields is especially notable due to the rate at which these fields have advanced in the past decade. The introduction of the Sanger sequencing method is undoubtedly a major milestone in the early phases of the genomics and transcriptomics eras. Developed by Frederick Sanger and colleagues in the 1980s,[3] the Sanger method was the prevailing sequencing method for genetic material for the following three decades.[4] However, advances in sequencing methods have accelerated the genomics field over the past ten years, which were sparked by the National Human Genome Research Institute (NHGRI) in 2004. Funds were made available to reduce costs and increase the speed of sequencing to reach the goal of sequencing the human genome for less than $1000 by 2014.[5] Next-Generation Sequencing (NGS) methods were introduced a few years later and included for example the 454 Genome Sequencer (Life Sciences, now Roche)[6] and the Sequencing by Oligo Ligation Detection (SOLiD) method (Applied Biosystems, now Life Technologies).[7] These techniques are now considered second generation sequencers, since they are being challenged by real time sequences, also called third generation sequencers, such as the PacBio RS platform (Pacific Biosciences).[8] Due to these improvements, we are capable of sequencing a large variety of materials, including bulk tissue, specific cell types, and even single cells.

The progression of NGS techniques has made vast demands on bioinformatic tools for management and data analysis. Databases, such as GenBank®, incorporate the latest data to keep members of the scientific community up to date and aid in managing "Omics" data. Additionally, a vast amount of bioinformatic tools has become available to facilitate data analysis. Various domains in the biomedical sciences have utilized these tools successfully to understand pathological conditions, for example the field of oncology, which is of interest to us in the context of resveratrol.

12.2. Cancer research and bioinformatics

A considerable amount of progress in understanding cancer has led to a decline in cancer mortality rates in the United States. Despite these advances, cancer is still the second leading cause of death in the United States.[9] The worldwide incidence and the mortality rates from cancer are rising and are expected to continue to rise over the next twenty years. According to the National Cancer Institute, 39.6 percent of American men and women will be diagnosed with cancer at one point in his or her lifetime (https://www.cancer.gov/about-cancer/understanding/statistics). These statistics not only demonstrate that cancer research is still necessary and relevant, but also show the importance of advancing research to find more efficient methods, chemicals, and drugs capable of preventing and treating cancer.

The use of bioinformatic tools has prospered in the field of oncology due to the potential of providing precision medicine. It is well-established that cancers arise through genomic alterations and while common driver mutations exist, such as TP53 and PTEN, we now know that the mutational landscape in cancers is often fairly distinct from patient to patient through NGS methods.[10] Targeted therapies that focus on particular tumor characteristics have been shown to increase patient survival and reduce side effects (reviewed in[11]). Such therapies are becoming the prevailing therapy in the treatment of breast cancer, melanoma and non-small cell lung cancer (NSCLC), among others (reviewed in[12]), and more targeted therapies are likely to be developed in the upcoming years.

Besides aiding in selecting patients for targeted therapies, bioinformatic tools are also employed to obtain more insights into the mechanisms of potential pharmaceuticals or chemopreventive agents. Bioinformatic resources are growing in importance in understanding and predicting drug targets, drug absorption and secretion, and drug toxicity.[13] Using this notion, we attempted to introduce bioinformatic tools that can be utilized to gain insight in the mechanisms of resveratrol as a potential chemopreventive agent. Using publicly available databases and bioinformatic tools, this chapter promotes the usage of bioinformatic tools, as this is becoming the mainstay of research.

12.3. Bioinformatic tools and resveratrol

12.3.1. *Introduction*

Due to the seemingly unrestricted availability of data and various data types, bioinformatic tools are continuously being developed to facilitate data analysis and mining. To bridge the gap between researchers and tool developers, a community-based search platform, OMICtools, was established in 2014.[14] This platform provides an overview of more than 4,400 web-accessible tools related to genomics, transcriptomics, proteomics and metabolomics. We focused on two types of tools: (i) databases that provide information on resveratrol and its interactors/target molecules, and (ii) bioinformatic tools that can be used to investigate interactions with its targets. To address the latter section, we have compiled a list of mostly cancer-related tools that provide information on four biological levels; DNA alterations (cBioPortal), methylation data (PubMeth), gene expression levels (Oncomine), and knock-down information (GenomeRNAi).

12.3.2. *Databases*

12.3.2.1. *DrugBank*

Established in 2006, DrugBank has grown to an extensive online database that comprises detailed drug data.[15] The database contains FDA-approved small molecules, FDA-approved biotech drugs, experimental drugs, and nutraceuticals. Each DrugCard contains information on pharmacological data, e.g. absorption and metabolism, chemical properties, current and performed clinical trials, pharmacoeconomics, and taxonomy. Additionally, three interactor types are provided where available, including drug targets, enzymes and carrier proteins. Based on these interactor types, BioInteractors are compiled, which are proteins in the DrugBank database that are similarly targeted by other drugs in the DrugBank database. DrugBank stands out from other drug databases due to the detailed biochemical information available for almost every compound in its directory. The utilities of DrugBank are wide-ranging; drugs can be queried by name, drug classes can be explored, chemical structures can be drawn and queried, protein sequences can be searched for interacting drugs, and selected text components can be downloaded.

Most of the mentioned information is available for resveratrol. For example, comprehensive chemical data and clinical trials are provided, but data on metabolism is sparse; half-life, clearance, and toxicity are "Not available". In terms of interactors, 25 resveratrol targets are currently listed, each containing general information such as function and gene names, and a reference for their interactions with resveratrol.

12.3.2.2. *PharmGKB*

PharmGKB is a comprehensive pharmacogenomics resource that collects data on genetic alterations, drugs, pathways, dosing guidelines, and drug labels.[16] PharmGKB offers many of the pharmacologic and clinical features of drug metabolism provided by DrugBank but extends clinical recommendations to suggest the use of certain drugs in patients with particular gene variants. For each drug, information is provided on prescriptions, drug labels, clinical annotations, and affected pathways. PharmGKB is aimed at supporting researchers in understanding how genetic variations contribute to individual differences as well as observed differences in isolated drugs reactions.

Limited information is available for resveratrol in PharmGKB, likely because resveratrol has not been extensively investigated in clinical trials. Basic information on resveratrol structure, alternative names, and molecular properties is provided. Additionally, a variety of links to other databases can be obtained. Although use of PharmGKB for examining resveratrol is currently limited, we highlight this database because the availability of integrated clinical information is crucial for future use of resveratrol and other compounds. The accumulation and union of different data types facilitate clinical implementations and highlight the utility of drugs in patients with certain genetic alterations or variations.

12.3.3. *Tools to investigate resveratrol interactors*

12.3.3.1. *cBioPortal*

12.3.3.1.1. Introduction

The cBioPortal (http://www.cbioportal.org) is an open cancer-genomics platform. The tool was developed at Memorial Sloan Kettering Cancer Center for cancer researchers to easily access molecular profiles and clinical attributes from large-scale cancer genomics projects.[17] The portal contains different data types per cancer study and includes: DNA copy-number data, mRNA and microRNA expression data, non-synonymous mutations, protein-level and phosphoprotein level data, DNA methylation data, and limited de-identified clinical data. This resource enables researchers to interactively explore genetic alterations across samples, genes, and pathways. This empowers researchers to translate these rich data sets into biologic insights and link them to clinical outcomes to facilitate novel discovery in biological systems. The database in cBioPortal contains tumor samples obtained from 20+ cancer studies.[17]

12.3.3.1.2. Resveratrol findings in cBioPortal

cBioPortal has been used to explore the link between resveratrol and prostate cancer.[18] In the study, 3 genes (*PTEN*, *TP53* and *CDKN1A*) were identified in the p53 signalling pathway using Kyoto Encyclopedia of Genes and Genomes (KEGG) pathway analysis. These genes were used in cBioPortal to explore the genetic alterations that occur in prostate cancer.[18] Alterations ranging from 1.9% to 72.1% were found for these pathways among 7 prostate cancer studies.[18–24] Upon closer examination of these studies, data from the Michigan study[21] showed that 72% of the investigated prostate cancer cases (44 individuals) have an alteration in at least one of the three genes queried.[18] In this study, no genetic alterations were found for *CDKN1A*. For *PTEN*, most alterations were classified as deep deletions with few cases of mutations. Gene changes found in *TP53* included deep deletions and missense/truncating mutations. The alterations in *PTEN* and *TP53* tended towards co-occurrence across samples, however no statistical significance ($p = 0.183$) was found by Mutual Exclusivity analysis.

cBioPortal was also utilized to explore the alteration of genes associated with resveratrol in colorectal cancer (CRC). A query of 2 selected genes, *APC* and *TP53*, was performed to evaluate their cancer genomic alterations and clinical profiles in CRC. Among data analyzed in the 4 CRC studies,[25–27] alterations in the specific gene sets/pathways ranged from 73.6% to 96.4%. A data set from The Cancer Genome Atlas (TCGA), the Provisional Study from

National Cancer Institute (NCI), contained the most number of cases (220 patients) and results showed that 73% and 54% of *APC* and *TP53*, respectively, had frequent alterations. Although alterations in these two genes demonstrate co-occurrence across samples, no statistical significance was revealed by Mutual Exclusivity analysis ($p = 0.385$).

12.3.3.1.3. Other applications of cBioPortal

cBioPortal contains different data types which provide additional functions that were not addressed in this section. For example, other data types (e.g. protein-level, post-translational modification data and microRNAs) can also be queried to investigate their over- or under expression in different cancer types. Overall, cBioPortal can help generate multiple visualization platforms when grouping alterations in data from any given type of cancer and using input from any given set of genes.[28–32]

12.3.3.2. *PubMeth*

12.3.3.2.1. Introduction

PubMeth is a cancer methylation database that includes genes reported to be methylated in cancers.[33] The database is constructed through text mining of Medline/PubMed abstracts. This approach increases speed and selectivity by providing aliases of genes that can be searched simultaneously. After text mining, records are manually annotated, which significantly raises the specificity and quality of the database. Text-mining is currently only available for PubMed abstracts, which will exclude epigenetic data presented in publication bodies. This feature assures high-quality and pertinent data, but limits the size of the database in terms of the number of genes that are available for epigenetic mining. PubMeth is freely available through http://www.pubmeth.org/. The goal of the database is to facilitate early detection of cancer and the development of personalized cancer treatments.[33]

A recent publication showed that resveratrol alters the cancer methylome in triple negative breast cancer cell lines.[34,35] Promoter hypermethylation decreased and DNA hypomethylation increased in specific genetic regions, including the restoration of hypomethylated and hypermethylated key tumor suppressor genes and oncogenes, respectively. In this regard, resveratrol might serve as a candidate drug for combination therapy in which this chemopreventive agent alters the methylation status of genes targeted by certain therapies and affects their efficacy. Therefore, acquiring insight into either the methylation pattern of resveratrol-interactors or the regions whereby resveratrol induces epigenetic alterations should provide better understanding for use of resveratrol in treatment, prevention or control of certain cancers.

12.3.3.2.2. Resveratrol findings in PubMeth

We have selected the resveratrol-interactor, prostaglandin-endoperoxide synthase 2 (PTGS2, aka cyclooxygenase-2), as an example to show the utility of PubMeth PTGS2, the rate-limiting enzyme in prostaglandin synthesis, is induced in many cells by inflammatory mediators. Up-regulation of PTGS2 is associated with tumor cell growth and metastasis, enhanced cellular adhesion, and inhibition of apoptosis. By contrast, induction of PTGS2 expression results in inhibition of growth of breast cancer and colon cancer cells.[36,37] Resveratrol-induced accumulation of PTGS2 results in its intranuclear colocalization with Ser15-phosphorylated p53 and p300 and facilitates apoptosis in breast cancer cells.[37]

With regard to the methylation status of PTGS2, twenty references are available that cover nine cancer types. Only prostate cancer showed a significant alteration in PTGS2 methylation status when comparing cancer to normal samples. PTGS2 was found to be methylated in prostate carcinomas (no subtype specified), but rarely or not methylated in control samples ($p = 0.0015$, two-tailed t-test). Interestingly, PTGS1, an isozyme of PTGS2, was one of the 16 CpG islands that display cancer-related hypermethylation in prostate adenocarcinoma. In addition, resveratrol directly binds PTGS2 and may regulate its effect on colony formation of human colon adenocarcinoma HT-29 cells depending on the presence or absence of other epigenetic targets and their co-association.[38]

Thus, not only are the epigenetic alterations that are direct targets of resveratrol of interest, but epigenetic targets *per se* are important to consider. The top 5 epigenetically altered genes in MDA-MB-231 breast cancer cells treated with resveratrol for 48 h are Zinc Finger Protein 217 (ZNF217), CPA4, AKT1, and C1orf54.[34] Interestingly, ZNF217 belongs to a gene region that is frequently amplified in human cancers.[39] Additionally, ZNF217 is a marker of poor prognosis in breast cancer and is proposed to be involved in epithelial-mesenchymal transition (EMT) and tumor invasion.[40] Presumably, hypermethylation of ZNF217 will reduce these processes and might slow down tumor growth as a consequence. Unfortunately, ZNF217 is currently not included in the PubMeth database, but this example highlights the importance of investigating resveratrol-induced epigenetic alterations in resveratrol research.

12.3.3.2.3. Other applications of PubMeth

We have provided an example of a specific gene search in this section, but PubMeth can also be queried by gene aliases, cancer type, or a combination of gene and cancer type. More than 300 cancer subtypes can be investigated and compared using these search options. Since PubMeth does not contain an exhaustive list of genes, the search option that combines gene and cancer type-specifications might not be suitable for less-studied genes or proteins. More proteins are likely to become available over time as the database expands.

12.3.3.3. *Oncomine*

12.3.3.3.1. Introduction

Oncomine is a web-based microarray gene expression data-mining platform that was established in 2004.[41] This platform contains more than 700 microarray datasets on cancer tissues (with healthy controls as reference) and serves to facilitate the mining of gene expression data by combining microarray data into one collective transcriptome data set. Microarray datasets are added to Oncomine through a microarray data pipeline that collects microarray studies from published data and draws relevant data sets directly from the Stanford Microarray Database and the NCBI Gene Expression Omnibus as they become available.[42] All datasets are curated before being added to the platform and analyses are standardized to ensure consistency in depiction of data.

Two versions of Oncomine are available; the Oncomine Research Edition is freely accessible to the academic and nonprofit cancer research communities, and the Oncomine Research Premium Edition is a subscription version that includes more extensive features, such as multi-gene search and export capabilities. Both versions can be accessed through www.oncomine.org.

12.3.3.3.2. Resveratrol findings in Oncomine

Depending on the research question, Oncomine can be queried through multiple methods. For example, individual genes, or multiple genes when using the Oncomine Research Premium Edition, can be searched and gene expression data, including statistical analyses for the queried gene, will be displayed for all available cancer types. This method was used in a study on quinone reductase 2 (QR2) in CRC.[43] QR2 is a known high-affinity interactor of resveratrol,[44] but little is known about its cellular role or function. Oncomine was used to evaluate if the expression of QR2 could provide information on the mechanisms of action of resveratrol in CRC.[45] The expression of QR2 was reported to be upregulated in certain CRC types, indicating that resveratrol might affect certain CRC tissues through QR2. Although not shown, other interactors of resveratrol could be mined in a similar manner to provide starting points for *in vitro* and *in vivo* studies.

Co-expression analyses are also provided for most datasets in Oncomine. Since functionally similar genes are often co-expressed,[46] this resource can be used to infer the function of lesser known genes, such as *QR2*. As an example, the top co-expressed genes of *QR2* reported in the PC/AA/C1/SB10C CRC cell lines (*RGS4*, *SYNGR3*, *CENPQ*, *MYBL2*, *HSPA4L*, *RFC5*, *MCM2*, *FEN1*, *LRP8*, *CHCT6*, *CDC6*, *C9orf40*, *DNMT1*, *PRIM2*, *HBS1L*, *RAD51*) (co-expression correlation >0.95) show an enrichment in DNA-dependent DNA replication genes ($p = 2.35E\text{-}08$, GO:0006261, geneontology.org), which hints to an involvement of QR2 in DNA replication processes.

Oncomine currently also contains three datasets of resveratrol-treated cell lines. The MCF7 (breast cancer), PC-3 (prostate cancer) and HL60 (promyelocytic leukemia) cell lines have been treated for 6 h with resveratrol at concentrations of 10 μM, 50 μM and 20 μM, respectively. A comparison of the overexpressed genes among these datasets shows that three genes (*ATF3*, *GADD45A*, and *PMAIP1*) are consistently overexpressed upon resveratrol treatment. These genes are involved in cellular stress responses, suggesting that resveratrol induces cell damage. Consistently under-expressed genes include *CENPE*, *DLGAP5*, *HMMR*, and *PLK1*. Three of these genes are involved in mitotic division, indicating that resveratrol might also induce cell cycle arrest.

12.3.3.3.3. Other applications of Oncomine

Oncomine provides several additional functions that were not addressed in this section. For example, entire pathways can be queried and their over- or under-expression in different cancer types can be investigated. Besides pathways, signatures of over- or under-expressed genes can also be compared among datasets to aid in identifying commonly altered pathways. Additionally, various filters can be used, such as microarray platforms or differential analysis (cancer vs. cancer, cancer vs. normal, or normal vs. normal), before or after queries to include only certain desired parameters. Also, cell lines and some primary cells can be queried for essential genes, which have been determined by short hairpin RNA (shRNA) studies. Lastly, each dataset in Oncomine contains references to publications and links to other raw data repositories if more in-depth information about the data or dataset is required.

12.3.3.4. *GenomeRNAi*

12.3.3.4.1. Introduction

The GenomeRNAi database contains RNA interference (RNAi) phenotype data for human and Drosophila genes. These data are obtained by extractions from 467 RNAi screens in humans and 207 RNAi screens in Drosophila. Two types of data are collected:

(i) information about the RNAi screen itself, and (ii) gene-reagent-phenotype associations. Each publication that is considered being included into the database can contain multiple screens when experiments are performed in different cell lines, or with different assays, or when it includes follow-up screens.

The database can be accessed at http://www.genomernai.org/ and is freely available. The goal of GenomeRNAi is to provide rich data sources for functional annotation across different species. This information can inform users about how to design new experiments with effective RNAi constructs or to investigate the efficacy of knocking down a gene of interest. Genes or proteins can be queried and all known RNAi designs are listed. Sequence information of each RNAi probe is provided, as well as on- and off-targets, and efficiency. Where available, phenotypic information is given, which indicates the effects of a specific RNAi construct.

12.3.3.4.2. Resveratrol findings in GenomeRNAi

Continuing with the example of using QR2 as a resveratrol interactor in the Oncomine section, we queried QR2 in the GenomeRNAi database. Twenty-four RNAi constructs have been designed and used to target QR2 using small interfering RNA (siRNA) and short hairpin RNA (shRNA) methods. While all constructs seem to be specific to QR2 (no known off-targets genes are reported), the knock-down efficiency ranges from 90% to 30%. Phenotypes that are observed when knocking-down QR2 include: (i) an increase in vaccinia virus (VACV) infections which are synthetically lethal with Ras; (ii) decreased hepcidin mRNA expression; (iii) decreased human papilloma virus 16 (HPV16) pseudovirus infections; and (iv) decreased viability. While these findings may appear genetically diffuse, the increase in virus infection and its synthetic lethality with Ras induction coupled with decreased viability might indicate that QR2 is involved in a process required for cell survival.

An example of a cancer-related resveratrol target is estrogen receptor alpha (ERα).[47] Twenty-three RNAi sequences are listed, again, both siRNA and shRNA constructs have been deposited. Associated phenotypes are: increased cell death in breast cancer cell lines, increased cell number in G1 with small nuclei, decreased viability, decreased viability in esophageal squamous lineage, decreased VACV infection, and decreased autoimmune regulator (AIRE) reporter expression. This clearly indicates a role for ERα in cell viability. Systematic investigations of resveratrol interactors, as exemplified here, will provide more information on the effects of resveratrol in cancer prevention and control as well as on its potential side effects.

12.3.3.4.3. Other applications of GenomeRNAi

Besides querying genes, phenotypes can be searched as well, which will list all related RNAi screens in the GenomeRNAi database. For example, querying apoptosis resulted in one RNAi screen that was aimed at identifying human kinases and phosphatases that regulate apoptosis and chemo-resistance. All identified targets are displayed in an interactive plot and targets can be continuously refined. Another option in GenomeRNAi is to search for specific RNAi identifiers or RNAi screens.

12.4. Conclusion and perspectives

Since the late 1990s, the yearly growth of citations on resveratrol in PubMed has been unprecedented. Resveratrol exhibits marked influences on diverse cellular processes and diseases, which can create a multifaceted challenge to dissect and prioritize mechanisms, targets, and gene-phenotype outputs. The challenge of data analysis can be overcome and countered by mining from high density, multi-dimensional biological datasets.

In recent years, the successful launch and execution of studies on high-throughput genomic screening studies as applied to cancer have frequently generated hundreds to thousands of novel and interesting gene hits. A web-based approach utilizing publicly available databases might effectively analyze candidate genes for subsequent linkage to experimentally verifiable pathways. Various bioinformatic tools have become available to aid in prioritizing these vast amounts of data. As we progress in our understanding and methodical capacities, additional tools will undoubtedly become available. The utilization of platforms such as OMICtools[14] should assist investigators in navigating towards practical and reliable analytical methods.

We attempted to facilitate the use of bioinformatic tools in resveratrol research by providing examples of virtual analyses that dissect some of the multi-modal publicly available databases. Although this chapter only features a fraction of the tools available, it may unclog conceptual bottlenecks or spark the use of complementary bioinformatic tools. In addition, acquiring information of novel targets of resveratrol and mining data on targets independent of resveratrol should unleash new ideas that form the basis for new testable hypotheses. The proposed strategy is significant because it will advance our understanding of the diverse bioactivities of resveratrol. It will also generate leads for specific biomarkers and therapeutic targets that might become essential in improving health benefits and disease interventions.

References

1. Hesper B, Hogeweg P. Bioinformatica: een werkconcept, vol. 1 Kameleon: (In Dutch.) Leiden: Leidse Biologen Club., 1970; 28–9.
2. Hogeweg P. The roots of bioinformatics in theoretical biology. *PLoS Comput Biol* 2011; **7**: e1002021.
3. Sanger F, Nicklen S, Coulson AR. DNA sequencing with chain-terminating inhibitors. *Proc Natl Acad Sci U S A* 1977; **74**: 5463–7.
4. van Dijk EL, Auger H, Jaszczyszyn Y, Thermes C. Ten years of next-generation sequencing technology. *Trends Genet* 2014; **30**: 418–26.
5. Schloss JA. How to get genomes at one ten-thousandth the cost. *Nat Biotechnol* 2008; **26**: 1113–5.
6. Margulies M, Egholm M, Altman WE, Attiya S, Bader JS, Bemben LA, Berka J, Braverman MS, Chen YJ, Chen Z, Dewell SB, Du L, *et al.* Genome sequencing in microfabricated high-density picolitre reactors. *Nature* 2005; **437**: 376–80.
7. Valouev A, Ichikawa J, Tonthat T, Stuart J, Ranade S, Peckham H, Zeng K, Malek JA, Costa G, McKernan K, Sidow A, Fire A, *et al.* A high-resolution, nucleosome position map of C. elegans reveals a lack of universal sequence-dictated positioning. *Genome Res* 2008; **18**: 1051–63.
8. Eid J, Fehr A, Gray J, Luong K, Lyle J, Otto G, Peluso P, Rank D, Baybayan P, Bettman B, Bibillo A, Bjornson K, *et al.* Real-time DNA sequencing from single polymerase molecules. *Science* 2009; **323**: 133–8.

9. Siegel RL, Miller KD, Jemal A. Cancer Statistics, 2017. *CA Cancer J Clin* 2017; **67**: 7–30.
10. Yates LR, Knappskog S, Wedge D, Farmery JHR, Gonzalez S, Martincorena I, Alexandrov LB, Van Loo P, Haugland HK, Lilleng PK, Gundem G, Gerstung M, *et al.* Genomic evolution of breast cancer metastasis and relapse. *Cancer Cell* 2017; **32**: 169–84.e7.
11. Masoud V, Pagès G. Targeted therapies in breast cancer: New challenges to fight against resistance. *World J Clin Oncol* 2017; **8**: 120–34.
12. Pagliarini R, Shao W, Sellers WR. Oncogene addiction: pathways of therapeutic response, resistance, and road maps toward a cure. *EMBO Rep* 2015; **16**: 280–96.
13. Lacombe D, Tejpar S, Salgado R, Cardoso F, Golfinopoulos V, Aust D, Folprecht G, Roth A, Stupp R. European perspective for effective cancer drug development. *Nat Rev Clin Oncol* 2014; **11**: 492–8.
14. Henry VJ, Bandrowski AE, Pepin AS, Gonzalez BJ, Desfeux A. OMICtools: an informative directory for multi-omic data analysis. *Database (Oxford)* 2014; **2014**.
15. Law V, Knox C, Djoumbou Y, Jewison T, Guo AC, Liu Y, Maciejewski A, Arndt D, Wilson M, Neveu V, Tang A, Gabriel G, *et al.* DrugBank 4.0: shedding new light on drug metabolism. *Nucleic Acids Res* 2014; **42**: D1091–7.
16. Whirl-Carrillo M, McDonagh EM, Hebert JM, Gong L, Sangkuhl K, Thorn CF, Altman RB, Klein TE. Pharmacogenomics knowledge for personalized medicine. *Clin Pharmacol Ther* 2012; **92**: 414–7.
17. Cerami E, Gao J, Dogrusoz U, Gross BE, Sumer SO, Aksoy BA, Jacobsen A, Byrne CJ, Heuer ML, Larsson E, Antipin Y, Reva B, *et al.* The cBio cancer genomics portal: an open platform for exploring multidimensional cancer genomics data. *Cancer Discov* 2012; **2**: 401–4.
18. Hsieh TC, Wu ST, Bennett DJ, Doonan BB, Wu E, Wu JM. Functional/activity network (FAN) analysis of gene-phenotype connectivity liaised by grape polyphenol resveratrol. *Oncotarget* 2016; **7**: 38670–80.
19. Barbieri CE, Baca SC, Lawrence MS, Demichelis F, Blattner M, Theurillat JP, White TA, Stojanov P, Van Allen E, Stransky N, Nickerson E, Chae SS, *et al.* Exome sequencing identifies recurrent SPOP, FOXA1 and MED12 mutations in prostate cancer. *Nat Genet* 2012; **44**: 685–9.
20. Baca SC, Prandi D, Lawrence MS, Mosquera JM, Romanel A, Drier Y, Park K, Kitabayashi N, MacDonald TY, Ghandi M, Van Allen E, Kryukov GV, *et al.* Punctuated evolution of prostate cancer genomes. *Cell* 2013; **153**: 666–77.
21. Grasso CS, Wu YM, Robinson DR, Cao X, Dhanasekaran SM, Khan AP, Quist MJ, Jing X, Lonigro RJ, Brenner JC, Asangani IA, Ateeq B, *et al.* The mutational landscape of lethal castration-resistant prostate cancer. *Nature* 2012; **487**: 239–43.
22. Taylor BS, Schultz N, Hieronymus H, Gopalan A, Xiao Y, Carver BS, Arora VK, Kaushik P, Cerami E, Reva B, Antipin Y, Mitsiades N, *et al.* Integrative genomic profiling of human prostate cancer. *Cancer Cell* 2010; **18**: 11–22.

23. Hieronymus H, Schultz N, Gopalan A, Carver BS, Chang MT, Xiao Y, Heguy A, Huberman K, Bernstein M, Assel M, Murali R, Vickers A, *et al.* Copy number alteration burden predicts prostate cancer relapse. *Proc Natl Acad Sci U S A* 2014; **111**: 11139–44.
24. Gao D, Vela I, Sboner A, Iaquinta PJ, Karthaus WR, Gopalan A, Dowling C, Wanjala JN, Undvall EA, Arora VK, Wongvipat J, Kossai M, *et al.* Organoid cultures derived from patients with advanced prostate cancer. *Cell* 2014; **159**: 176–87.
25. Brannon AR, Vakiani E, Sylvester BE, Scott SN, McDermott G, Shah RH, Kania K, Viale A, Oschwald DM, Vacic V, Emde AK, Cercek A, *et al.* Comparative sequencing analysis reveals high genomic concordance between matched primary and metastatic colorectal cancer lesions. *Genome Biol* 2014; **15**: 454.
26. Comprehensive molecular characterization of human colon and rectal cancer. *Nature* 2012; **487**: 330–7.
27. Seshagiri S, Stawiski EW, Durinck S, Modrusan Z, Storm EE, Conboy CB, Chaudhuri S, Guan Y, Janakiraman V, Jaiswal BS, Guillory J, Ha C, *et al.* Recurrent R-spondin fusions in colon cancer. *Nature* 2012; **488**: 660–4.
28. Keshava Prasad TS, Goel R, Kandasamy K, Keerthikumar S, Kumar S, Mathivanan S, Telikicherla D, Raju R, Shafreen B, Venugopal A, Balakrishnan L, Marimuthu A, *et al.* Human Protein Reference Database–2009 update. *Nucleic Acids Res* 2009; **37**: D767–72.
29. Matthews L, Gopinath G, Gillespie M, Caudy M, Croft D, de Bono B, Garapati P, Hemish J, Hermjakob H, Jassal B, Kanapin A, Lewis S, *et al.* Reactome knowledgebase of human biological pathways and processes. *Nucleic Acids Res* 2009; **37**: D619–22.
30. Schaefer CF, Anthony K, Krupa S, Buchoff J, Day M, Hannay T, Buetow KH. PID: the Pathway Interaction Database. *Nucleic Acids Res* 2009; **37**: D674–9.
31. Cerami EG, Gross BE, Demir E, Rodchenkov I, Babur O, Anwar N, Schultz N, Bader GD, Sander C. Pathway Commons, a web resource for biological pathway data. *Nucleic Acids Res* 2011; **39**: D685–90.
32. Gao J, Aksoy BA, Dogrusoz U, Dresdner G, Gross B, Sumer SO, Sun Y, Jacobsen A, Sinha R, Larsson E, Cerami E, Sander C, *et al.* Integrative analysis of complex cancer genomics and clinical profiles using the cBioPortal. *Sci Signal* 2013; **6**: pl1.
33. Ongenaert M, Van Neste L, De Meyer T, Menschaert G, Bekaert S, Van Criekinge W. PubMeth: a cancer methylation database combining text-mining and expert annotation. *Nucleic Acids Res* 2008; **36**: D842–6.
34. Medina-Aguilar R, Pérez-Plasencia C, Marchat LA, Gariglio P, García Mena J, Rodríguez Cuevas S, Ruíz-García E, Astudillo-de la Vega H, Hernández Juárez J, Flores-Pérez A, López-Camarillo C. Methylation landscape of human breast cancer cells in response to dietary compound resveratrol. *PLoS One* 2016; **11**: e0157866.
35. Medina-Aguilar R, Pérez-Plasencia C, Gariglio P, Marchat LA, Flores-Pérez A, López-Camarillo C, García Mena J. DNA methylation data for

identification of epigenetic targets of resveratrol in triple negative breast cancer cells. *Data Brief* 2017; **11**: 169–82.
36. Williams JL, Nath N, Chen J, Hundley TR, Gao J, Kopelovich L, Kashfi K, Rigas B. Growth inhibition of human colon cancer cells by nitric oxide (NO)-donating aspirin is associated with cyclooxygenase-2 induction and beta-catenin/T-cell factor signaling, nuclear factor-kappaB, and NO synthase 2 inhibition: implications for chemoprevention. *Cancer Res* 2003; **63**: 7613–8.
37. Tang HY, Shih A, Cao HJ, Davis FB, Davis PJ, Lin HY. Resveratrol-induced cyclooxygenase-2 facilitates p53-dependent apoptosis in human breast cancer cells. *Mol Cancer Ther* 2006; **5**: 2034–42.
38. Zykova TA, Zhu F, Zhai X, Ma WY, Ermakova SP, Lee KW, Bode AM, Dong Z. Resveratrol directly targets COX-2 to inhibit carcinogenesis. *Mol Carcinog* 2008; **47**: 797–805.
39. Tabach Y, Kogan-Sakin I, Buganim Y, Solomon H, Goldfinger N, Hovland R, Ke XS, Oyan AM, Kalland KH, Rotter V, Domany E. Amplification of the 20q chromosomal arm occurs early in tumorigenic transformation and may initiate cancer. *PLoS One* 2011; **6**: e14632.
40. Vendrell JA, Thollet A, Nguyen NT, Ghayad SE, Vinot S, Bièche I, Grisard E, Josserand V, Coll JL, Roux P, Corbo L, Treilleux I, *et al.* ZNF217 is a marker of poor prognosis in breast cancer that drives epithelial-mesenchymal transition and invasion. *Cancer Res* 2012; **72**: 3593–606.
41. Rhodes DR, Yu J, Shanker K, Deshpande N, Varambally R, Ghosh D, Barrette T, Pandey A, Chinnaiyan AM. ONCOMINE: a cancer microarray database and integrated data-mining platform. *Neoplasia* 2004; **6**: 1–6.
42. Rhodes DR, Kalyana-Sundaram S, Mahavisno V, Varambally R, Yu J, Briggs BB, Barrette TR, Anstet MJ, Kincead-Beal C, Kulkarni P, Varambally S, Ghosh D, *et al.* Oncomine 3.0: genes, pathways, and networks in a collection of 18,000 cancer gene expression profiles. *Neoplasia* 2007; **9**: 166–80.
43. Doonan BB, Schaafsma E, Pinto JT, Wu JM, Hsieh TC. Application of open-access databases to determine functional connectivity between resveratrol-binding protein QR2 and colorectal carcinoma. *In Vitro Cell Dev Biol Anim* 2017; **53**: 575–8.
44. Buryanovskyy L, Fu Y, Boyd M, Ma Y, Hsieh TC, Wu JM, Zhang Z. Crystal structure of quinone reductase 2 in complex with resveratrol. *Biochemistry* 2004; **43**: 11417–26.
45. Patel KR, Brown VA, Jones DJ, Britton RG, Hemingway D, Miller AS, West KP, Booth TD, Perloff M, Crowell JA, Brenner DE, Steward WP, *et al.* Clinical pharmacology of resveratrol and its metabolites in colorectal cancer patients. *Cancer Res* 2010; **70**: 7392–9.
46. Romijn EP, Christis C, Wieffer M, Gouw JW, Fullaondo A, van der Sluijs P, Braakman I, Heck AJ. Expression clustering reveals detailed co-expression patterns of functionally related proteins during B cell differentiation: A proteomic study using a combination of one-dimensional gel electrophoresis,

LC-MS/MS, and stable isotope labeling by amino acids in cell culture (SILAC). *Mol Cell Proteomics* 2005; **4**: 1297–310.

47. Lappano R, Rosano C, Madeo A, Albanito L, Plastina P, Gabriele B, Forti L, Stivala LA, Iacopetta D, Dolce V, Andò S, Pezzi V, *et al.* Structure-activity relationships of resveratrol and derivatives in breast cancer cells. *Mol Nutr Food Res* 2009; **53**: 845–58.

Index

Lightning Source UK Ltd.
Milton Keynes UK
UKHW021244181218
334207UK00002B/25/P

9 789813 270909